AF323626

7
Topics in Heterocyclic Chemistry

Series Editor: R. R. Gupta

Topics in Heterocyclic Chemistry
Series Editor: R. R. Gupta

Recently Published and Forthcoming Volumes

Heterocycles from Carbohydrate Precursors

Volume Editor: El Sayed H. El Ashry

With contributions by

E. S. H. El Ashry · J. A. S. Cavaleiro · M. Chmielewski
M. A. F. Faustino · J. G. Fernández-Bolaños · L. Fišera
B. Furman · B. Grzeszczyk · F. Hernández-Mateo · Z. Kałuża
K. M. Khan · Y. El Kilany · Ó López · N. M. Nahas · A. El Nemr
S. Perveen · F. Santoyo-González · A. Stencel
J. P. C. Tomé · W. Voelter

 Springer

The series *Topics in Heterocyclic Chemistry* presents critical reviews on "Heterocyclic Compounds" within topic-related volumes dealing with all aspects such as synthesis, reaction mechanisms, structure complexity, properties, reactivity, stability, fundamental and theoretical studies, biology, biomedical studies, pharmacological aspects, applications in material sciences, etc. Metabolism will be also included which will provide information useful in designing pharmacologically active agents. Pathways involving destruction of heterocyclic rings will also be dealt with so that synthesis of specifically functionalized non-heterocyclic molecules can be designed.

The overall scope is to cover topics dealing with most of the areas of current trends in heterocyclic chemistry which will suit to a larger heterocyclic community.

As a rule contributions are specially commissioned. The editors and publishers will, however, always be pleased to receive suggestions and supplementary information. Papers are accepted for *Topics in Heterocyclic Chemistry* in English.

In references *Topics in Heterocyclic Chemistry* is abbreviated *Top Heterocycl Chem* and is cited as a journal.

Springer WWW home page: springer.com
Visit the THC content at springerlink.com

Library of Congress Control Number: 2007928119

ISSN 1861-9282
ISBN 978-3-540-72956-3 Springer Berlin Heidelberg New York
DOI 10.1007/978-3-540-72957-0

Springer is a part of Springer Science+Business Media

springer.com

Cover design: *Design & Production* GmbH, Heidelberg
Typesetting and Production: LE-TEX Jelonek, Schmidt & Vöckler GbR, Leipzig

Printed on acid-free paper   02/3100 YL – 5 4 3 2 1 0

Topics in Heterocyclic Chemistry
Also Available Electronically

For all customers who have a standing order to Topics in Heterocyclic Chemistry, we offer the electronic version via SpringerLink free of charge. Please contact your librarian who can receive a password or free access to the full articles by registering at:

springerlink.com

If you do not have a subscription, you can still view the tables of contents of the volumes and the abstract of each article by going to the SpringerLink Homepage, clicking on "Browse by Online Libraries", then "Chemical Sciences", and finally choose Topics in Heterocyclic Chemistry.

You will find information about the

- Editorial Board
- Aims and Scope
- Instructions for Authors
  Sample Contribution

at springer.com using the search function.

Preface

This book is a volume in the series *Topics in Heterocyclic Chemistry*. It covers the key methods used for designing synthetic approaches to heterocycles from carbohydrates and the value and scope of these methods. Carbohydrates are widely distributed in nature and constitute the largest part of renewable biomasses. Moreover, many carbohydrates and their derivatives are commercially available at relatively cheap prices. Consequently their utilization is highly encouraged and economically they are of great significance. Moreover, carbohydrates are highly functionalized compounds that can be readily derivatized and/or cyclized to provide heterocyclic compounds.

This book provides a modern account and an up-to-date description of the advancement in the synthesis of diverse heterocycles from carbohydrates. Carbohydrates can be considered as a source of chiral centers in addition to the variable modification thereof. Herein the elaboration of the carbohydrate molecules for providing different heterocycles is the main objective and team efforts from leaders of the topics has been gathered in this volume.

This book is designed to be suitable for students and researchers. It is highly recommended as a reference book and for teaching the fascinating topics related to carbohydrates, heterocycles and organic synthesis. In addition to its importance in academia, it is also an excellent source for information about the variety of methods used in the synthesis of heterocycles important to industry.

The first chapter, by El Ashry, El Kilany and Nahas, describes the manipulation of carbohydrate carbon atoms in the synthesis of heterocycles, whereby a wide range of heterocycles are presented to illustrate how different functionalities, either existing or generated on the carbohydrate molecules, can be manipulated for the construction of heterocycles. The second chapter, by Fernández-Bolaños and López, deals with the synthesis of heterocycles from glycosylamines and glycosyl azides. The derived heterocycles could have significant biological activity and be related to the naturally occurring heterocycles; thus a plethora of heterocycles, such as N-glycosyl heterocycles, polyhydroxy pyrrolidines, oxapyrrolidines, piperidines and azepanes, have been prepared. The third chapter, also by Fernández-Bolaños and López, includes the synthetic versatility of carbohydrate isothiocyanates as building blocks in the preparation of a plethora of heterocycles. The fourth chapter, by Furman, Kałuża, Stencel, Grzeszczyk and Chmielewski, deals with the formation of β-lactams from carbohydrates; β-lactams are a well-known class of antibiotics. In the fifth chapter, Santoyo-González and Hernandez-Mateo report on azide-alkyne 1,3-dipolar cycloadditions, a valuable tool in carbohydrate chemistry. The sixth chapter, by Cavaleiro, Tomé, and Faustino, has reported the synthesis of glycoporphyrins. Although few examples of glycoporphyrins have been isolated from a natural source, the attachment of saccharide units to porphyrin macrocycles gives derivatives that might be of great significance for certain medicinal and other applications. The seventh chapter, by El Nemr and El Ashry, reports on new developments in the synthesis of anisomycin and its analogues; anisomycin is a natural product that has interesting biological activity against certain pathogenic protozoa, strains of fungi and in the treatment of certain diseases, such as amoebic dysentery and tricomonas vaginitis. The eighth chapter, by Fisera, deals with the 1,3-dipolar cycloadditions of sugar derived nitrones and their utilization in the synthesis of heterocycles, which allow the synthesis of tailor-made products of possible biological interest. The final chapter, by Khan, Perveen and Voelter, reports on the anhydro sugars as useful tools for chiral heterocycle syntheses.

The discussions with and suggestions from Professors H. El Khadem, R. Schmidt, V. Whitmann, J. G. Fernández-Bolaños, R.A. Field, J. Caveilaro, A.M. Campos, A. Lobo, A. Vieira and M. Chmielewski are highly appreciated. The continued encougement and facilities from Alexander von Humboldt Foundation in Germany are gratefully acknowledged. Finally, thanks to Prof. Gupta for including this topic in this series.

Alexandria, May 2007   El Sayed H. El Ashry

Contents

Contents of Topics in Current Chemistry Volume 215

Glycoscience

Volume Editor: Stütz, A. E.
ISBN: 978-3-540-41383-7

Top Heterocycl Chem (2007) 7: 1–30
DOI 10.1007/7081_2006_047
© Springer-Verlag Berlin Heidelberg
Published online: 1 February 2007

Manipulation of Carbohydrate Carbon Atoms for the Synthesis of Heterocycles

E. S. H. El Ashry[1,2] (✉) · Y. El Kilany[3] · N. M. Nahas[3]

[1]The International Center for Chemical Sciences, HEJ Research Institute,
University of Karachi, 75270 Karachi, Pakistan
Eelashry60@hotmail.com

[2]Chemistry Department, Faculty of Science, Alexandria University, Alexandria, Egypt

[3]Chemistry Department, Girls Campus, Umm-Alqura University, Mecca, Saudi Arabia

Abstract This work provides an overview of the role of carbohydrates as precursors for the synthesis of heterocycles. It is limited to heterocycles where one or more of their carbon atoms are from carbohydrate reactants. A part of the sugar may remain linked to the heterocycles thus providing nucleoside analogues. The following heterocycles were included: furans, pyrrols, thiophenes, pyrazoles, imidazoles, oxazoline, dioxolanes, thiazolidines, triazoles, oxadiazoles, oxadiazolines, thiadiazoles, thiadiazolines, pyridines, pyridazines, pyrimidines, oxazines, pyrazines, quinoxalines, triazines, seven-membered rings and their fused ring systems.

Keywords Carbohydrates · Sugars · Heterocycles · Nucleosides

1
Introduction

The synthesis of heterocycles from carbohydrates dates back to the early work on carbohydrates. A series of publications on "Heterocycles from carbohydrate precursors" was initiated in 1976 [1] and a complementary one on "The scope of reactions of hydrazines and hydrazones" was also published. Carbohydrate hydrazones and osazones have recently been reviewed as organic raw materials for nucleosides and heterocycles [2]. The heterocycle skeleton may include a carbon back bone and heteroatoms or the carbon back bone only from the sugars. The remaining unutilized sugar moieties can provide acyclonucleosides [3–5]. Recently, a book on the synthesis of naturally occurring nitrogen heterocycles from carbohydrates was published by El Ashry and El Nemr [6]. Other relevant reviews published include N-heterocycles from saccharide derivatives by El Khadem [7], glycosylthio-heterocycles [8], and chiral enaminones and hydrazones derived from chiral pool starting materials such as aldose [9].

2
Value of Carbohydrates as Raw Materials

The use of carbohydrates as organic raw materials has been comprehensively reviewed by Lichtenthaler [10–13]. The major bulk of the renewable biomass is represented by carbohydrates in the form of polysaccharides [10–15]. In addition to their utilization in food products their other uses are confined mainly to the textile, paper and coating industries [12]. Monomeric sugars or their oligosaccharides—that can be derived from polysaccharides—are excellent precursors for the synthesis of organic compounds and are available in large quantities at a low price. Moreover, the produced organic compounds are eco-friendly chemicals for use in the pharmaceutical and agrochemical industries [13].

A method to produce sugars from a biomass by hydrolysis and then to subject the sugars to dehydration to form heterocyclic compounds, useful as liquid fuels, is described in a patent [16]. The formation of heterocycles from carbohydrates is an interesting subject in food research [17, 18]. Flavor contribution and formation of heterocyclic oxygen-containing key aroma compounds in thermally processed foods were reviewed [17, 18]. Furanone, pyranone, and other heterocyclic colored compounds from sugar-glycine model Maillard systems have also been analyzed [19].

3
The Synthesis of Heterocycles from Carbohydrates

Herein an overview of the most relevant approaches for the synthesis of unnatural heterocycles of biological and industrial potential from carbohydrates is presented; the natural heterocycles have been previously reviewed [6]. This review is limited to heterocycles with one or more of their carbon skeletons derived from carbohydrate precursors—those formed by cycloaddition reactions are not included. Also, carbohydrates with strained ring systems, oxiranes, aziridines, and thiiranes have already been reviewed and are not included herein [20]. The synthetic approaches for the reviewed heterocycles are divided according to the size of the heterocyclic rings and the number of hetero atoms in the ring. The bicyclic ring systems are included under the smaller ring of their skeleton.

3.1
Five-Membered Heterocycles with One Hetero Atom

There are three heterocyclic ring systems that can be included under this heading. These rings include oxygen, nitrogen, or sulfur hetero atoms while the most studied ones are those with oxygen.

3.1.1
Furans

Furfural **1** can be obtained from pentosans via their hydrolysis to D-xylose which undergoes dehydrative cyclization by the action of acid [11]. Similarly, acid-induced dehydrative cyclization of inulin hydrolyzates or D-fructose gave 5-hydroxymethyl furfural **2** [21]. The chemistry of **1** and **2** has been well developed because the aldehydic functionality can be readily transformed into other functional groups (Fig. 1). Moreover, the additional functionality in **2** has provided variable polyesters, polyamides, and resins [22–24].

The 5-(α-D-glucosyloxy-methyl)furfural (**4**) became readily available from isomaltulose (**3**) by dehydrative cyclization of its fructose part by the action of acid [25]. Its aldehydic group has been also transformed into a variety of functional groups of which compounds such as **5** and **6** are representative; they can be used as nonionic surface active agents and exhibit liquid crystalline properties. Such compounds are characterized by having a hydrophilic glucose part and a hydrophobic fat-alkyl moiety separated by a quasi-aromatic spacer [26–28]. The diastereoisomers of ranunculin were

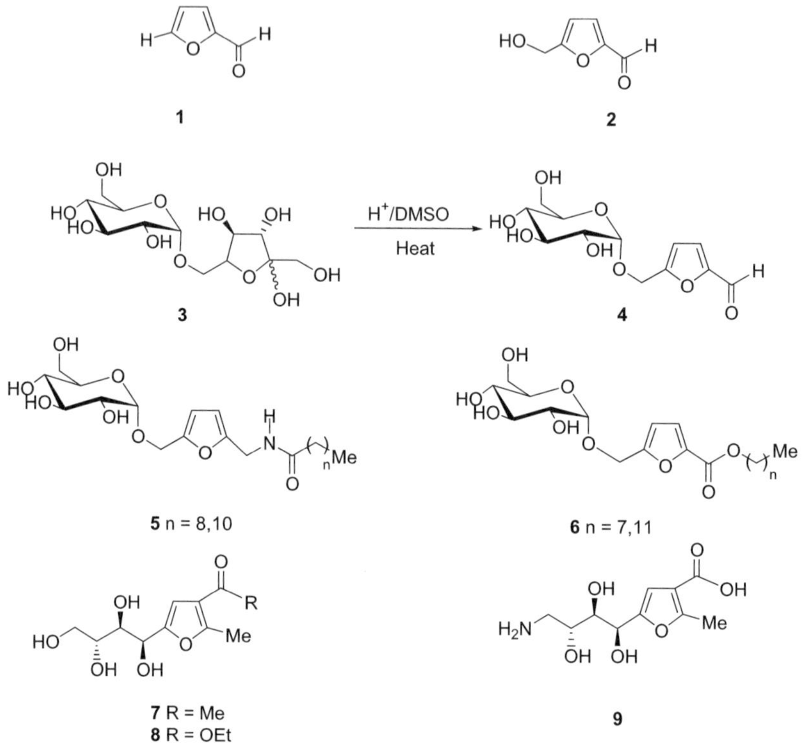

Fig. 1 Furan derivatives

synthesized by coupling of acetobromoglucose with (5S)-5-(hydroxymethyl)-2-(5H)-furanone, using a Koenigs–Knorr reaction in the presence of silver oxide, followed by acid hydrolysis, or by starting with acetylated (5-formyl 2-furyl)methyl-D-glucopyranosides and subsequent conversion [29, 30].

Reaction of D-glucose with 2,4-pentanedione or ethylacetoacetate in the presence of $ZnCl_2$ gave the respective D-arabino-tetrahydroxybutyl furans **7** and **8** [31]; two sugar carbon atoms were utilized in building the furan ring. The alditolyl residues have been oxidatively cleaved to aldehyde or carboxylic acid and can be chemically modified to **9** to prepare hetarylene-carbopeptoid libraries [32].

3.1.2
Pyrrols

On the other hand, five-membered heterocycles with one nitrogen in the ring are less readily available directly from carbohydrates. Thus, the pyrrole **11** has been formed from ammonium mucate of **10** under thermal conditions [33]. The respective pyrrole analogs of **7** and **8** with the tetrahydroxybutyl residues were obtained from D-glucosamine by reaction under mild basic conditions (Na_2CO_3/MeOH) with acetylacetone or ethylacetoacetate to give **12** and **13**, respectively [34, 35]. D-Fructose gave upon heating with acetylacetone in the presence of ammonium carbonate in DMSO, the pyrrole derivative **12** [35]. Such pyrrole derivatives can be degraded to simple functionalized pyrroles and cyclized to give C-nucleoside analog **14** (Fig. 2). The furan derivatives **2** and **4** can be converted to the respective pyrrole derivatives [36].

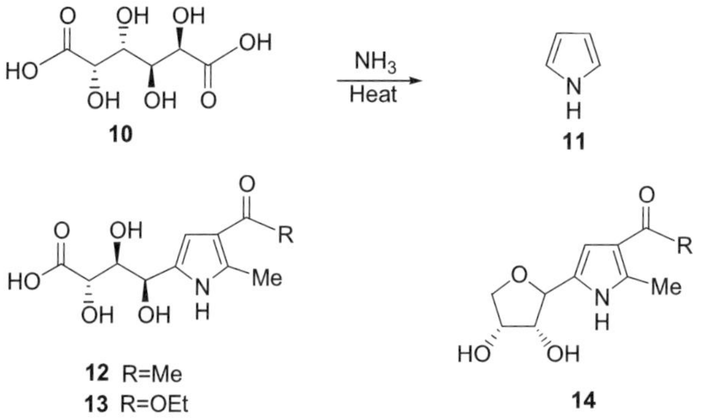

Fig. 2 Pyrrole derivatives

3.1.3
Thiophenes

Thiophene derivatives were also prepared from the respective furans **2** [37]. Thus, the 2,5-dihydroxymethyl thiophenes were obtained from **2** by reduction, protection of the hydroxyl groups, oxidation with *m*-chloroperbenzoic acid to the respective *cis*-hexenedione whose reduction with titanium trichloride and subsequent reaction with Lawesson reagent gave the protected thiophene derivative.

3.2
Five-Membered Ring Heterocycles with Two Hetero Atoms

Hetero atoms in the five-membered heterocycles can exist at 1,2 or 1,3 positions of the ring. The first of which is only represented by pyrazoles, whereas all possible variations may exist for the 1,3-hetero atoms.

3.2.1
Pyrazoles

El Khadem et al. [38–41] prepared the fist pyrazole derivative **16** by the action of acetic anhydride on D-glucose phenylosazone **15**, readily available from reaction of D-glucose with phenylhydrazine. The reaction was extended to other osazones of sugars and isomaltulose which gave the respective pyrazole **19** (Fig. 3). The cyclization was also extended to the mixed bishydrazones. Moreover, the isolation of intermediates from the reaction led to a proposal for a mechanism for the formation of the pyrazoles [42, 43]. Pyrazoles can be dehydrazonated, then deacetylated to give **17** which can also be degraded to give 3,5-difunctionalized pyrazole derivatives **18** [44, 45]. Recently, we applied the use of microwave irradiation for the synthesis of the pyrazole **16** and its derivatives (El Ashry et al., unpublished results) [46–48].

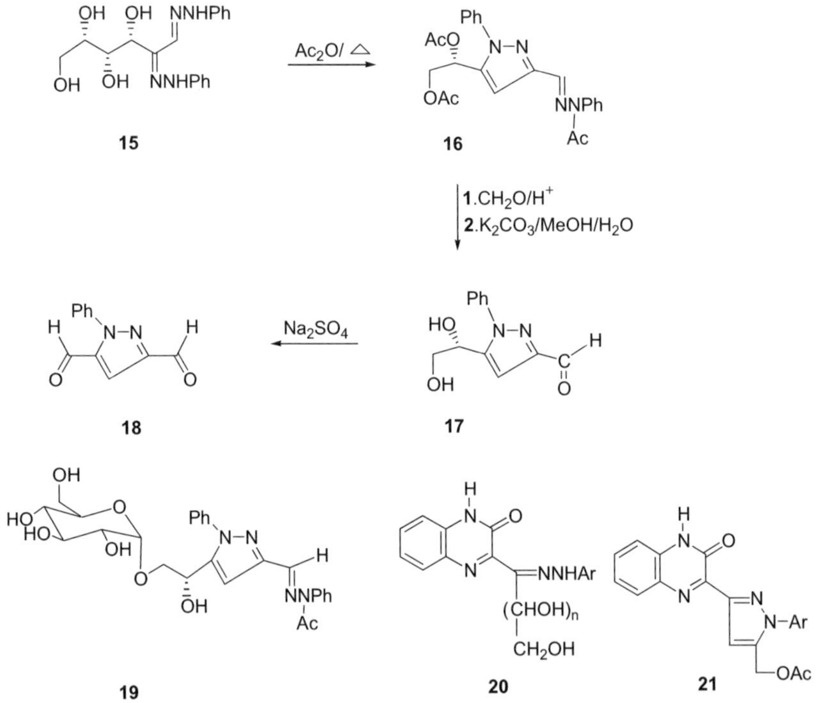

Fig. 3 Pyrazoles from osazones and hydrazones

Although the action of acetic anhydride on sugar arylhydrazones gave only the respective per-acetyl derivatives, the hydrazones **20**, derived from ascorbic acids, gave the pyrazole derivative **21** which can also be prepared under microwave irradiation (El Ashry et al., unpublished results) [49, 50].

The hydrazones **22** derived from partially protected D-ribose or D-glucose exist in equilibrium with the corresponding cyclic structures (Fig. 4). Subsequent cyclization of the respective hydrazine residue with 1,3-dicarbonyl compounds or unsaturated nitriles gave the pyrazoles **23** and **24** [51–55]. Tosylation of the isopropylidene derivative **23** gave the pyrazolium tosylate **26** via the intramolecular cyclization of **25** [54].

Dehydro-ascorbic acid osazones can be rearranged by ring opening of the lactone which upon cyclization gave the respective pyrazolones which can be converted into different functionalized derivatives [49, 50, 56–60].

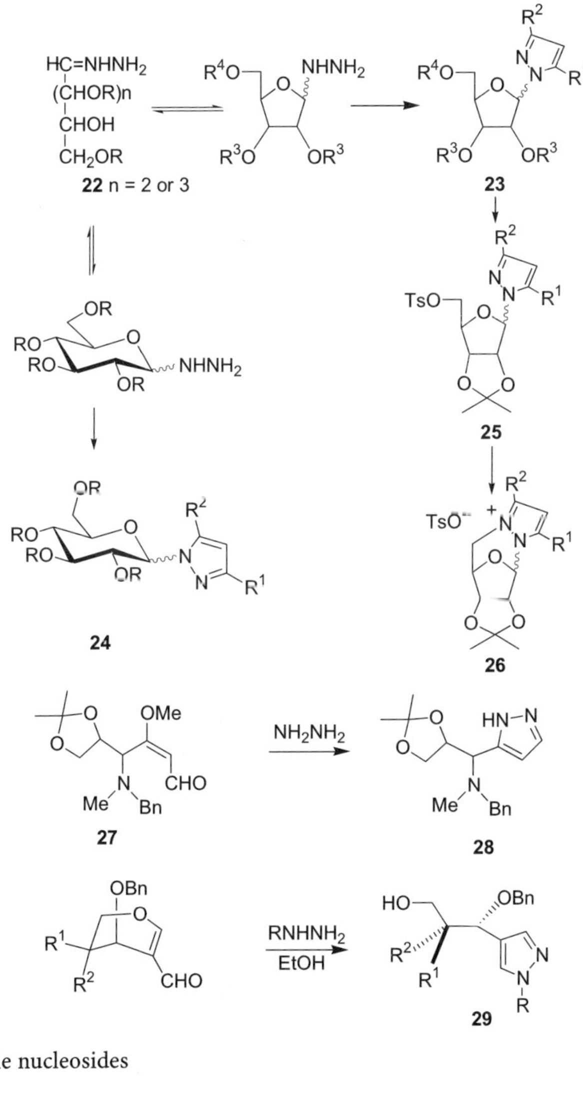

Fig. 4 Pyrazole nucleosides

Degradation of D-glactose phenylhydrazone with alkali gave 1-phenyl-pyrazole [61]. Periodate oxidation of D-glucose phenylosazone gave mesoxaldehyde bis(phenylhydrazone) which can be transformed into pyrazole derivatives [62, 63]. Microwave irradiation effected also a similar transformation (El Ashry et al., unpublished results).

Condensation of syn- or anti-**27** with hydrazine afforded new pyrazole derivatives **28** with a stereodefined and protected amino diol side chain [64]. The preparation of push-pull substituted unsaturated monosaccharide derivatives and their use in the synthesis of nucleoside analogs have been reviewed [65]. Thus, the 2-formyl pentose glycals were transformed to the corresponding acyclo-C-nucleosides **29** [66]. Similarly, the benzylated 2-formylglycals reacted with hydrazine derivatives to afford the substituted 1,2,4-tri-O-benzyl-1C-(1H-pyrazol-4-yl)-D-tetritols; the deprotection of which was achieved with Pd/H$_2$ to yield the 1C-(1-methyl-1H-pyrazol-4-yl)-D-tetritols [67]. 3-O-Benzyl-6-deoxy-1,2-O-isopropylidene-α-D-xylo-hept-5-ulofuranurono-nitrile was reacted with *N,N*-dimethylformamide dimethylacetal in THF to furnish the (E)-3-O-benzyl-6-deoxy-6-dimethyl-aminomethylene-1,2-O-isopropylidene-α-D-xylo-hept-5-ulofuranurono-nitrile as a major product, and on treatment with carbon disulfide and methyl iodide under basic conditions afforded 3-O-benzyl-6-deoxy-1,2-O-isopropylidene-6-[bis(methylsulfanyl)methylene]-α-D-xylo-hept-5-ulofuranurono-nitrile. Further reaction with hydrazines yielded the reversed pyrazole-C-nucleoside analogs [68].

3.2.2
Imidazoles

A convenient access to the synthesis of 4-hydroxymethyl imidazole from D-fructose has been performed upon heating with formaldehyde and concentrated ammonia in the presence of CuCO$_3$/Cu(OH)$_2$ [69, 70]. This could be attributed to a retro-aldol fission of D-fructose to glyceraldehyde and dihydroxyacetone that could have taken place under such conditions. Consequently, the expected product **31** which could result from D-fructose was obtained in a poor yield [71]. The retro-aldol fission and the draw back from forming copper salts of the products, which required decomposition with hydrogen sulfide, can be prohibited to a certain extent by heating D-fructose with formamidinium acetate in liquid ammonia in a pressure vessel to give **31** [72–74]; the same procedure can be applied for preparing hydroxymethyl imidazole from dihydroxy acetone. Moreover, heating D-fructose with formamidinium acetate in the presence of boric acid and hydrazines as dehydrating agents gave **31** [11]. The reaction can be applied to D-xylose and isomaltulose to give **30** (R = H) [11]. Similarly, the 2-methyl-imidazole **30** (R = Me) was obtained upon using ethylacetimidate (Fig. 5). The use of 3-O-benzyl-D-glucose led to the respective benzyl derivative whose reaction

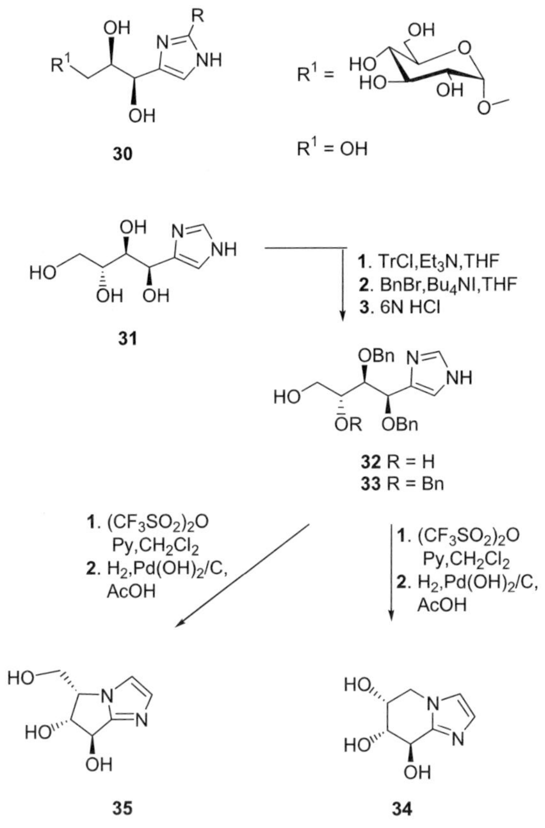

Fig. 5 Imidazole nucleosides and their annulation

with CBr_4 in the presence of Ph_3P gave, after debenzylation, sugar-annulated imidazoles [72–74]. The stereoisomeric imidazolo-pentoses were also synthesized [72–74].

The imidazole derivative **31** was transformed to bicyclic analogs by selective tritylation at the nitrogen and the primary hydroxyl groups followed by benzylation and then detritylation to give **33** [71]. When the benzylation was carried out at lower temperatures followed by detritylation, the dibenzyl derivative **32** was obtained. Both **32** and **33** were readily cyclized by the action of triflic anhydride in pyridine and chloroform to give **34** and **35**, respectively as imidazole analogues of 6-epicastanospermine and 3,7a-diepialexine.

Another synthetic approach for the annulated imidazoles was done by starting with L-arabinose that converted to the aldehyde **36** whose condensation with different glyoxals gave the linear imidazole sugars, which was followed by removal of the trityl group to give **37** (Fig. 6). An intramolecular nucleophilic substitution took place upon mesylation of **37** followed by heating to give the respective protected derivative of **38** whose catalytic

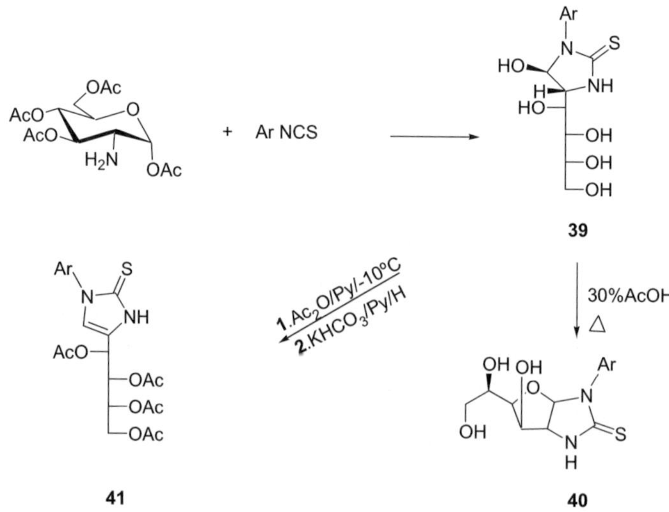

Fig. 6 Imidazolo-pyridine

debenzylation gave **38**. Various bicyclic analogues were also achieved by the chemical modification of the imidazole ring [76].

Chiral imidazolidin-2-thione or 2-one were prepared from D-glucoseamine by reaction with o,o′-disubstituted arylisothiocyanates to give **39** which can be cyclized to **40** [76]. Thus, a variety of derivatives were prepared carrying different aryl groups and as a consequence of hindered rotation, some of them exist as stable rotamers (Fig. 7). Treatment of **39** with acetic anhydride in pyridine at low temperature gave the respective per-O-acetyl derivative imidazolidine thione, which upon elimination of acetic acid gave the imidazoline thion **41** as a mixture of atropisomers [76].

Fig. 7 Imidazole nucleosides

The imidazole ring can be introduced by the addition of 2-lithio1-[(dimethylamino)methyl]-1H-imidazole to 2,3,4,6-O-benzyl-glucono-1,4-lactone to give **42** [77]; the reduction of which gave **43** and **44**. Mesylation of the latter followed by NaH/DMF gave **45** in 38% yield. Regioselective aroylation of **43** and **44** followed by NaH/DMF gave a mixture of anomers **46** and **47** (80%) presumably via an elimination-addition mechanism (Fig. 8).

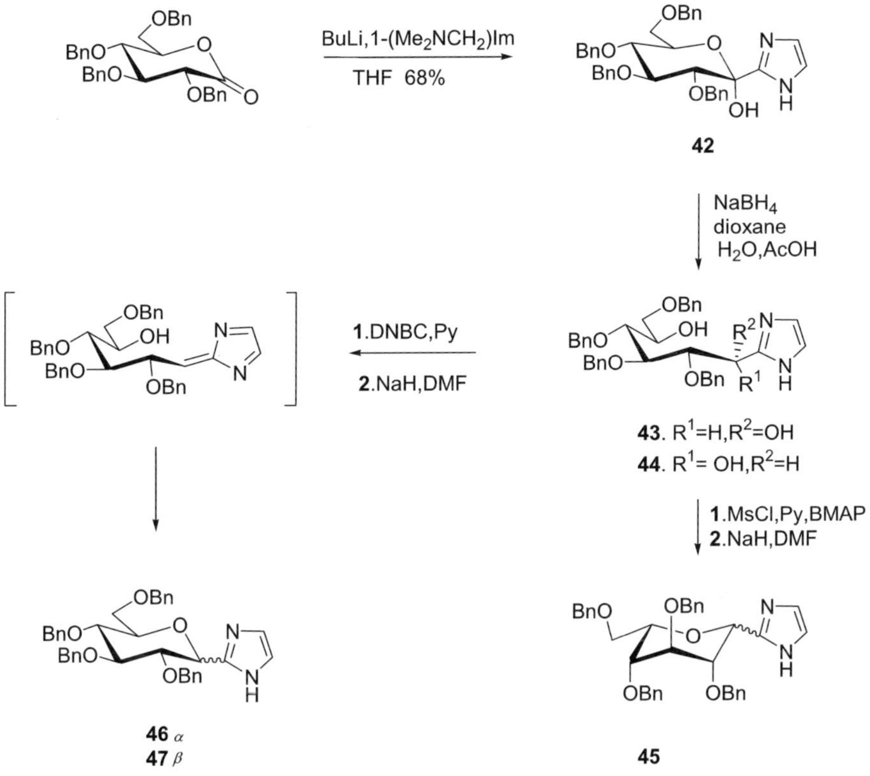

DMAP=4-Dimethylaminopyridine

3,5-DNBC=3,5-Dinitrobenzoyl chloride

Fig. 8 Imidazole nucleosides from lactones

The *N*-allyloxy-carbonyl derivative of 2-methyl-(1,2-dideoxy-α-D-glucopyrano)[2,1-d]-1-imidazolines, inhibitors of β-*N*-acetylglucosaminidases, have been formed as a byproduct during a study on tetra-O-acetyl-2-allyloxy-2-carbonylamino-β-D-glucose synthesized as a glycosyl donor and when no available acceptor was present in the reaction media of acetonitril in the presence of trimethysilyl triflate it formed almost quantitatively [78, 79].

3.2.3
Oxazolines, Thiazolines, and Dioxolanes

The oxazoline **49** was formed from the glycosyl halide **48** by abstraction of the anomeric leaving group to give the respective oxazolinium ion and then loosing a proton (Fig. 9). They are important intermediates for the synthesis of oligosaccharides and this led to variant substitution on the ring [80, 81]. Glucooxazolines have been prepared in a two-step process involving iodoamidation of protected D-glucals and subsequent basic cyclization [80, 81]. The thiazoline analogues of **50** were also prepared by reaction of 2-acetamido-2-deoxy-1,3,4,6-tetra-O-acetyl-β-D-glucopyranose with Lawesson's reagent; here selective thionation of the amide was followed by cyclization through displacement of anomeric acetate to give the thiazoline triacetate **50** [83]. On the other hand, acylated bromosugars **51** can be readily transformed in the presence of alcohol and base to the cyclic orthoesters **52**, precursors for glycoside-bond formation [84]. Reaction of aldehydo sugar acetates with thioglycollic acid gave the respective oxathiolanones [85]. Synthesis of pseudo-C-nucleoside including the thiazole and thiadiazolidinone rings was accomplished starting from 3-O-benzyl-1,2-O-isopropylidene-D-ribo-pentodialdehydo-1,4-furanose [86].

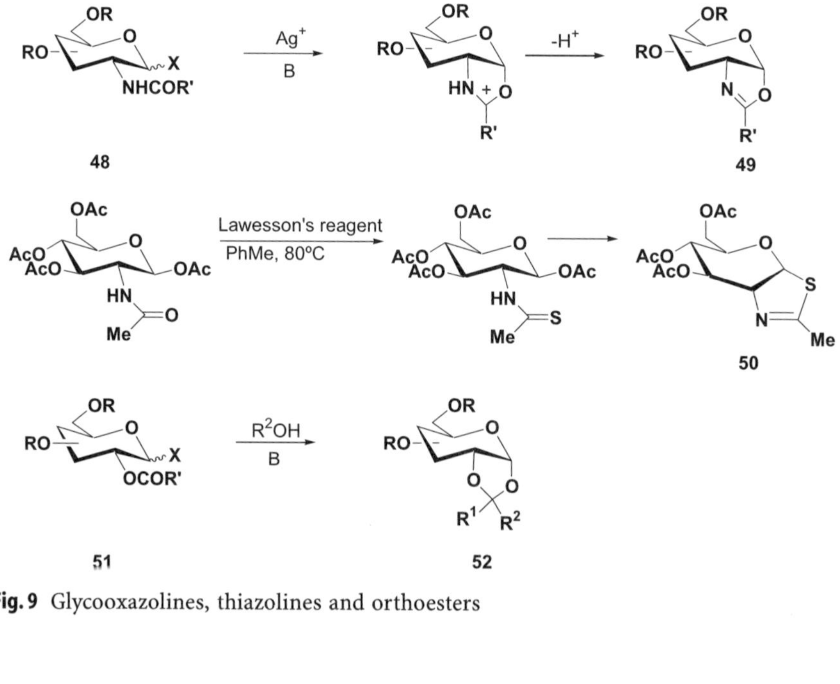

Fig. 9 Glycooxazolines, thiazolines and orthoesters

3.2.4
Thiazolidines

Reaction of L-cysteine with monosaccharides gave the corresponding 2-(poly-hydroxyalkyl)-4-thiazolidinecarboxylic acids **53**, from which the peracetyl derivatives and the peracetyl methyl esters were prepared. The 2-(D-gluco-pentahydroxypentyl)-benzothiazoline, the 5-chloro derivative, and the D-galacto analogue were also prepared by heating 2-amino- or 2-amino-4-chlorothiophenol and sugar in pyridine [86]. The condensation of D-penicillamine and penta-O-acetyl-aldehydo-D-galactose gave the two isomeric products of **54** [87–89].

On the other hand, reaction of D-glucorono-3,6-lactone **55** with L-cysteine ester **56** gave stereoselectively the bicyclic thiazolidine lactam **57**, a precursor for polyol peptidomimetics [90, 91]. The thiazolidine **58** was found to be

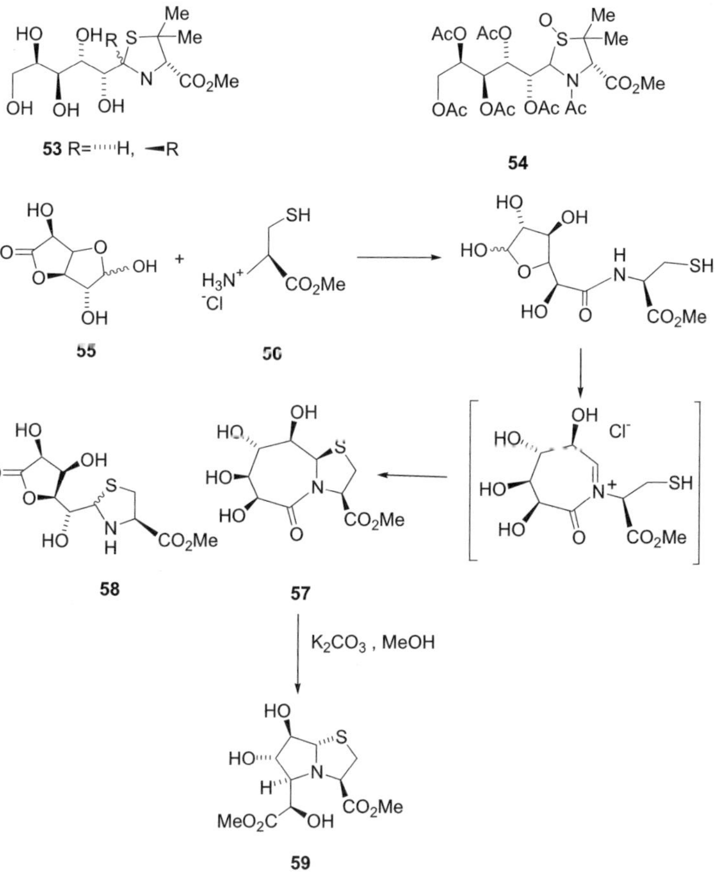

Fig. 10 Thiazolidines

formed quantitatively in pyridine, which is in equilibrium with the starting materials, but in a solvent mixture of water/pyridine (9 : 1) the bicyclic scaffold **57** was formed slowly (Fig. 10). Unexpected ring contraction of 7,5-fused bicyclic thiazolidinelactams yields hexahydropyrrolo [2,1-b]thiazoles in one step, whereby their stereocenters are inverted. A mechanism for the reaction was proposed [90, 91].

The chiroptical properties of optically active thiazolidines derived from aldoses and natural mercapto aminoacids was studied [92]. PMR parameters for thiazolidine-4(R)-carboxylate derivatives were obtained by computer-assisted analysis of their spectra. The polyacetoxy-alkyl side chains have planar zig-zag conformations. The configurations at C-2 in the diastereoisomers were ascertained on the basis of the $J_{NH,CH}$ coupling constants [93]. The conformation and stereochemistry of diastereomeric sulfoxides of methyl 3-acetyl-5,5-dimethyl-2-(D-galactopentaacetoxypentyl)-1,3-thiazolidine-4-carboxylate 1-oxides were performed by ^{1}H- and ^{13}C-NMR spectral analysis [94].

3.3
Five-Membered Heterocycles with Three Hetero Atoms

There are three main classes of heterocycles that can be classified under this heading. These are heterocycles with three nitrogen atoms and with two nitrogen atoms in addition to one oxygen or sulfur atom.

3.3.1
Triazoles

The triazoles exist either as 1,2,3- or 1,2,4-triazoles. The different 1,2,3-triazoles from carbohydrates such as **60** and **61** have been reviewed [2]. Thus, cyclization of sugar arylosazones by a boiling solution of copper sulfate gave the corresponding arylosotriazoles; the cyclization can also be affected under a variety of conditions [95]. Sugar bis(benzoyl hydrazones) gave **61** upon oxidation with iodine and mercuric oxide [96]. Degradation of the sugar chain opened the path to different types of heterocycles.

The 1,2,4-triazole nucleosides such as **62** were prepared from the 1-hydrazino-glycopyranosyls or furanosyls using different cyclizing agents [51, 97, 98].

Reaction of sugars with benzamidrazone iodide followed by cyclization with acetic anhydride gave cyclic or acyclic nucleoside analogues of 5-methyl-3-phenyl-1,3,4-triazoles depending on the sugar [99]. Thus, D-glucose, D-galactose, and D-arabinose gave **63**, **64**, and **65**, respectively (Fig. 11).

Dehydrative cyclization of glyconolactones with thiocarbohydrazide gave 4-amino-5-mercapto-1,2,4-triazol-3-yl acyclo C-nucleosides **66** [100–104]. Similarly, the double-headed acyclonucleosides **68** were prepared from galac-

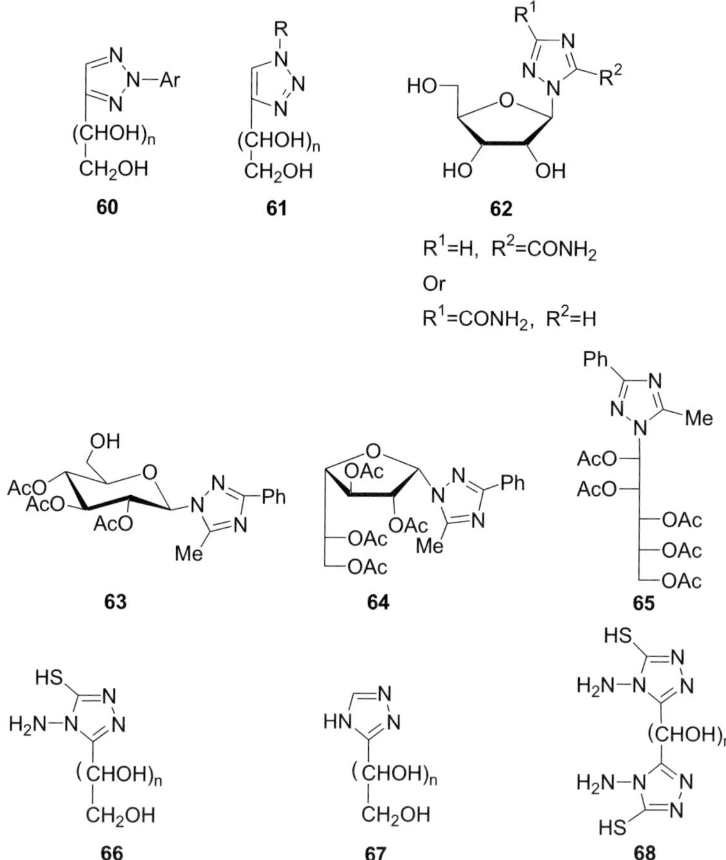

Fig. 11 Triazole analogues

taric acid and tartaric acid [105, 106]. Nitrous acid transformed **66** to **67**. Microwave irradiation enhanced these reactions to give the same products in a higher yield with much shorter reaction times [107].

Acetylation of N^1-(aldopyranosylamino)guanidines with D-gluco, D-galacto, and L-arabino configurations gave acylaminoguanidines that cyclized under mild conditions, boiling in ethanol or treatment with cold 0.1 M sodium methylate solution, to afford 3-amino-N1-glycopyranosyl-5-methyl-1H-1,2,4-triazoles [108].

Reaction of heterocycles with a hydrazine residue at the α-position with sugars gave the respective hydrazones whose oxidative heterocyclization gave the 1,2,4-triazolo ring fused to the heterocycle. A variety of fused heterocycles such as pyridines, quinolines, triazines, and fused ring systems were prepared as acyclic nucleosides [2–5, 109–111].

3.3.2
Oxadiazoles and Oxadiazolines

Oxadiazole acyclonucleosides **69** were prepared from the respective sugar acylhydrazones by oxidation with iodine and mercuric oxide [112–114]. On the other hand the oxadiazoline analogues **71** were obtained by the action of acetic anhydride on the acyl hydrazones [115–119]. Thus, cyclization of 2,3,4,5,6-penta-O-acetyl-aldehydo-D-galactose aroylhydrazones with boiling acetic anhydride yielded 3-acetyl-5-aryl-2-(D-galacto-1,2,3,4,5-pentaacetoxy-pentyl)-1,3,4-oxaidazoline. The spiro oxadiazoline **73** was similarly prepared from the benzoylhydrazone of the 3-oxo derivative of diacetone glucose [120].

3.3.3
Thiadiazoles and Thiadiazolines

The thiadiazoles **70** were prepared from sugar thiobenzoylhydrazones and thiosemicarbazones by the action of ferric chloride (Fig. 12). Acetylation of **70** afforded the 2-(*N*-acetylarylamino)-5-polyacetoxyalkyl-1,3,4-thiadiazoles and periodate oxidation gave the thiadiazolecarbaldehydes [121–123]. The respective thiadiazoline **72** was prepared by the action of acetic anhydride on the corresponding sugar thiosemicarbazone [123, 124]. Dehydrative cyclization of bis-aryl thiosemicarbazide derivatives of galactaric acid gave thiadiazole or triazole derivatives by the action of phosphorousoxy chloride and sodium ethoxide, respectively. The thiadiazole derivatives were also obtained by the reaction of D-glucono-1,5-lactone with S-methylhydrazine carbodithioate and cyclization with acetic anhydride [125–128]. The thiadiazolotriazoles were prepared by the condensation of 4-amino-3-aryl-1,2,4-triazole-5-thiols with D-gluconic and galactaric acids [125–128]. The syntheses of 2,3-dihydro-1,3,4-thiadiazoles and 1,3,4-thiadiazoles from 1,2-O-isopropylidene-α-D-xylo-1,5-pentadialdo-1,4-furanose thiosemicarbazone and 1,2:3,4-di-O-isopropylidene-α-D-galacto-1,6-hexodialdo-1,5-pyranose thiosemicarbazone were reported [129, 130]. The diastereomerically pure (3′R) 2-acetamido-4-*N*-acetyl-3′-spiro-[1′,2′:5′,6′-di-O-isopropylidene-α-D-glucofuranosyl]-1,3,4-thiadiazoline was also synthesized [131]. Cyclocondensation of 2-phenylthio-

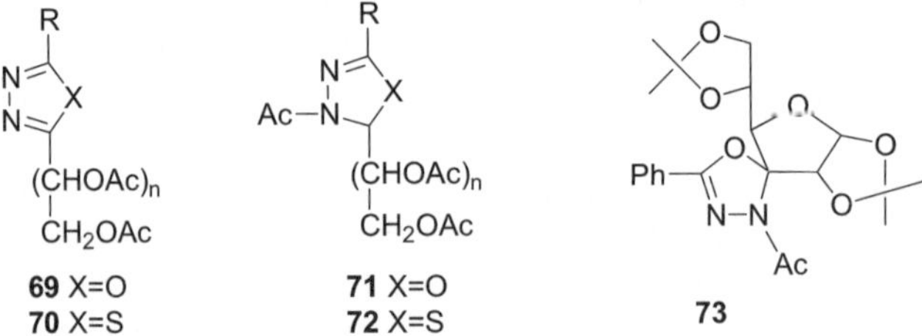

Fig. 12 Oxa- and thia-diazoles and diazolines

benzhydrazide with monosaccharides gave 2-polyhydroxyalkyl-1,3,4-thiadi-azoline as shown by spectroscopic evidence and by single X-ray analysis [132].

3.4
Five-Membered Heterocycles with Four Hetero Atoms

The number of five-membered heterocycles with four hetero atoms is limited to those with nitrogen atoms. Thus, the tetrazole acyclonucleosides were formed from sugars N,N'-diphenylformazanes by the action of N-bromosuccinimide or lead tetraacetate [133, 134]. The synthesis of the tetra-O-benzoyl-D-lyxononitrile and its conversion to (D-lyxo-tetritol-1-yl)-tetrazole derivatives upon reaction with sodium azide was reported [133, 134]. A pseudo C-nucleoside including a tetrazole ring was achieved by conversion of 3-O-benzyl-1,2-O-isopropylidene-D-ribo-pentodialdehydo-1,4-furanose to the respective nitrile and then reaction with sodium azide [86].

3.5
Six-Membered Heterocycles with One Heteroatom

Heterocycles with one oxygen or nitrogen atom are the only ones that may be included under this heading. Six-membered oxygen heterocycles such as glycals, 2-hydroxyglycal esters, and kojic acid are well-known classes of compounds and readily obtained from sugar derivatives [11]. They were reviewed elsewhere and consequently no more details will be given here.

3.5.1
Pyridines

The 3-pyridinols **74** can be prepared from furfural by reductive amination to furfurylamine and oxidation with hydrogen peroxide [135]. The respective 6-hydroxymethyl derivative **75** was prepared from 4-hydroxymethylfurfural by reductive amination followed by exposure to bromine in water/methanol to give a 2,5-dibromo derivative whose hydrolysis, elimination of water, and cyclization gave 6-hydroxymethyl-3-pyridinol [136]. Similarly, glucosylated furfuryl amine derived from isomaltulose was converted to **76** (Fig. 13).

Reaction of levoglucosenone with amides of α-nitrocarboxylic acids resulted in the formation of the tetrahydropyridones **77** and with acetoacetic acid amides, mixtures of the keto and enol tautomers were obtained [137].

Reaction of the aldehyde D-aldurono-2,5-lactone with L-cysteine methyl ester gave the bicyclic thiazolidino-pyridone **78** as a single diastereoisomer with the new bridgehead stereocenter in the S-configuration. Regioselective activation of the α-hydroxy group to the triflate whose subsequent azide displacement, reduction, and Boc protection gave the epimeric dipeptide mimetics. Conversion of this to the tripeptide **80** gave in addition a product

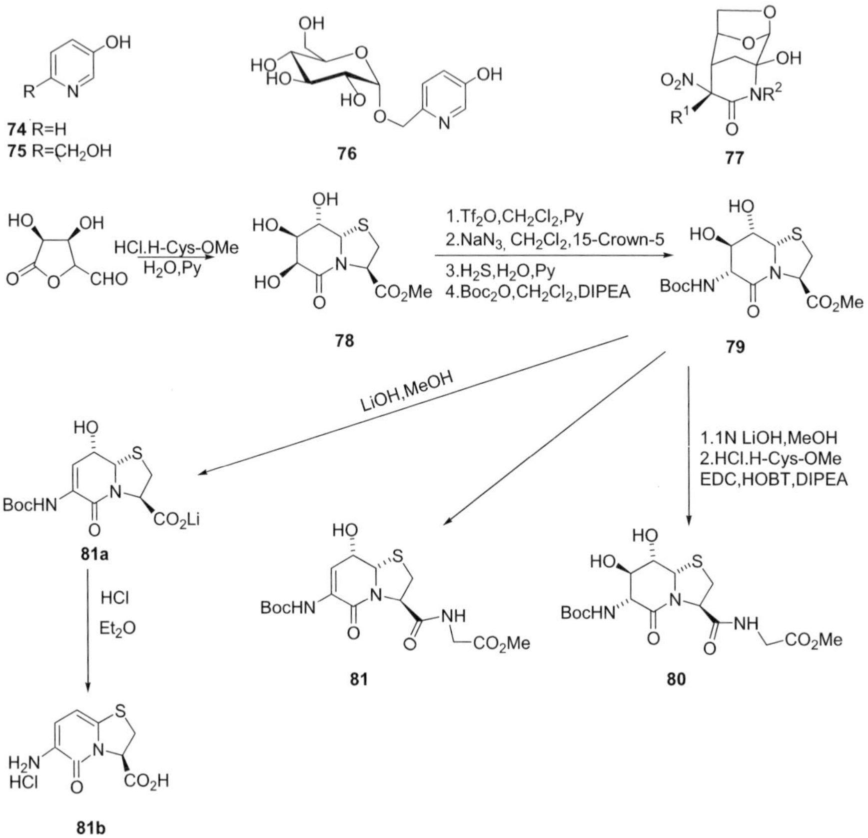

Fig. 13 Pyridine analogues

81 resulting from the elimination of one molecule of water, which can undergo a loss of another molecule of water to give a diene analogue [138, 139]. Hydrolysis of the Boc in **79** gave **81a** which underwent elimination of another molecule of water to give **81b**.

The reaction of 2-formylglycals with malononitrile afforded push-pull butadienes with a sugar moiety. The treatment of these branched-chain sugars with ammonia yielded nicotinonitrile acyclo-C-nucleosides. Furthermore, a one-step ring transformation of 2(3)-formylglycals with *N*-aryl-acetoacetanilides gave pyridone acyclo-C-nucleosides [140–142].

Treatment of 2-formylglycals **82**, in presence of piperidine and acetic acid, with cyanoacetamide provided 2(3)-[2-(aminocarbonyl)-2-cyanovinyl]-1,5(2,6)-anhydro-2(3)-deoxyhex-1-enitols, which are converted into polyhydroxyalkylated 2-aminonicotinamides and pyridinecarbonitriles, **83** Fig. 14 [140–142]. Similarly, reaction of **82** with cyanomethyl imidazole gave the imidazopyridine **84** [140–142].

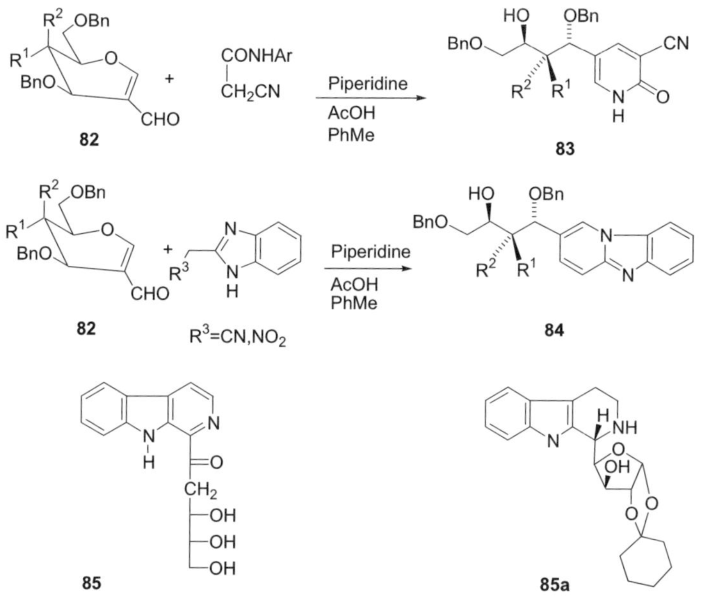

Fig. 14 Pyridines from 3-formyl glycals

The B-carboline derivative **85** was isolated as a natural product from a hybrid plant cell culture. It was synthesized from tryptamine and (E)-4,5,6-tri-O-acetyl-2,3-dideoxy-D-erythro-hex-2-enose. Condensation of tryptamine with 1,2-O-cyclohexylidene D-xylo-pentodialdose gave tetra-hydro analogue **85a** [143–145].

3.6
Six-Membered Heterocycles with Two Hetero Atoms

The presence of two hetero atoms in a six-membered ring increased the possibility of the variant heterocycles which are presented in this section. Location of the hetero atoms can be at the 1,2-, 1,3-, or 1,4-positions. Moreover they can be similar or different.

3.6.1
Pyridazines

Oxidative ring opening of the acetyl or benzyl derivatives of the furan **86** by metachloroperbenzoic acid gave the respective hexene dione **87**; *cis-trans* isomerization did not take place because of the acidic conditions (Fig. 15). Reaction of **87** with hydrazine furnished the pyridazine **88** [36]. The glucosyl derivative **90** was prepared from the respective furan **86** by the action of bromine water to give **89** which cyclized upon reaction with hydrazine to

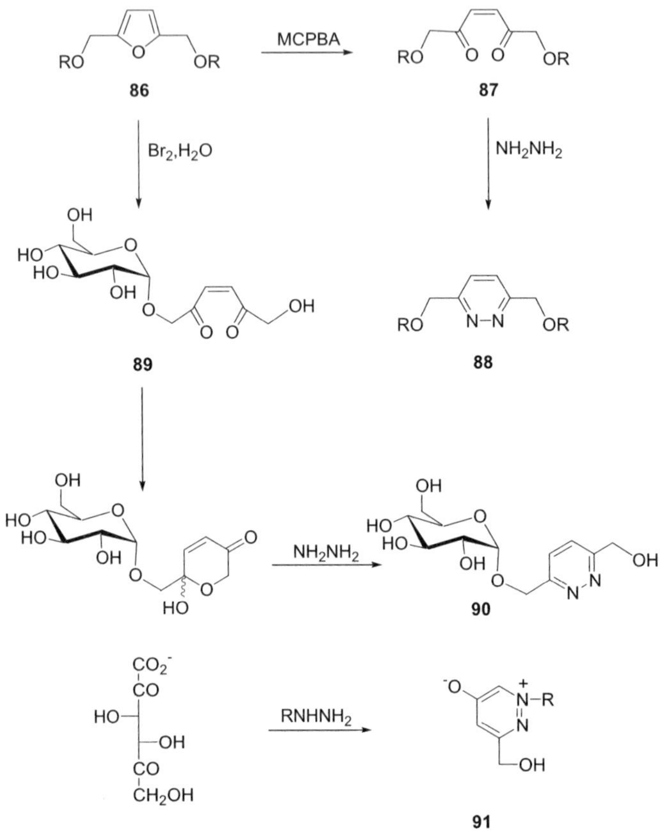

Fig. 15 Pyridazines from furans

give **90** [36]. Calcium 2,5-diketogluconate, prepared by microbial oxidation of D-glucose, was reacted with substituted hydrazines to give the N-substituted 5-oxidopyridazinium derivatives **91** [146]. An access to enantiomerically pure pyridazines and oxazines from epoxy pyranosides via keto ester functionalization was also reported [146–148].

3.6.2
Pyrimidines and their Fused Ring Systems

Reactions of 2-formyl-galactal, presented as an unsaturated sugar derivative with push-pull functionalization, with guanidinium and amidinium salts, respectively were carried out under basic conditions to furnish the substituted 5-(1,2,4-tri-O-benzyl-D-lyxo-1,2,3,4-tetrahydroxy-butyl)pyrimidines **92** (Fig. 16). Treatment of the 2-formyl pentose glycals with 2-aminobenzimidazole and 3-amino-1,2,4-triazole, respectively afforded 3-(1,2,4-tri-O-benzyl-D-lyxo-1,2,3,4-tetrahydroxy-butyl)benzo[4,5]imidazo[1,2-a]pyrimidine **94**

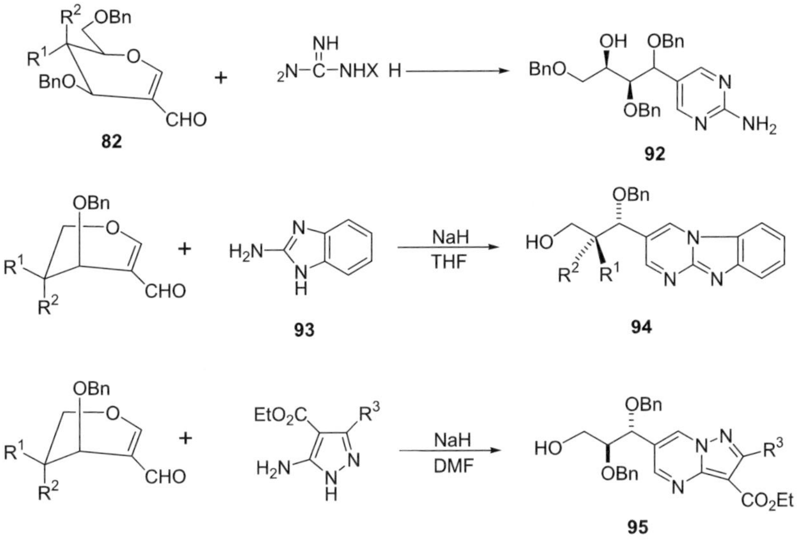

Fig. 16 Pyrimidines from 2-formyl-glycals

and 6-(1,2,4-tri-O-benzyl-D-lyxo-1,2,3,4-tetrahydroxy-butyl)-[1,2,4]triazolo-[1,5-a]pyrimidine [149].

1-Deoxy-3,4:5,6-di-O-isopropylidene-D-arabino-hex-2-ulose was reacted with N,N-dimethylformamide di-methyl acetal and tert-butoxy[bis(dimethyl-amino)]methane, respectively to furnish (E)-1,2-dideoxy-1-dimethylamino-4,5:6,7-di-O-isopropylidene-D-arabino-hept-1-en-3-ulose. This push-pull activated heptenulose underwent ring closure reactions with various N,N'-nucleophiles like hydrazine hydrate, amidinium, guanidinium, and isothiouronium salts in the presence of bases to yield pyrazole and pyrimidine derivatives, all of which derivatized with a di-O-isopropylidenated tetritol side chain. The isopropylidene groups were removed with aqueous TFA in the case of pyrimidine analogs [150]. Reaction of 2-formy glycals with 2-aminoimidazole 93 and 2-aminopyrazole gave 94 and 95, respectively [66].

The protected 2-formyl-L-arabinal reacted with thiourea and cyanamide in the presence of sodium hydride to afford, via ring transformations, the 5-[1R,2S-1,2-bis(benzyloxy)-3-hydroxypropyl]-1,2-dihydropyrimidines. Similarly, treatment with 3-amino-2H-1,2,4-triazole yielded 6-[1R,2S-1,2-bis(benzyloxy)-3-hydroxypropyl][1,2,4]-triazolo[1,5-a]pyrimidine [151]. Reaction of 27 with 2-aminoimidazole and 2-aminobenzimidazole gave 96 and its bezo-analogues [64].

The synthesis of the mesoionic thiazolopyrimidine acyclonucleosides 97 incorporating the 2,3-dihydroxypropyl moiety was carried out starting by reaction of 2-bromothiazole with excess 1-amino-2,3-propanediol acetonide via an aromatic nucleophilic substitution reaction to yield 1-(2-thiazolylamino)-

2,3-propanediol acetonide, which was condensed with substituted bis(2,4,6-trichlorophenyl) malonic esters to form a series of protected acyclonucleosides **98**. The deprotection of the isopropylidene group of **98** was done by using p-toluenesulfonic acid catalyst in methanol to give **97** [152].

Reaction of levoglucosenone with urea, thiourea, or *N*-cyano- or *N*-nitroguanidine resulted in the formation of the pyrimidine systems **99** in a stereospecific manner [153, 154]. Its reaction with α-aminoazoles yielded azolo[1,5-a]pyrimidine systems fused with a carbohydrate fragment. The reaction occurs much more smoothly than in the case of other α,β-unsaturated ketones. The reactions of levoglucosenone with β-dicarbonyl compounds (dimedone, barbituric acid) in the presence of a base resulted in pyran ring closure [153, 154].

The major species present in the aqueous solution of hexoses with 1,3-diaminopropane was acyclic hexahydro-pyrimidines [155].

3.6.3
Oxazines

Dehydrative cyclization of the condensation product of 2,3,4,5-tetra-O-acetylgalactaroyl chloride with anthranilic acid gave 1,2,3,4-tetra-O-acetyl-1,4-bis(4H-benzoxazin-4-one-2-yl)-galacto-tetritol **100** (X = O) (Fig. 17). Its

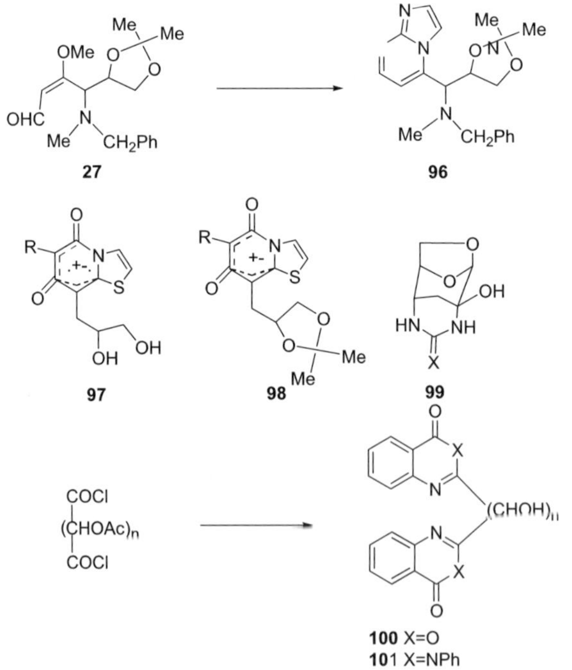

Fig. 17 Pyrimidines and oxazines

reaction with aniline in the presence of PCl_3 afforded 1,4-bis(3-phenylquin-azolin-4-one-2-yl)-1,2,3,4-tetra-O-acetyl-galacto-tetritol **101** (X = NPh) [156].

3.6.4
Pyrazines and Quinoxalines

Synthesis of the tricyclic pyrano[2,3-b]quinoxalines (Fig. 18) **102**, with a hydroxymethyl function at the 2-position was described by reaction of o-phenylenediamine with phenylhydrazone derivatives of sugars [157].

The reaction of kojic acid with one and two equivalents of 1,2-diamino-4,5-dimethylbenzene afforded 2-hydroxy-5,6-dihydro-8,9-dimethylpyrido[1,2-a]quinoxaline **103** and 2,3,10,11-tetramethyl-5,6-dihydro-13H-quinoxalino[2′,

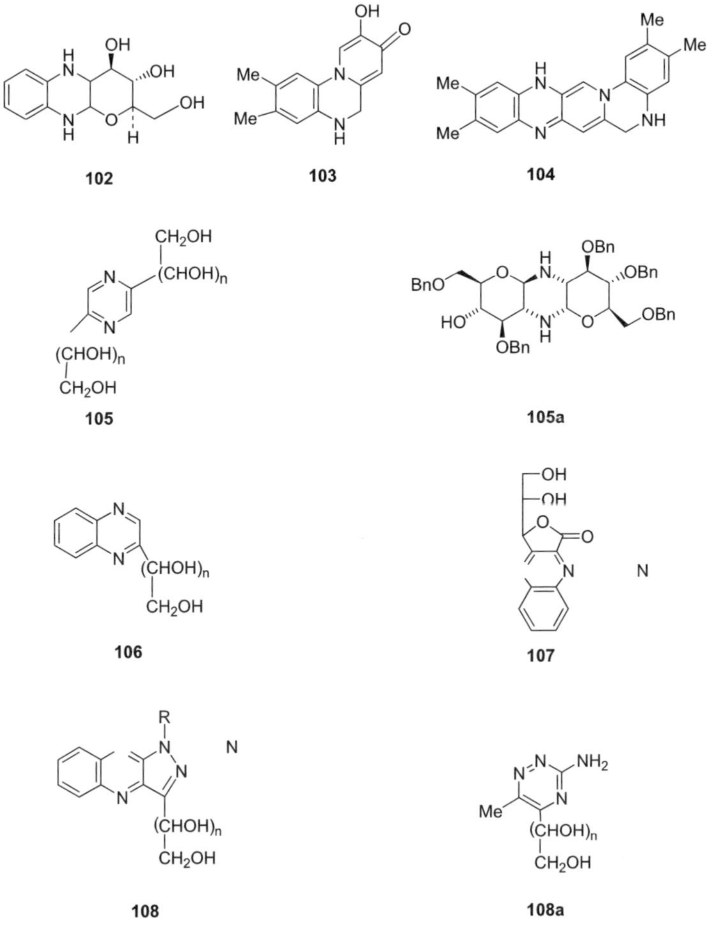

Fig. 18 Pyrazines, quinoxalines and triazines

3′:4,5]pyrido[1,2-a]quinoxaline **104**, respectively [158]. The pyrazine **105** was obtained from the condensation of two molecules of D-glucoseamine [159–161]. Dimerization of 2-amino-3,4,6-tri-O-benzyl-2-deoxy-D-glucose gave fructosazine and bis-tetrahydro-pyrano-piperazine **105a** derivatives in a ratio 1 : 5, respectively [159–161]. The quinoxalines **106** can be prepared from the reaction of o-phenylene diamine with osuloses [2, 7]. Different quinoxaline derivatives such as **107** were prepared from L-ascorbic acid and its analogues with o-phenylene diamines [49, 50]. In the presence of phenylhydrazine the product was a quinoxalinone derivative **20** whose cyclization gave the pyrazoloquinoxaline **108** [49, 50].

3.6.5
Triazines

The triazine derivatives **108a** were prepared by condensing aminoguanidine with the respective dicarbonyl derivatives [162].

3.7
Seven-Membered Heterocycles

The hydrophilic diazepinones were prepared from α-glucopyranosyl-hydroxymethyfurfural **35**, oxidation of which with m-chloroperbenzoic acid or singlet

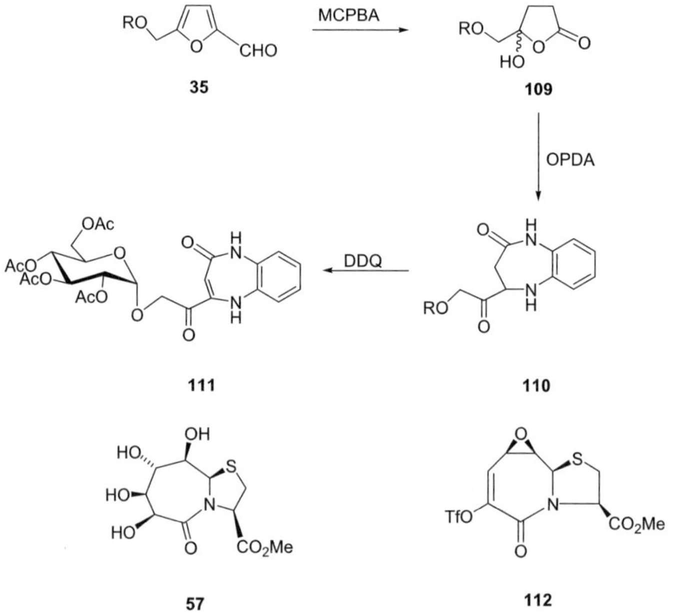

Fig. 19 Seven-membered heterocycles

oxygen gave the hydroxyl butenolide **109**. Its reaction with o-phenylene di-amine generated the benzodiazepinones **110** which upon dehydrogenation gave **111** (Fig. 19). Deacetylation of **110** and **111** gave the deacetylated derivatives which are soluble in water [36].

Recently, the reaction of L-cysteine methyl ester with uronic acids produced interesting types of heterocyclic compounds. A seven-membered bi-cyclic ring system fused with the thiazoline ring **57** was discussed earlier in this article, because the reaction was related to the reaction of the sugar to give a thiazoline ring [90, 91]. The regioselective conversion of the secondary hydroxyl group of D-glucoronic acid without the requirement of O-protecting groups allowed the conversion of **57** to **112** [163].

A recent publication described an interesting synthesis for glyco-conju-gated hexahydroazepindiones from saccharide imides via the Norrish type II

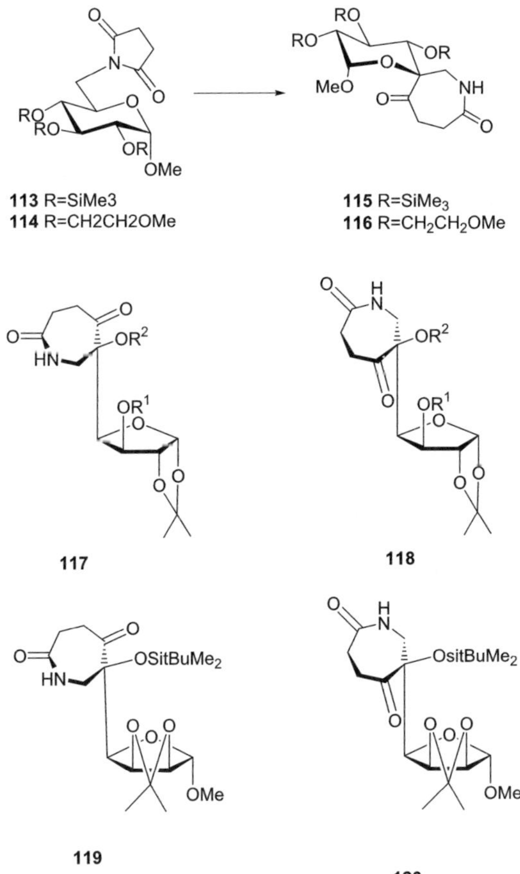

Fig. 20 Hexahydroazepindiones

reaction [164]. Here the primary succinimido derivatives such as **113** and **114** were obtained by nucleophilic displacement of the protected 6-tosylates of D-glucose, D-mannose, and D-galactose in both their pyranoside and furanoside forms with potassium succinimide. Then the imido sugars were photochemically transformed into hexahydroazepindione derivatives **115** and **116**, respectively by means of the Norrish type II reaction (Fig. 20). The isomeric **117** and **118** as well as **119** and **120** were prepared in a similar way. The regiochemistry of the photoreaction was controlled by configurational, conformational, and electronic features.

Acknowledgements This work was partially supported by HEC in Pakistan and AvH in Germany. The authors thank Nighat Rizvi for the drawing of the figures.

References

1. El Ashry ESH, El Ashry Y (1976) Chem Ind (London) 372
2. El Ashry ESH, Rashed N (2000) Curr Org Chem 4:609
3. El Ashry ESH, El Kilany Y (1996) Adv Heterocycl Chem 67:391
4. El Ashry ESH, El Kilany Y (1997) Adv Heterocycl Chem 68:1
5. El Ashry ESH, El Kilany Y (1998) Adv Heterocycl Chem 69:129
6. El Ashry ESH, El Nemr A (2005) Synthesis of Naturally Occurring Nitrogen Heterocycles from Carbohydrates. Blackwell, Oxford, UK
7. El Khadem H (1970) Adv Carbohydr Chem 25:351
8. El Ashry ESH, Awad LF, Atta AI (2006) Tetrahedron 62:2943
9. Svete J (2004) Monatsh Chem 135:629
10. Lichtenthaler FW (1991) Carbohydrates as Organic Raw Materials. VCH, Weinheim
11. Lichtenthaler FW (2002) Acc Chem Res 35:828
12. Lichtenthaler FW, Mondel S (1997) Pure Appl Chem 69:1853
13. Lichtenthaler FW, Peters S (2004) CR Chimie 7:65
14. Descotes G (1993) Carbohydrates as Organic Raw Materials II. VCH, Weinheim
15. Van Bekkum H, Roper H, Voragen A (1996) Carbohydrates as Organic Raw Materials III. VCH, Weinheim
16. Lightner GE (2003) US Pat Appl Publ US 2003032819 Application: US 2001-923644 20010807. CAN 138:173289
17. Schieberle P, Hofmann T (2002) ACS Symposium Ser 826:207
18. Milic LB, Djilas SM, Canadanovic-Brunet JM (1993) Food Chem 46:273
19. Ames JM, Bailey RG, Mann J (1999) J Agric Food Chem 47:438
20. Cousins GS, Hoberg JO (2000) Adv Strain Inter Org Mol 8:113
21. Kuster BFM (1990) Starch/Stärke 42:314
22. Gandini A, Belgacem MN (1997) Prog Polym Sci 22:1203
23. Koch H, Pein J (1985) Polym Bull (Berlin) 13:525
24. Koch H, Pein J (1985) Starch/Starke 35:304
25. Lichtenthaler FW, Klimesch R, Moiler V, Kunz M (1993) Liebigs Ann 975
26. Lichtenthaler FW, Martin D, Weber T, Schiweck H (1989) Ger Offen 3.936.522
27. Lichtenthaler FW, Martin D, Weber T, Schiweck H (1993) Liebigs Ann 967
28. Hanemann T, Haase W, Lichtenthaler FW (1997) Liquid Cryst 22:47
29. Cottier L, Descotes G, Soro Y (2005) J Carbohydr Chem 24:55

30. Cottier L, Descotes G (1991) Trends Heterocycl Chem 2:233
31. Garcia-Gonzales F (1956) Adv Carbohydr Chem 11:97
32. Moreno-Vargas AJ, Fuentes J, Fernandez-Bolanos J, Robina I (2001) Tetrahedron Lett 42:1283
33. McElvain SM, Bolliger K (1941) Org Synth Collect Vol 1:473
34. Garcia-Gonzales F, Gómez-Sanchez A (1965) Adv Carbohydr Chem 20:303
35. Rozanski A, Bielawski K, Boltryk J, Bartulewicz DR (1991) Akad Med Bialystok 35–36:57
36. Merino P, Franco S, Merchan FL, Tejero T (2000) Recent Res Dev Syn Org Chem 3:65
37. Lichtenthaler FW, Brust A, Cuny E (2001) Green Chem 3:201–209
38. El Khadem H, Mahammed-Aly MM (1963) J Chem Soc 4929
39. El Khadem H, El-Shafei, Mohamed-Aly MM (1964) J Org Chem 29:1565
40. El Khadem H (1964) J Org Chem 29:3072
41. El Khadem H, El-Shafei, Abdel Rahman MMA (1965) Carbohydr Res 1:31
42. Somogyi L (1986) Carbohydr Res 152:316
43. Somogyi L (1985) Carbohydr Res 144:71
44. Diehl V, Cuny E, Lichtenthaler FW (1998) Heterocycles 48:1193
45. Oikawa M, MOiler C, Kunz M, Lichtenthaler FW (1998) Carbohydr Res 309:269
46. El Ashry ESH, Ramadan E, Kassem AA, Haggar M (2005) Adv Heterocycl Chem 88:1
47. El Ashry ESH, Kassem AA, Ramadan E (2006) Adv Heterocycl Chem 90:1
48. El Ashry ESH, Kassem AA (2006) ARKIVOC, ix:1
49. El Ashry ESH (1982) Nitrogen derivatives of L-ascorbic acid in ascorbic acid, chemistry, metabolism and uses. In: Seib PA, Tolbert M (eds) Advances in Chemistry Series. Am Chem Soc, Washington, DC, p 43 200:179
50. El Ashry ESH, Mousaad A, Rashed A (1992) Adv Heterocycl Chem 53:233
51. Schmidt RR, Heermann D (1981) Chem Ber 114:2825.
52. Schmidt RR, Kary J, Guilliard W (1977) Chem Ber 110:2433
53. Schmidt RR, Guilliard W, Heermann D (1981) Liebigs Ann Chem 2309
54. Schmidt RR, Guilliard W, Karg J (1977) Chern Ber 110:2445
55. Kett WE, Bately M, Redmond JW (1997) Carbohydr Res 299:129
56. El Khadem HS, El Ashry ESH (1968) J Chem Soc 2248
57. El Khadem HS, El Ashry ESH (1973) J Heterocycl Chem 10:1051
58. El Ashry ESH, El Kilany Y, Singab F (1988) J Chem Soc Perkin Trans 1 133
59. El Ashry ESH, El Kilany Y, Singab F (1988) J Chem Soc Perkin Trans 1 139
60. Abdel Hamid H, Rashed N, El Kilany Y, El Ashry ESH (1992) Carbohydr Res 224:41
61. Simon H, Moldenhauer W (1967) Chem Ber 100:3121
62. Diels O, Meyer R, Onnen O (1936) Ann 525:94
63. EI Khadem H, Abdel Rahman MMA (1966) Carbohydr Res 3:25
64. Al-Harrasi A, Reissig H (2005) Synlett 1152
65. Peseke K, Feist H, Quincoces J (2001) Targets Heterocyc Syst 5:299
66. Bari A, Feist H, Michalik D, Michalik M, Peseke K (2004) Synthesis 2863
67. Montero A, Michalik M, Feist H, Reinke H, Rudloff I, Peseke K (2004) J Carbohydr Chem 23:313
68. Hashmi IA, Feist H, Michalik M, Reinke H, Peseke K (2006) J Carbohydr Chem 25:19
69. Weidenhagen R, Hermann R (1937) Ber Dtsch Chem Ges 70:570
70. Toter R, Darby WJ (1964) Org Synth Collect 3:460
71. Frankowski A, Seliga C, Bur D, Streith J (1991) Helv Chim Acta 74:934
72. Streith J, Boiron A, Frankowski A, Le Nouen D, Rudyk H, Tschamber T (1995) Synthesis 944
73. Tschamber T, Siendt H, Boiron A, Gessler F, Streith J (2001) Eur J Org Chern 1335

74. Dziun P, Schunack W (1973) Arch Pharm (Weinheim) 306
75. Dubost E, Le Nouen D, Streith J, Tarnus C, Tschamber T (2006) Eur J Org Chem 610
76. Avalos M, Babiano R, Cintas P, Hursthouse MB, Jimenez JL, Light ME, Palacios JC, Silvero G (2005) Tetrahedron 61:7931
77. Graier T, Vasella A (1995) Helv Chim Acta 78:1738
78. Kato M, Uno T, Hiratake J, Sakata K (2005) Bioorg Med Chem 13:1563
79. Heinemann F, Hiegemann M, Welzel P (1992) Tetrahedron 48:3781
80. Banoub J, Boulanger P, Lafont D (1992) Chem Rev 92:1167
81. DeCastro M, Marzabadi CH (2005) J Carbohydr Chem 24:179
82. Wittmann V, Lennartz D (2002) Eur J Org Chem 1363
83. Knapp S, Vocadlo D, Gao Z, Kirk B, Lou J, Withers SG (1996) J Am Chem Soc 118:6804
84. Lemieux RU, Morgan AR (1965) Cand J Chem 43:2199
85. Somogyi L (1987) Carbohydr Res 166:166
86. Rauter AP, Padilha M, Figueiredo JA, Ismael M, Justino J, Ferreira H, Ferreira MJ, Rajendran C, Wilkins R, Vaz PD, Calhorda MJ (2005) J Carbohydr Chem 24:275
87. Bognar R, Kolodynska Z, Somogyi L, Gyorgydeak Z, Szilagyi L, Nemes-Nanasi EL (1969) Acta Chim Acad Sci Hung 62:65
88. Bognar R, Somogyi L, Gyorgydeak Z (1970) Liebig Ann Chem 738:68
89. Bognar R, Gyorgydeak Z, Szilagyi L, Czira G, Tamas J (1977) Liebigs Ann Chem 1536
90. Geyer A, Moser F (2000) Eur J Org Chem 1113
91. Agoston K, Geyer A (2004) Tetrahedron Lett 45:1895
92. Gyorgydeak Z, Levai A, Snatzke G (1987) Croat Chem Acta 60:185
93. Szilagyi L, Gyorgydeak Z (1976) Carbohydr Res 48:159–169
94. Kover KE, Gyorgydeak Z (1992) Magn Res Chem 30:137
95. El Khadem HS (1965) Adv Carbohydr Chem 20:139
96. El Khadem H, Shaban MAE (1966) Carbohydr Res 2:178
97. Schmidt RR, Guilliard W, Herrmann D (1981) Liebigs Ann Chem 2309
98. Jung KH, Schmidt RR, Heermann D (1981) Chem Ber 114:2834
99. El Ashry ESH, Awad LF, Winkler MJ (2000) J Chem Soc Perkin Trans 1 829
100. Awad LF, El Ashry ESH (1998) Carbohydr Res 321:9
101. Awad LF, El Ashry ESH (1999) Nucleos Nucleot 18:557
102. Balbaa M, Mansour H, El-Sawy H, El Ashry ESH (2002) Nucleos Nucleot Nucl 21:695
103. El Ashry ESH, Awad LF (2001) Nucleos Nucleot Nucl 20:103
104. El Ashry ESH, Awad LF (2001) Nucleos Nucleot Nucl 20:901
105. El Ashry ESH, Rashed N, Mosaad A (1987) J Carbohydr Chem 6:599
106. Moustafa AH, Haggam RA, Younes ME, El Ashry ESH (2005) Nucleos Nucleot Nucl 24:1885
107. El Ashry ESH, Awad LF, Abdel Hamid HM, Atta AI (2005) Nucleos Nucleot Nucl 24:427
108. Gyorgydeak Z, Holzer W, Thiem J (1997) Carbohydr Res 302:229
109. El Ashry ESH, Abdel-Ghanie M (2004) Nucleos Nucleot Nucl 23:567
110. Rashed N, Ibrahim E-SI, El Ashry ESH (1994) Carbohydr Res 254:295
111. Rashed N, Shoukry M, El Ashry ESH (1994) Bull Chem Soc Jpn 67:149
112. El Khadem H, Shaban MAE, Nassr MAM (1970) Carbohydr Res 13:470
113. El Khadem H, Shaban MAE, Nassr MAM (1972) Carbohydr Res 23:103
114. Nassr MAM (1983) Org Prep Proc Int 15:329
115. Somogyi L (1988) Carbohydr Res 182:19
116. Somogyi L (1987) Carbohydr Res 165:318
117. El Kilany Y, Awad L, Mackawy K, El Ashry ESH (1988) Alex J Pharm Sci 2:139

118. El Ashry ESH, El Kilany Y, Abdallah AA, Mackawy K (1983) Carbohydr Res 113:273
119. El Ashry ESH, Nassr MMA, Abdel Rahman MMA, Rashed N, Mackawy K (1980) Carbohydr Res 82:149
120. El Ashry ESH, Abdel Rahman MA, El Kilany Y, Rashed N (1987) Carbohydr Res 163:123
121. Holmberg B (1954) Ark Kemi 7:513
122. El Ashry ESH, Nassr MMA, El Kilany Y, Mousaad A (1986) J Prak Chem 328:1
123. El Ashry ESH, Nassr MMA, El Kilany Y, Mousaad A (1987) Bull Chem Soc Jap 60:3405
124. Somogyi L (1979) Carbohydr Res 311:325
125. Nasr AZ (1996) Alexandria J Pharm Sci 10:139
126. Shaban MA, Nasr AZ, Taha MAM (1995) Pharmazie 50:534
127. Shaban MA, Nasr AZ, Taha MAM (1995) J Carbohydr Chem 14:985
128. Taha MAM (1995) Alexandria J Pharm Sci 7:79
129. Alho MAM, D'Accorso NB (1997) J Heterocycl Chem 34:1415
130. Alho MAM, D'Accorso NB (2000) J Heterocycl Chem 37:811
131. Alho MAM, D'Accorso NB (2002) J Heterocycl Chem 39:137
132. Argay G, Csuk R, Gyorgydeak Z, Kalman A, Snatzke G (1995) Tetrahedron 51:12911
133. D'Accorso NB, Thiel IME (1993) An Quim 89:266
134. Zemplen G, Mester L, Eckhart E (1953) Chem Ber 86:472
135. Elming N, Clauson-Kaas N (1956) Acta Chem Scand 10:1603
136. Muller C, Diehl V, Lichtenthaler FW (1998) Tetrahedron 54:10703
137. Samet AV, Kislyi VP, Chernysheva NB, Reznikov DH, Ugrak BI, Semenov VV, Zelinskogo ND (1996) Izves Akad Nauk Seriya Khim 409
138. Tremmel P, Geyer A (2005) Eur J Org Chem 3475
139. Tremel P, Brand J, Knapp V, Geyer A (2003) Eur J Org Chem 878
140. Rudloff I, Michalik M, Montero A, Peseke K (2001) Synthesis 1686
141. Rudloff I, Peseke K, Reinke H (1998) J Prak Chem/Chem-Zeitung 340:334
142. Montero A, Feist H, Michalik M, Quincoce J, Peseke K (2002) Synthesis 664
143. Kitajima M, Shirakawa S, Abdel-Moty SGA, Takayama H, Sakai S, Ami N, Stöckigt J (1996) Chem Pharm Bull 44:2195
144. Lehnhoff S, Ugi I (1995) Heterocycles 40:801
145. Manna RK, Jaisankar P, Giri VS (1995) Synth Commun 25:3027
146. Imada K (1973) J Chem Soc Chem Commun 796
147. Voelter W, Kahn KM, Shekhani MS (1996) Pur Appl Chem 68:1347
148. Al-Qawasmeh RA, Al-Tel T, Abdel-Jalil RJ, Voelter W (1999) Chem Lett 541
149. Montero A, Feist H, Michalik M, Quincoces J, Peseke K (2002) J Carbohydr Chem 21:305
150. Methling K, Kopf J, Michalik M, Reinke H, Buerger C, Oberender H, Peseke K (2003) J Carbohydr Chem 22:537
151. Bari A, Feist H, Michalik M, Peseke K (2005) Molecules 10:837
152. Stoelting DT, Mbagwu GO, Scott T, Long M, Sharpe EL (2002) J Heterocycl Chem 39:719
153. Samet AV, Yamskov AN, Ugrak BI, Vorontsova LG, Kurella MG, Semenov VV, Zelinskogo ND (1996) Izves Akad Nauk Ser Khim 415
154. Samet AV, Yamskov AN, Ugrak BI, Semenov VV, Zelinskogo ND (1997) Izves Akad Nauk Ser Khim 46:532
155. Lammers H, Peters JA, van Bekkum H (1994) Tetrahedron 50:8103
156. El Ashry ESH, Rashed N, Mousaad A (1987) J Carbohydr Chem 6:599
157. Goswami S, Adak AK (2005) Tetrahedron Lett 46:221

158. El Ashry ESH, El Kilany Y, Mousaad A (1986) Curr Sci 55:891
159. Rohovec J, Kotek J, Peters JA, Maschmeyer T (2001) Eur J Org Chem 3899
160. Sumoto K, Irie M, Mibu N, Miyano S, Watanabe K, Yamaguchi T (1991) Chem Pharm Bull 39:792
161. Kerns RJ, Toida T, Linhardt RJ (1996) J Carbohydr Chem 15:581
162. Hirsch J, Petrakova E, Feather MS (1995) J Carbohydr Chem 14:1179
163. Agoston K, Geyer A (2005) Chem-A Eur J 11:6407
164. Stark M, Thiem J (2006) Carbohydr Res 341:1543

Top Heterocycl Chem (2007) 7: 31–66
DOI 10.1007/7081_2007_053
© Springer-Verlag Berlin Heidelberg
Published online: 12 May 2007

Synthesis of Heterocycles from Glycosylamines and Glycosyl Azides

José G. Fernández-Bolaños (✉) · Ó López

Departamento Química Orgánica, Facultad de Química, Universidad de Sevilla, Apartado 553, 41071 Sevilla, Spain
bolanos@us.es

Abstract Glycosylamines and glycosyl azides are valuable and versatile intermediates in carbohydrate chemistry; furthermore they are easily available in a highly stereoselective fashion. Glycosylamines and their derivatives glycosyl enamines and glycosyl imines can be transformed into *N*-glycosyl heterocycles. Glycosyl enamines are also used in the preparation of polyhydroxy pyrrolidines, oxapyrrolizidines, piperidines and azepanes, an intramolecular nucleophilic displacement of a sulfonyloxy group by a stabilized amide anion being the key step. Glycosyl azides can be converted into a plethora of heterocyclic derivatives by different reactions such as the Staudinger reaction, thermolysis and 1,3-dipolar cycloadditions. Much effort has been devoted to the synthesis of glycosyl triazoles by using the Dimroth reaction, ketene aminals, phosphorus ylides, azide–alkyne Huisgen cycloaddition and the especially successful Cu(I)-mediated Huisgen cycloaddition.

Keywords Glycosyl azides · Glycosylamines · Glycosyl enamines · Glycosyl heterocycles · Triazoles · Click chemistry

Abbreviations

BzCl	Benzoyl chloride
CAN	Cerium(IV) ammonium nitrate
CSA	Camphor-10-sulfonic acid
DBU	1,8-Diazabicyclo[5.4.0] undec-7-ene
DMAD	Dimethyl acetylenedicarboxylate
DME	1,2-Dimethoxyethane
DMF	*N,N*-Dimethylformamide
DMPU	*N,N'*-Dimethyl propylene urea
DPPE	1,2-Bis(diphenylphosphino)ethane
EDPA	Ethyldiisopropylamine
Glyc	Glycosyl
HCV	Hepatitis C virus
HMPT	Hexamethylphosphoramide
MW	Microwave
MsCl	Mesyl chloride
NOE	Nuclear Overhauser effect
Piv	Pivaloyl
PMB	*p*-Methoxybenzyl
Py	Pyridine
quant	Quantitative
rt	Room temperature
TBAF	Tetrabutylammonium fluoride
TBDMS	*tert*-Butyldimethylsilyl
TFA	Trifluoroacetic acid
Tf$_2$O	Triflic anhydride
THF	Tetrahydrofurane
TMSOTf	Trimethylsilyl triflate
TsOH	*p*-Toluenesulfonic acid

1
Introduction

Glycosylamines and glycosyl azides are valuable and versatile intermediates in carbohydrate chemistry. Thus, glycosylamines have been used extensively in the synthesis of glycopeptides [1–4], *N*-glycoconjugates [5, 6] and *N*-glycosylamidines, useful as glycosidase inhibitors [7, 8]. Glycosylamines have also proved to be efficient auxiliaries in the stereoselective Strecker synthesis of α-amino nitriles and the Ugi reaction yielding α-amino acid amides [9].

Glycosylamines can be prepared directly from unprotected sugars, using ammonia in methanol [7, 10–12], aqueous ammonium hydrogen carbonate [13, 14] or ammonium carbamate in methanol [15–18]. Procedures based on the use of glycals [19], α-hydroxy nitriles [20], or the Burgess reagent, recently proposed by Nicolaou [21], can be mentioned among the methods to introduce the amino group after the hydroxyl group protection.

Glycosyl azides [22–24] have been used as precursors of the corresponding glycosylamines by reduction with hydrogen in the presence of different catalysts, with propane-1,3-dithiol, complex hydrides, boranes and phosphines (Staudinger reduction) [4, 22, 25]. Besides that, glycosyl azides can be easily transformed into glycosyl amides [26–28], carbodiimides [29] and ureas [30]. The dipolar nature of the azido group allows glycosyl azides to be excellent substrates in cycloaddition reactions leading to heterocycles derivatives [25, 31].

2
Glycosyl Enamines as Key Intermediates in the Synthesis of Heterocycles

2.1
Synthesis of Polyhydroxy Pyrrolidines

Reaction of glycosylamines or aminosugars 1 with diethyl (ethoxymethylene)malonate to give the corresponding enamine derivatives 2 was found [32–34] to be quite an efficient procedure for protecting amino moieties in carbohydrates, stable in both acidic and basic conditions (Scheme 1). Deprotection is carried out in moist dichloromethane with chlorine or bromine, to give hydrochloride or hydrobromide 3, usually crystalline.

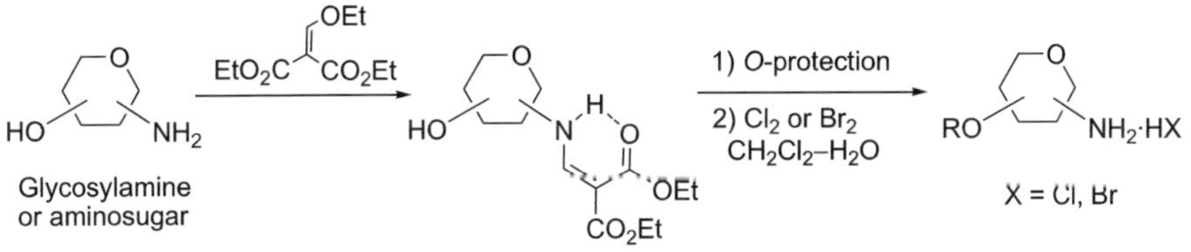

Scheme 1 Preparation of sugar-derived enamines as protective groups

Since then, this protective group has mainly been used in the preparation of sugar-derived isothiocyanates and isocyanates as intermediates in the preparation of thioureas [35–39], ureas [39] or pseudo-nucleosides [40].

Fuentes and co-workers reported [41] the transformation of 4-*O*-mesyl-*N*-diethoxycarbonylvinyl-β-D-glycopyranosylamines of *gluco*, *xylo* and *ribo* configurations 4 into 1,4-anhydro-iminosugars (2-oxa-7-azabicyclo[2.2.1]-heptane derivatives) 6 upon treatment with an equimolecular amount of sodium methoxide in hexamethylphosphoramide (Scheme 2). The reaction took place through the stabilized intermediate amide anion 5, which displaces the mesyloxy group by nucleophilic attack on C-4 of the sugar.

IUPAC recommends that the name azasugar should be used for compounds having a carbon replaced with nitrogen, whereas the name imino-

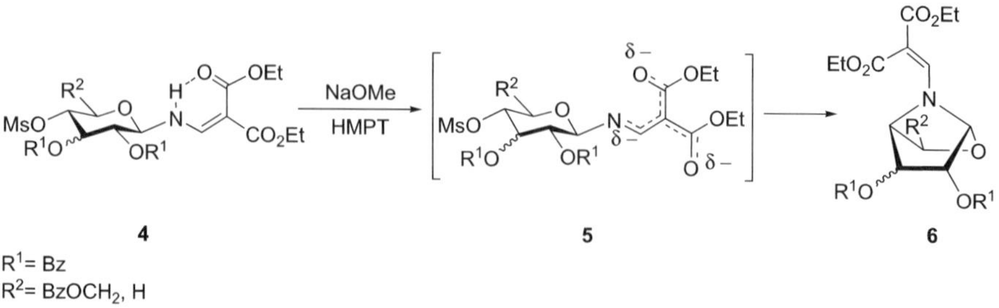

R^1= Bz
R^2= BzOCH$_2$, H

Scheme 2

sugar should be used for compounds where the ring oxygen has been replaced with nitrogen [42].

When starting from the per-*O*-mesylated derivative **7** the reaction only proceeded through C-6 to form the corresponding 1,6-anhydro-iminosugar **8** (Scheme 3) and no reaction was observed through C-4 [41].

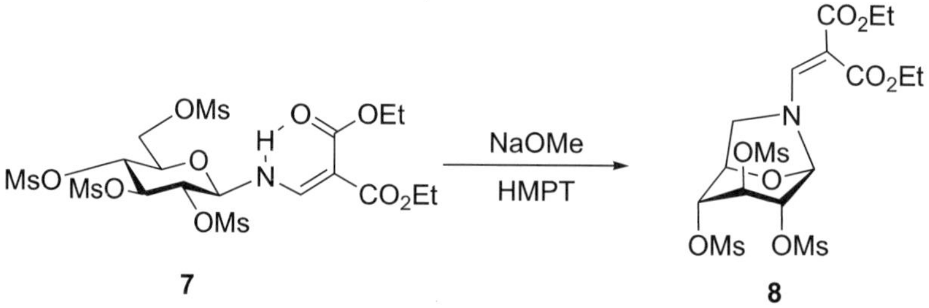

Scheme 3

These anhydro-iminosugars proved to be valuable synthetic intermediates for the preparation of some derivatives of synthetic and biological interest. Thus, the opening of the pyrrolidine ring of **9** in the presence of aqueous trifluoroacetic acid led to the stereoselective migration of the protected amino moiety from the anomeric position to C-4, as depicted in Scheme 4 [43]. By this procedure 4-aminoaldoses such as **10** can be obtained, in an efficient alternative to nucleophilic displacement of 4-sulfonyloxy groups by azide.

When a mixture of acetic anhydride and trimethylsilyl triflate was used, the opening of the C1 – O bond took place to yield the corresponding acylated polyhydroxy pyrrolidines **11** as mixtures of α- and β-anomers [43], compounds of interest as potential glycosidase inhibitors. However, when the opening of the pyranose ring is carried out in methanol containing lithium perchlorate, followed by addition of a 2 : 1 mixture of TFA and trifluoroacetic anhydride, only **12** was obtained as a single diastereoisomer (Scheme 4).

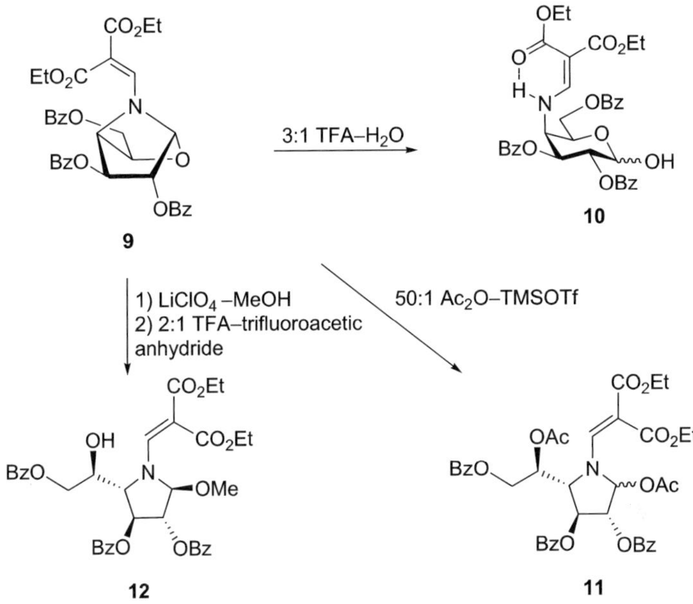

Scheme 4

Furthermore, reduction of 1,4-anhydro-iminosugar **13** with sodium cyano-borohydride in acetic acid and deprotection in basic medium (Scheme 5) led to an equilibrium between *N*-acetylpyrrolidine **14** and 2-oxapyrrolizidine

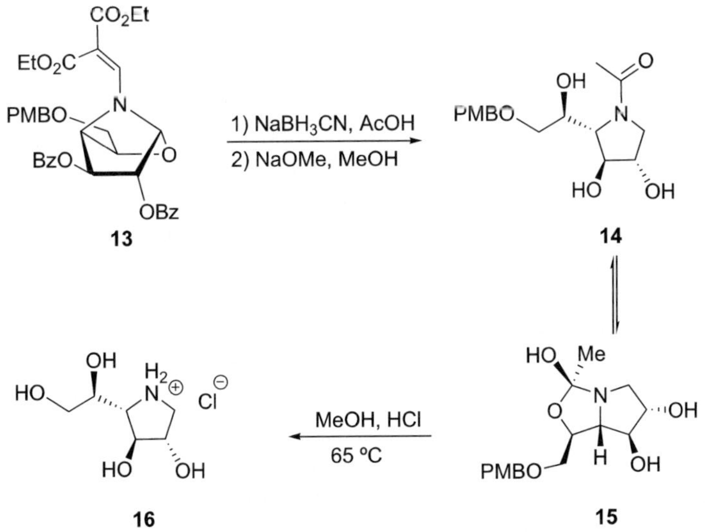

Scheme 5

15 (minor product) [44]. Simultaneous *O*- and *N*-deprotection in acidic conditions led to hydrochloride **16**, a galactosidase mutase and galactan biosynthesis inhibitor [45].

2.2
Synthesis of Polyhydroxy 2-Oxapyrrolizidines

2-Oxapyrrolizidine **18** was obtained [43] from 1,4-anhydro-iminosugar **17** by pyranose ring opening followed by addition of the free 5-OH group to the double bond of the enamine moiety (Scheme 6). These compounds can be considered analogues of polyhydroxylated pyrrolizidine alkaloids, which display important biological properties [46–49].

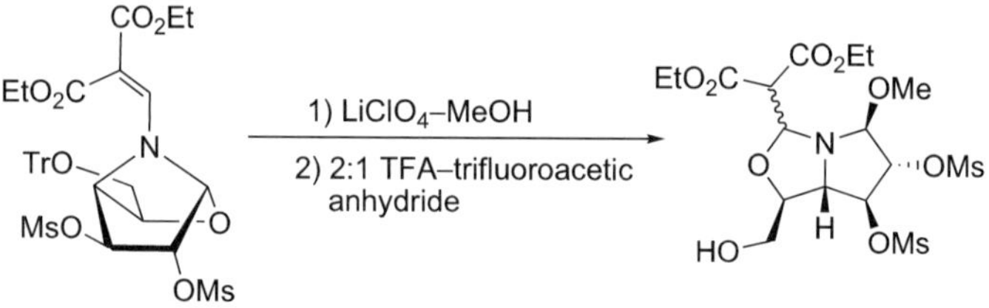

Scheme 6

Fuentes and co-workers have also carried out [50, 51] the transformation of glycosylenamines into bicyclic and monocyclic iminosugar thioglycosides. Thus, starting from glycosylenamines of D-*gluco*, D-*xylo* and L-*rhamno* configurations bearing a mesyloxy group on C-4, the corresponding 1,4-anhydro-iminosugars were prepared according to the conditions described in Scheme 7. When these bicyclic compounds were treated with ethanethiol in the presence of *p*-toluenesulfonic acid, nucleophilic attack of the thiol on C-1 led to a transient pyrrolidine derivative that spontaneously cyclized by an internal cycloaddition of the hydroxyl group on the carbon–carbon double bond (Scheme 7) to give 2-oxa-pyrrolizidines.

For example, bicyclic pyrrolizidine **21** was obtained in good yield from β-D-glucopyranosylenamine via 1,4-anhydro-iminosugar **13**, as depicted in Scheme 7. This compound existed as a 1 : 4 mixture of the 5*R* and 5*S* diastereoisomers, as the attack of ethanethiol takes place preferably on the opposite side of the bulky OBz group. The stereoselectivity of the cyclization step in **19** was 100% as only **20** with the 3*S* configuration was detected.

This kind of compound combines two structural motifs that are of interest, both from the synthetical and biological point of view; they are iminocyclitols and thioglycosides. The latter type of compounds have been successfully used in glycosylation reactions [52–55], as well as exhibiting interesting biological activities [56, 57].

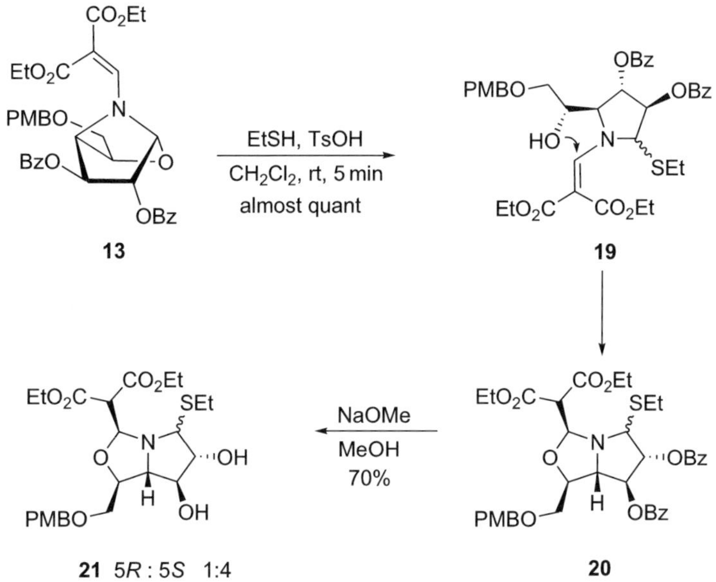

Scheme 7

2.3
Synthesis of Polyhydroxy Piperidines

Fuentes's group reported the synthesis of polyhydroxy piperidines (Scheme 8) starting from partially protected β-D-ribopyranosylenamine **22** [58]. Thus, mesylation on the free hydroxyl group led to the corresponding 5-mesyloxy and 5-chloro derivatives **23** and **24**, the latter obtained by nucleophylic substitution on the mesyloxy group when longer reaction times and more concentrated solutions were used.

Basic treatment of the enamino derivative bearing a leaving group on C-5, afforded, as expected, the corresponding bicyclic derivative **25**. This compound was subjected to a ring opening reaction in acidic medium using a series of thiols as nucleophiles. This gave piperidine-derived iminosugars **26a–d** bearing a thioglycosidic aglycon in the pseudo-anomeric position.

It is remarkable that only the β-anomer, that is, the 2S stereoisomer was obtained in all cases, thus proving that reaction proceeded via a pure S_N2 mechanism. The configuration of the pseudo-anomeric position was assigned by NOEs experiments [58].

In contrast, attempts at opening 1,5-anhydro-iminosugar **29** using ethanethiol and TsOH did not lead to the expected piperidine (Scheme 9), but afforded α-L-furanoid thioglycoside **30** [50]. Bicyclic derivative **29** was prepared from mannopyranosyl enamine **27** by selective monobenzoylation on the primary hydroxyl group, and acetal protection of the *cis*-arranged hydroxyl

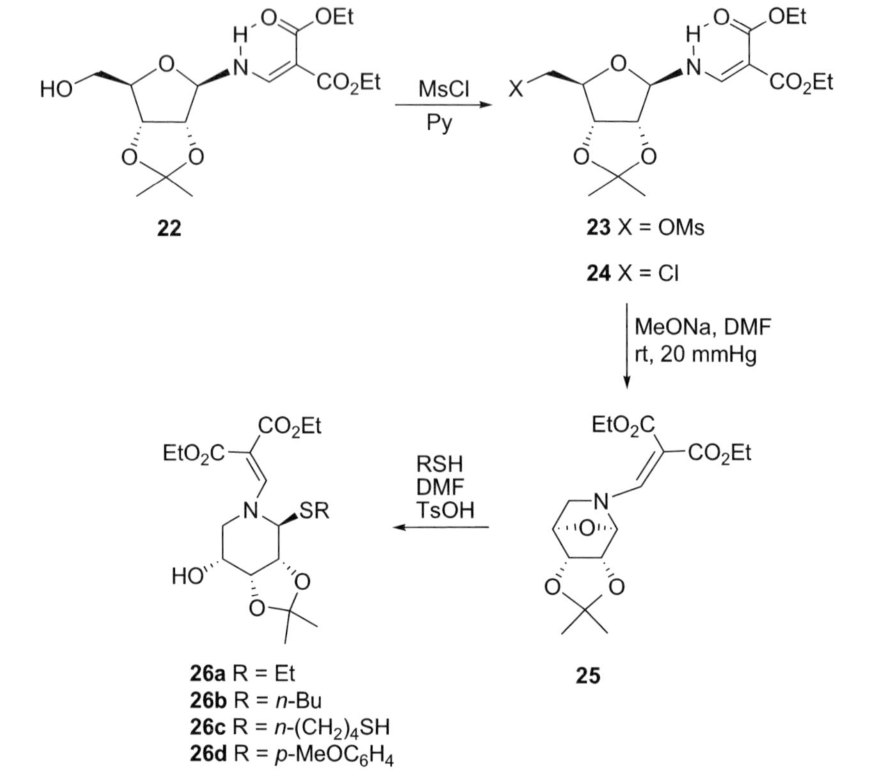

Scheme 8

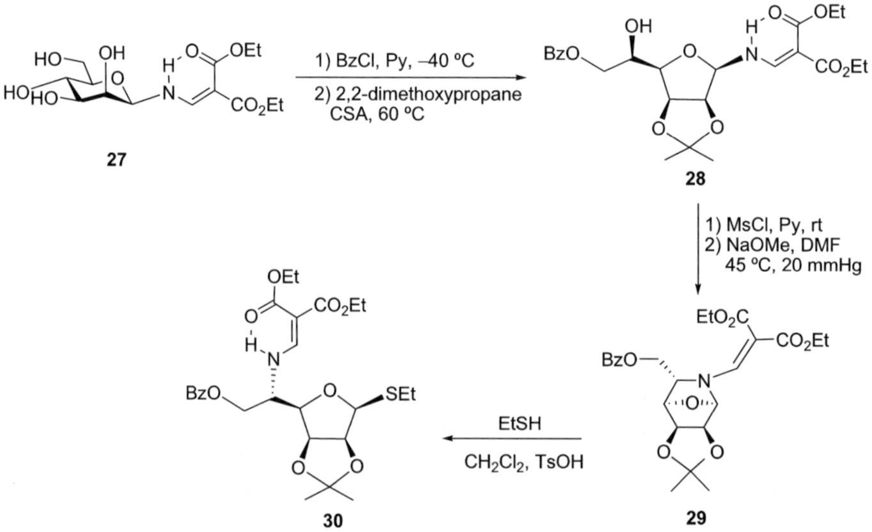

Scheme 9

groups on C-2 and C-3 to give furanosylenamine **28**. Mesylation of the free hydroxyl group, followed by basic treatment allowed the preparation of **29**; its ring opening took place via cleavage of the C1 – N bond rather than the C1 – O bond.

In a similar way, a series of furanoid thioglycosides were obtained when starting from β-L-rhamnopyranosylenamine [50].

2.4
Synthesis of Polyhydroxy Azepanes

Another interesting application of glycosylenamines is their conversion into polyhydroxy azepanes [59, 60]. This family of iminosugars is reported to exhibit important inhibitory properties against glycosidases [61–63]. Thus, 1,6-anhydro-iminosugars of D-*gluco*, D-*manno*, and D-*galacto* configurations **34** were prepared in good yields (80–93%) from β-D-glycopyranosylenamines **31** [59]. Displacement of the mesyloxy group by the amide anion using sodium methoxide in DMF, followed by *O*- and *N*-deprotection of **32**, and reduction of **33** gave polyhydroxy azepanes **34**, as depicted in Scheme 10. Compounds **34** were tested against a series of glycosidases, but turned out to be either poor inhibitors or totally inactive.

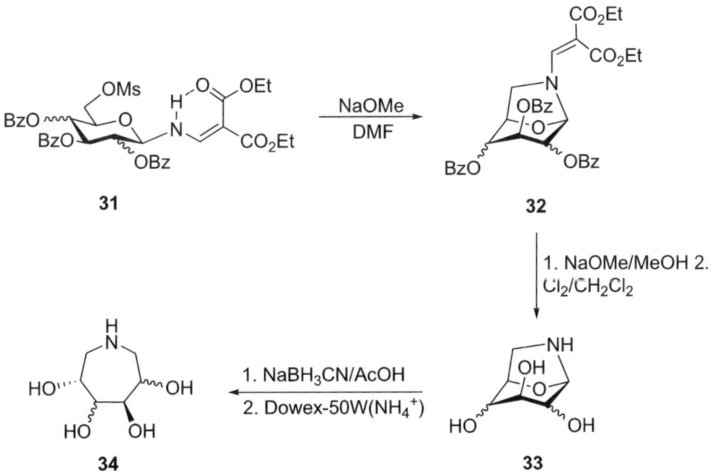

Scheme 10

Synthesis of polyhydroxy azepane **35** with hemiaminal structure was carried out [60] by removal of the enamino group of the 1,6-anhydro-iminosugar **32** of D-*gluco* configuration by treatment with chlorine followed by ring opening with aqueous acetic acid (Scheme 11). Attempts to carry out the ring opening in the presence of the enamino moiety were unsuccessful, and the starting materials were stable even in acidic medium in boiling methanol.

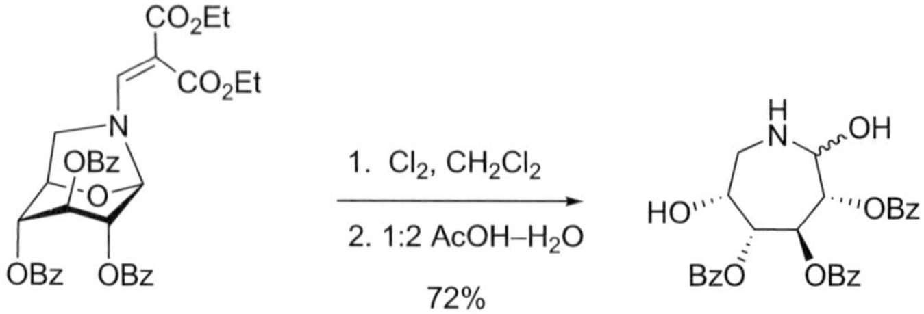

Scheme 11

32 (D-*gluco*)   **35**

All these reactions confirm the high versatility of glycosylenamines, which can be easily transformed into 1,4- and 1,6-anhydro-iminosugars, bicyclic compounds that can be used in the preparation of pyrrolidines, pyrrolizidines, piperidines and azepanes.

2.5
Synthesis of Pyrimidine Nucleosides

Recently, Avalos and co-workers [64] have described a novel one-pot reaction to transform glycopyranosyl enamines of *gluco* and *galacto* configuration, **36a** and **36b**, into uracil nucleosides **37a,b**, carrying an ethoxycarbonyl group at C-5 of the base (Scheme 12). The coupling of enamines with chlorosulfonyl isocyanate took place in refluxing toluene within 48–72 h, the labile chlorosulfonyl group being removed during the work-up. The reaction was accelerated by irradiation with microwaves (1 h), although the yields were lower.

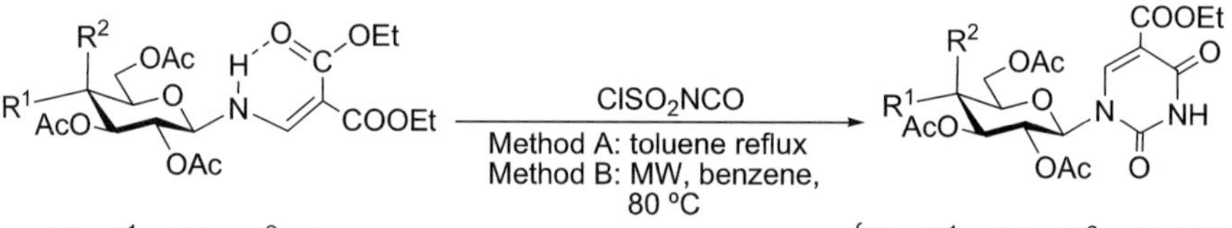

36a R^1 = OAc, R^2 = H
36b R^1 = H, R^2 = OAc

ClSO₂NCO
Method A: toluene reflux
Method B: MW, benzene, 80 °C

Method A
37a R^1 = OAc, R^2 = H 44%
37b R^1 = H, R^2 = OAc 70%

Method B
37a R^1 = OAc, R^2 = H 30%
37b R^1 = H, R^2 = OAc 58%

Scheme 12

A mechanistic approach for these transformations (Scheme 13) was suggested involving the zwitterionic intermediate **41** as the key step. The polar intermediate could be generated either by a [2 + 2] cycloaddition followed by opening of the β-lactam ring of **40** [65], or by direct nucleophilic attack on the electrophilic carbon of the heterocumulene. Migration of the ethoxycar-

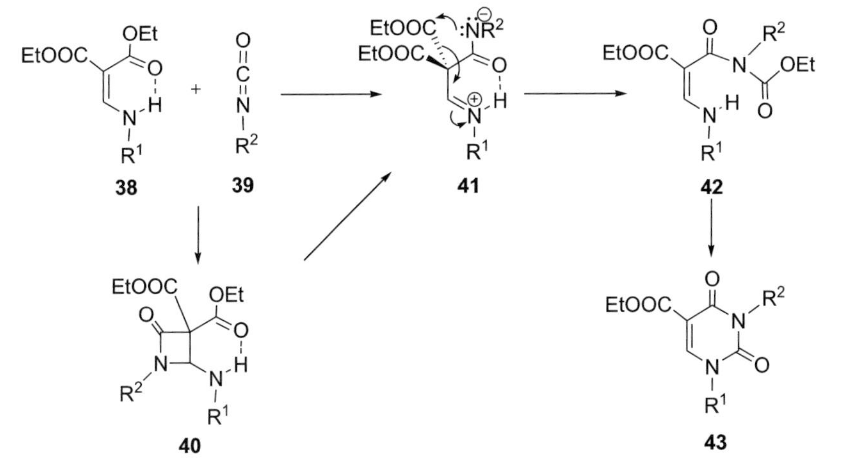

Scheme 13

bonyl group of **41**, followed by ring closure of **42** would lead to the uracil derivative **43**.

3
Glycosyl Imines and Amines for the Synthesis of Glycosyl Heterocycles

Stereoselective synthesis of chiral piperidones from an immobilized glycosyl-amine has been developed by Kunz et al. [66]. Thus, O-protected galactopyra-nosyl azide **44**, coupled to polystyrene through a hexamethylene spacer unit, was quantitatively reduced to amine **45** with 1,3-propanedithiol (Scheme 14). The immobilized galactosylamine was transformed into the corresponding galactosyl imines **46** with high E-isomeric purity by acid-catalysed conden-sation with aldehydes. Subsequent domino Mannich–Michael condensation of the glycosyl imines with Danishefsky's diene gave 2-substituted piperidi-

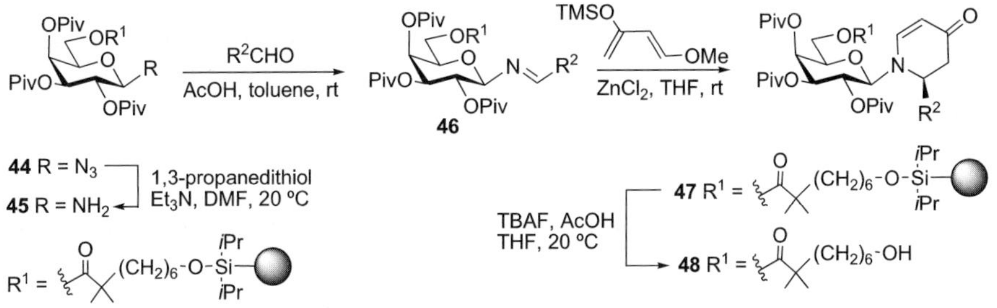

Scheme 14

none derivative **48** in high diastereoselectivity. This reaction was previously studied in solution by Kunz et al. starting from 2,3,4,6-tetra-*O*-pivaloyl-β-D-galactopyranosylamine [67, 68] in order to prepare *N*-glycosyl heterocycles. Furthermore, reaction of **47** with $R_2Cu(CN)Li_2$ (Lipshutz cuprates) in the presence of $BF_3 \cdot OEt_2$ gave polymer-bound galactosyl 2,6-substituted piperidinones in excellent *cis* selectivity [66].

The preparation of *N*-glycosyl monocyclic β-lactams from *O*-protected β-D-galatopyranosyl **49** [69] and β-L-arabinopyranosylamine [70] has been described (Scheme 15). The key step was the Staudinger [2+2] cycloaddition [71, 72] of the Schiff base **50** of glycosylamine and cinnamaldehyde with phthalimidoketene, generated in situ from phthalimidoacetyl chloride and triethylamine. This cycloaddition yielded β-lactam **51** as a single isomer, as previously reported for D-glucosamine propanedithioacetal [73, 74]. Some of the novel *N*-glycosylated β-lactams were found to be active against *Staphylococcus citrus*, *Klebsiella pneumoniae*, *Escherichia coli* and *Bacillus subtilis* [69].

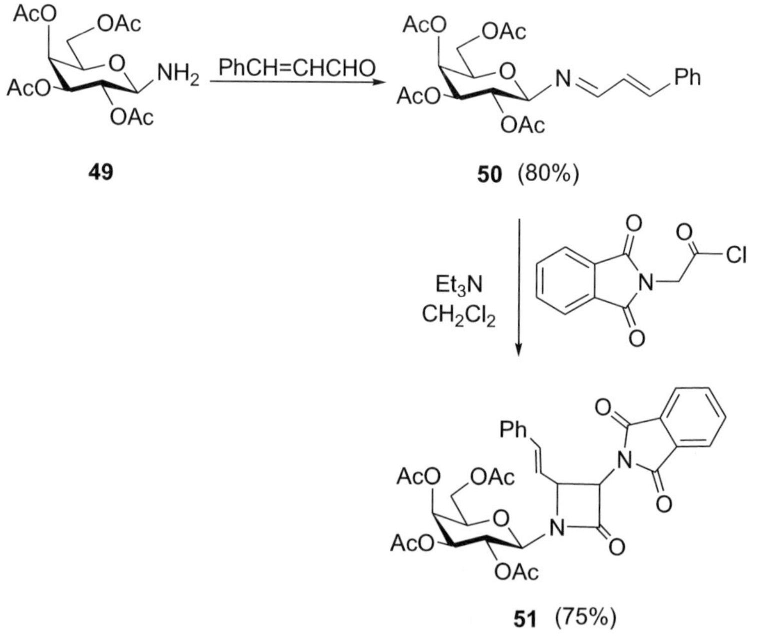

Scheme 15

Thiem et al. have described [75, 76] the synthesis of hexopyranosyl imides of *gluco* and *manno* configuration starting from *O*-protected glycosylamines available by reduction of glycosyl azides. Reaction of **52**, **53** with succinic anhydride followed by a cyclization reaction of the corresponding

N-glycosylsuccinamidic acids **54**, **55** under acetylation conditions, afforded N-glycosyl imides **56**, **57** as a mixture of separable anomers (Scheme 16).

Photolysis of N-glycosyl imides by irradiation with UV light of 254 nm wavelength to give highly functionalized heterocycles has also been studied by Thiem's group [75, 76]. The regiochemistry of this reaction is controlled by the anomeric configuration of the glycosyl imides. Abstraction of H-2 or H-5 of the β- or α-anomers of glucopyranosyl imide **56** by the excited carbonyl function gives place to 1,4- or 1,5-diradicals **58**, **61**, respectively (Scheme 16). Recombination of the 1,4-diradical **58** gives the unstable tricyclic derivative **59**, which is transformed into a 1,2-fused bicyclic glucopyranoso-azepanedione **60**. Similarly, the 1,5-diradical **61** gives the stable tricyclic derivative **62**. In contrast, photolysis of the α-D-mannopyranosyl imide **57** gives a mixture of a bicyclic and a tricyclic derivative, as both H-2 and H-5 could be abstracted by the excited carbonyl function [75].

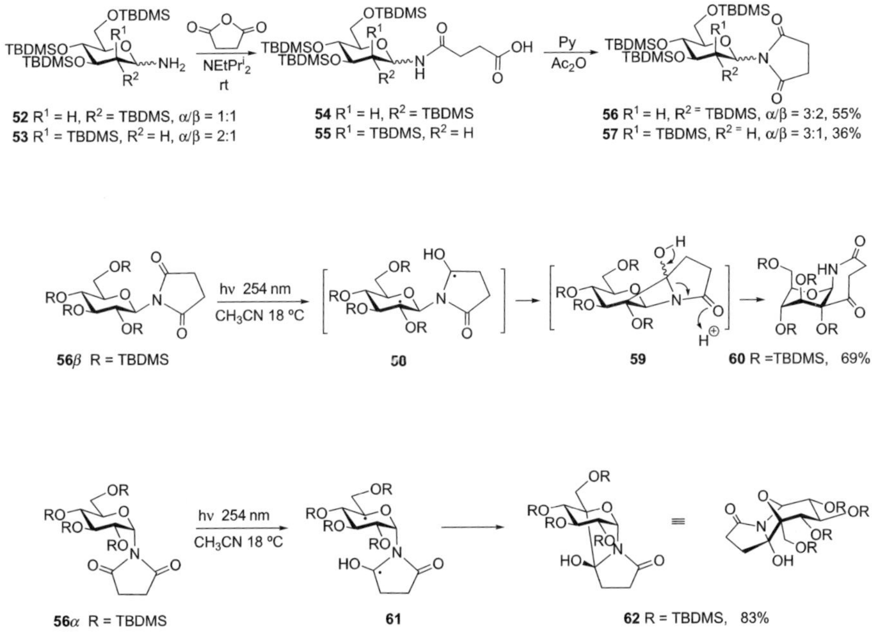

Scheme 16

N-Glucopyranosy urea **65**, prepared by coupling O-protected glucosylamine **63** with isocyanate **64** (Scheme 17), could be transformed easily into 3-glycosylated 5,6-dihydropyrimidine **66** by refluxing in chloroform in the presence of pyridine [77, 78]. In contrast, the mannopyranosyl analogue of **65** did not react to form the six-membered heterocyclic ring system under identical reaction conditions, and only decomposition products were observed [78]. An intramolecular ring closure to give N-glucopyranosyl

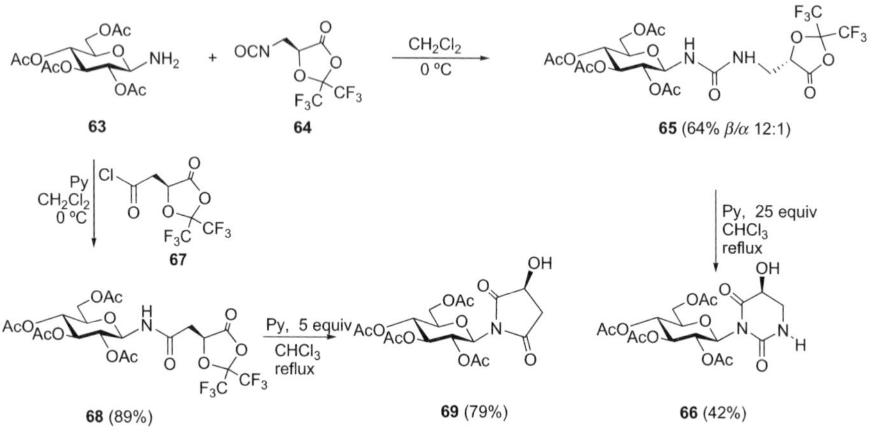

Scheme 17

3-hydroxysuccinimide **69** was observed (Scheme 17) when *N*-glucosylated malic acid derivative **68** was treated with pyridine in chloroform under reflux [79].

4
Glycosyl Azides as Key Intermediates in the Synthesis of Heterocycles

4.1
Staudinger Reaction of Glycosyl Azides

Pinter and co-wokers [80] described the synthesis of bicyclic *trans*-fused glucopyranoso-oxazolidinone **73** and the *galacto* isomer by reaction of the corresponding glycopyranosyl phosphinimines with carbon dioxide (Scheme 18). A one-pot procedure to prepare **73** from glucopyranosyl azide **70** in high yield (97%) has been described recently [81].

Pinter and co-wokers also studied the reaction of α-D-glucopyranosyl and α-D-xylopyranosyl azides **75** and **76** with PPh$_3$ and CO$_2$ (Scheme 19). This gave a mixture of the expected bicyclic *cis*-fused glycopyranoso-oxazolidinones **77** and **78**, together with the α-D-furanose analogues **79** and **80**, and unexpected 1,3-*cis* bicyclic carbamates [82].

Pinter's and Ichikawa's groups have used **73** (Scheme 18) for the synthesis of *N*-glycosyl ureas **74** in high yield in DMF [83] and in water [81, 84, 85] in spite of the relative sensitivity of **73** to water. Following the same methodology, cellobiosyl and lactosyl azides **81** and **82** were transformed (Scheme 20) into the corresponding disaccharide oxazolidinones **83** and **84**, which were used for the preparation of urea-tethered pseudo-oligosaccharides **85** and **86** [81].

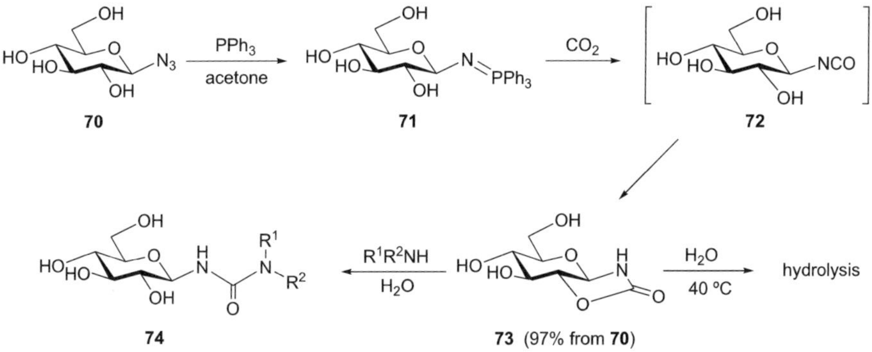

Scheme 18

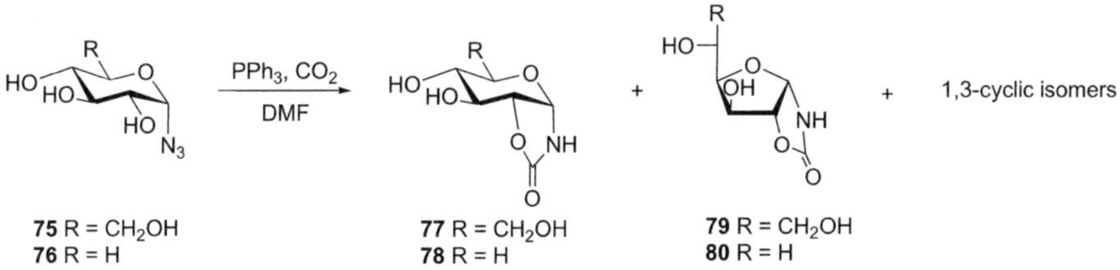

Scheme 19

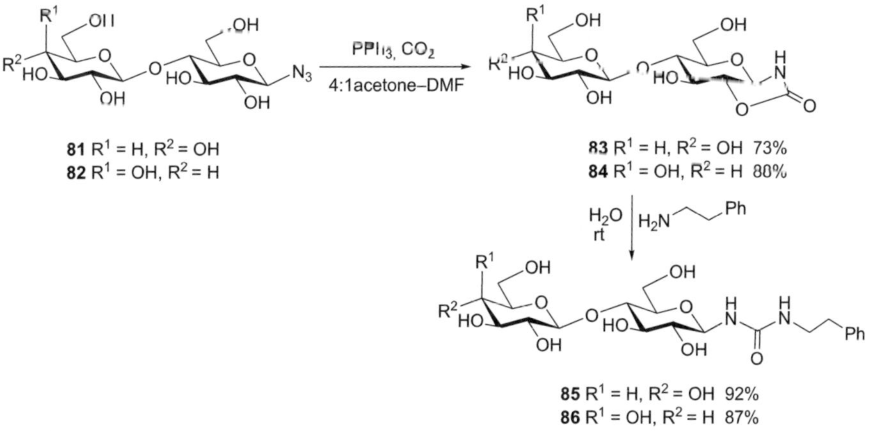

Scheme 20

No evidence of an isocyanate/carbamate equilibrium has been detected for **73**; in contrast, such equilibrium was reported for its isothiocyanate isoster [86, 87]. PM3 semi-empirical quantum chemical computations, undertaken to elucidate the mechanism involved in the synthesis of **73**, suggested [88] a glycosyl isocyanate intermediate **72**, the **72**→**73** step being an exothermic process ($\Delta E = -23.2\,\text{kJ mol}^{-1}$).

The twisted structure of the oxazolidinone ring of **73**, preventing delocalization of the lone pair electrons on nitrogen into the carbonyl group, has been argued to explain the increased reactivity of **73** towards nucleophilic attack of amines or thiols [81]. However, the formation of *N*-glycosyl ureas or thiolcarbamates through a transient glycosyl isocyanates should not be discarded.

Oxazoline **92** has been prepared [89] by treatment of acetylated glucopyranosyl azides of either β- or α-configurations **87** and **88** with PPh₃ in refluxing 1,2-dichloroethane (Scheme 21) by a mechanism involving α/β-anomerization of the intermediate phosphinimines **89** and **91** followed by ring closure of the α-anomer **91**. Coupling of in-situ generated oxazoline **92** with different acylating reagents gave mixtures of α- and β-glucopyranosyl amides, the α/β ratio being dependent on the acylating reagent and metal salts employed. The best results in terms of yield and stereoselectivity were obtained with thiopyridyl ester, in the presence of $CuCl_2$ to give the α-glucopyranosyl amide **93** (α/β ratio 98 : 2) in excellent yield.

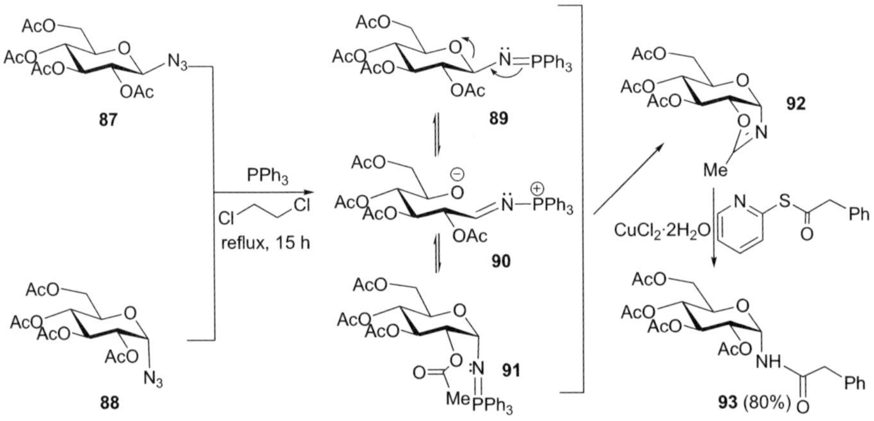

Scheme 21

Glycosyl azides bearing an additional azido group at C-1, such as **94** and its *manno* and *galacto* isomers, were transformed into bicyclic glycopyranoso[3,4-*d*]-1,2,3-triazoles [90] by an anomalous Staudinger reaction (Scheme 22). Thus, reaction of per-*O*-acetylated glucopyranosylidene 1,1-diazide **94** [91] with triphenylphosphine led to resonance-stabilized *v*-triazolo-pyranosyl phosphinimine derivative **95** in high yield (86%). A mechanism involving β-elimination of acetic acid and cycloaddition of the azido anion to the C-2,C-3 double bond and elimination of a second molecule of acetic acid was suggested [92]. The glycopyranosylidene diazides could

94 → **95** (86%)

96 (48%)

Scheme 22

easily obtained from the corresponding 1,1-dihalogenated precursors [91]. Transformation of the fused heterocyclic iminophosphorane **95** on treatment with aqueous ethanolic ammonia gave v-triazole-carboxamide **96**, bearing a chiral trihydroxypropyl side-chain [91].

4.2
Thermolysis of Glycosyl Azides

A synthetic way to incorporate monosaccharides (D-glucose, D-galactose) and oligosaccharides (lactose, maltose, maltotriose) into [60]fullerene by thermolysis of per-O-acetyl glycosyl azides in the presence of fullerene (Scheme 23) was reported by Kobayashi [93]. This procedure, based on the cycloaddition reaction of alkyl azides with fullerenes [94], gave a mixture of two inseparable stereoisomers of N-β-glycopyranosyl azafulleroids in moderate 13–28% yields.

The reaction of per-O-acetyl-β-D-glucopyranosyl azide **87** with [60]fullerene gave **97** as a mixture of two isomers in a molar ratio of ca. 2 : 1

87 → **97**   28% (57% based on consumed C_{60})

Scheme 23

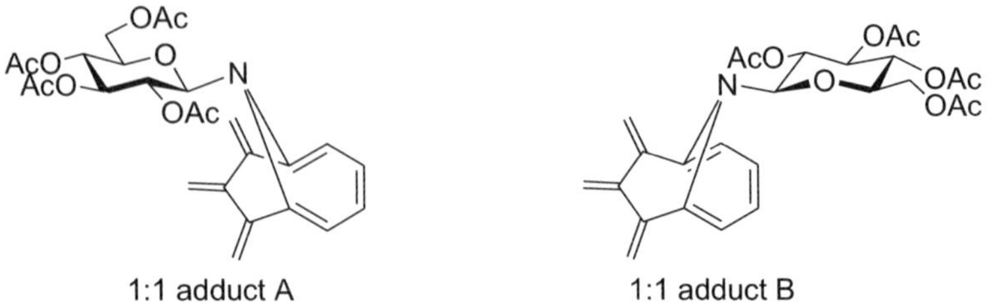

Fig. 1 Suggested structures of adducts A and B

(adducts A and B in Fig. 1). It was suggested that the pyramidal inversion at the nitrogen atom was "frozen" by steric or electronic restrictions to afford two *N*-invertomers of azafulleroids.

Thermolysis of both anomers of acetylated 5-thio-D-xylopyranosyl azide **98** and **99** by refluxing in chlorobenzene afforded tetrahydrothiazepine **100** by loss of N_2 and ring expansion [95] (Scheme 24). No differences in substrate reactivity and substrate selectivity could be detected for either anomer. A concerted process in which the sulfur atom assists the departure of molecular nitrogen was suggested. In a similar way, methyl 1-azido-D-glycopyranosides rearranged by photolysis with a stereocontrolled migration of the endocyclic oxygen atom to afford an oxazepine derivative [96]. In contrast, photolysis of thiosugar derivatives **98** and **99** led to decomposition compounds [95].

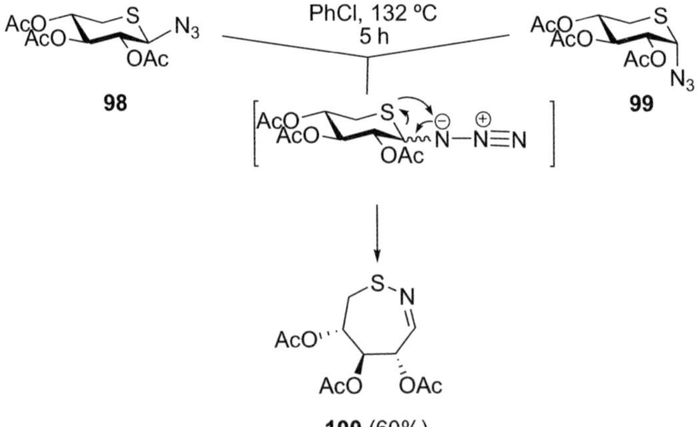

Scheme 24

4.3
Synthesis of Tetrazoles

Amide-linked *N*-glycosyl furane **103** and its thiophene analogue have been prepared (Scheme 25) by the Staudinger–Aza–Wittig reaction using glycosyl azides with DPPE, 1,2-bis(diphenylphosphino)ethane, in the presence of acid chlorides [26]. This type of glycosylamide heterocycles have been shown to possess interesting biological activities as endothelial cell growth inhibitors [97, 98]. DPPE has been shown to be a convenient substitute for triphenylphosphine and for polymer-supported triphenylphosphine [28] in Staudinger-type reactions in terms of cost, reaction time and easy removal of by-products [98]. The imidoyl chloride intermediate **102** could be isolated and transformed with high yield into the glycosyl tetrazole **106** by nucleophilic attack of azide, subsequent cyclization of **104** and loss of the chloride of **105** [26].

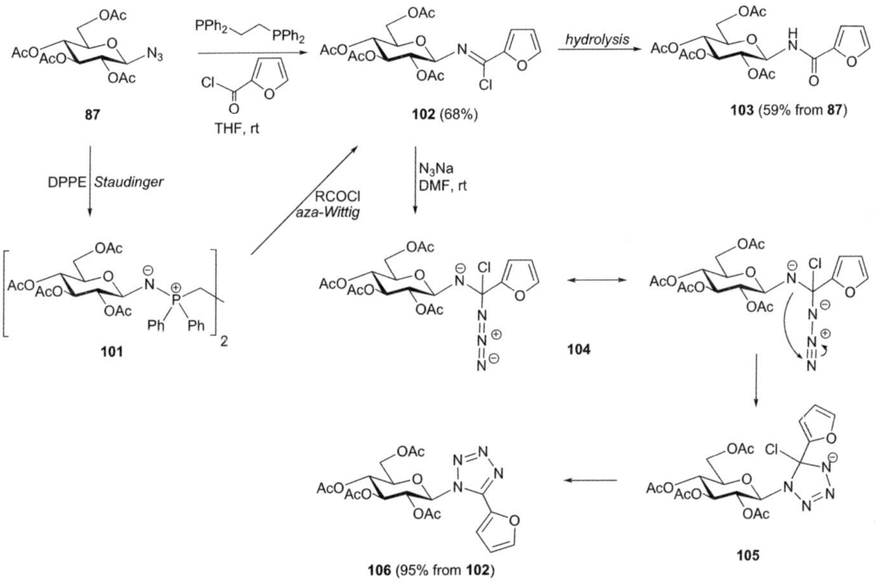

Scheme 25

The attempt at preparing the 6-cyano derivative **108** by displacement of the tosyloxy group at C-6 by reaction of glycosyl azide **107** with KCN/DMF at 60 °C gave an unexpected tricyclic derivative **110** [99] in a 58% yield (Scheme 26). Its formation could be explained by 1,3-dipolar cycloaddition between the 6-cyano and the 1-azido groups of **108**, followed by a tetrazole ring-opening [100] of tetracyclic **109**. The last step, involving breakage of the

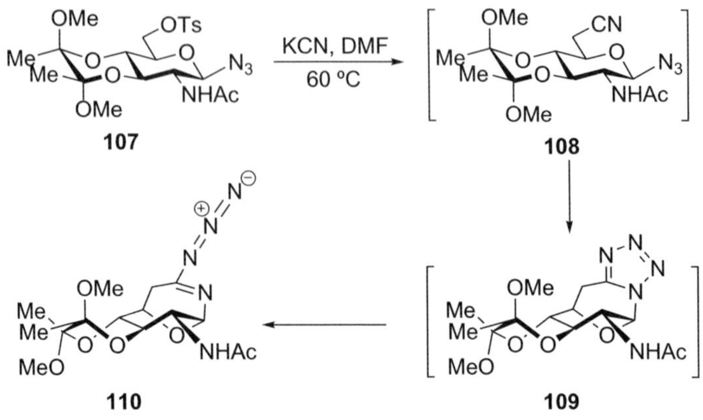

Scheme 26

N – N bond, was related to the high conformational strain of **109** [99]. No evidence of the configuration of the acetal carbons of the 1,4-dioxane ring was reported.

4.4
Synthesis of Triazoles

Triazoles comprise an interesting family of heterocyclic compounds displaying important biological activities such as insecticidal [101], anticancer [102–104], antiviral [104], HIV protease inhibitors [105] or antibacterial [106].

4.4.1
Dimroth Reaction

8-Aza-3-deazaguanine nucleosides **113a–e** have been prepared [107] from O-protected glycosyl azides **111a–e** by a Dimroth reaction [108]. Treatment of **111a–e** with dimethyl 3-oxoglutarate in DMSO in the presence of K_2CO_3 (Scheme 27) gave glycosyl triazoles **112a–e** as single regioisomers, followed by selective manipulation of the 1,2,3-triazole C-4 and C-5 substituents. The Dimroth reaction has been interpreted as either an azido cycloaddition to the enolate intermediate of the CH-active compound or a nucleophilic addition of a carbanion to the azido group [108]. 8-Aza-3-deazaisoguanine nucleosides have been prepared following a similar pathway [109].

4.4.2
Synthesis of Triazoles from Cyclic Acyl Ketene Aminals

Huang et al. have described the reaction of acylated β-D-ribofuranosyl [103] and β-D-glucopyranosyl [110, 111] azides **111a** and **87**, respectively, with het-

Scheme 27

erocyclic ketene aminals (cyclic 1,1-enediamines) such as **114** or **116** to give
N-glycosyl triazoles **115** and **117** in moderate to excellent yields (Scheme 28).
Some of the prepared glucopyranosyl triazoles were found to be biologically active, showing antitumor and antiviral activities [103, 111]. It was suggested that these cyclic 1,1-enediamines behave mostly as nucleophiles rather
than 1,3-dipolarophiles [112]. Therefore, the glycosyl triazoles are formed by
nucleophilic attack of the α-carbon of the ketene aminal to the terminal nitrogen atom of the azide, followed by the subsequent cyclocondensation and
aromatization with loss of a water molecule [112].

Scheme 28

4.4.3
Synthesis of Triazoles from Phosphorus Ylides

Several D-ribofuranosyl [113] and L-arabinofuranosyl 1,2,3-triazole derivatives [114] such as **119** have been prepared by cycloaddition of the corresponding glycosyl azide to 2-ethoxycarbonyl-2-oxoethylidene-triphenylphosphorane, followed by reaction with methanolic ammonia or primary or secondary amines (Scheme 29). Some of prepared nucleosides proved to be potent inhibitors of HIV-1 replication. Phosphorus ylides containing an α-oxo group in the chain behave in a similar way as activated acetylenes with electron-donating groups [108]. Presumably the azide adds to the enolate form of the ylide, followed by elimination of Ph_3PO [108].

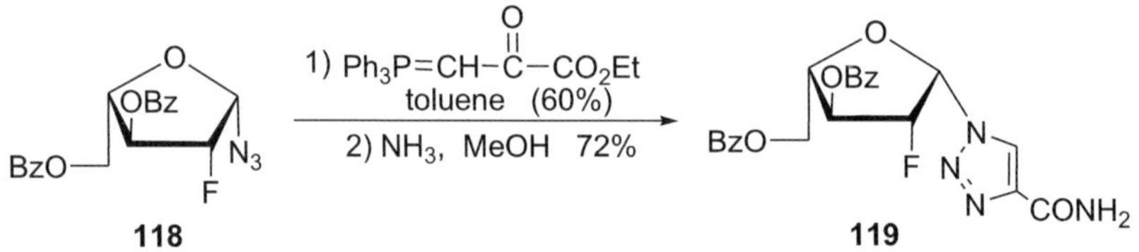

Scheme 29

4.4.4
Thermal 1,3-Dipolar Reactions

Thermal 1,3-dipolar reaction of azido and alkyne derivatives, known as Huisgen cyclization [115], leads to 1,2,3-triazoles as a mixture of 1,4- and 1,5-regioisomers, as activation energies for both processes are quite close to each other (Scheme 30) [116]. Only when terminal electron-deficient alkynes are used good regioselectivity is obtained in thermal conditions [117], the 1,4-adduct being the major regioisomer.

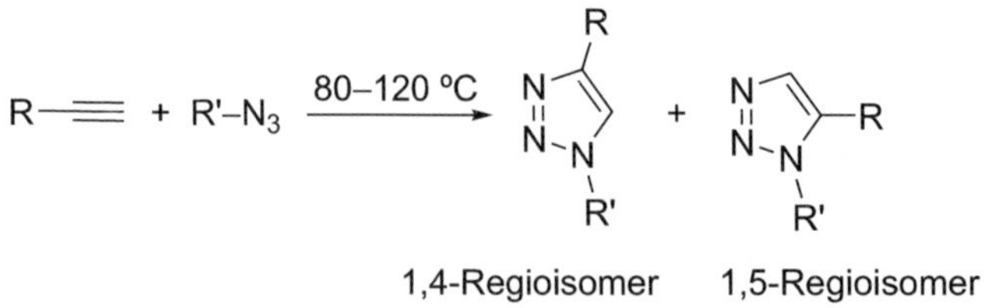

Scheme 30

Despite being quite energetic species, azides and alkynes are relatively low reactive species, for kinetic reasons [118], making the title cycloaddition slow and temperature-demanding. On the other hand, these functional groups are compatible with a wide range of temperature, pH and solvents.

As an example, Hager and co-workers [119] carried out the thermal reaction between per-*O*-acetylated β-D-galactopyranosylazide and the non-terminal alkyne 3,3,3-trifluoropropynylbenzene, to give a mixture of the expected regiosiomers.

Glycosyl azide **121** bearing an additional cyano group at C-1, easily prepared from the 1-bromo-glycosyl cyanide **120** [120], was transformed (Scheme 31) into the triazole-dicarboxylic acid derivative **122** by 1,3-dipolar cycloaddition of the azido moiety to dimethyl acetylenedicarboxylate (DMAD) [121]. In a similar way, acetylated α- and β-cellobiosyl, lactosyl, maltosyl and melibiosyl azides, and the commercially available acetylated β-D-glucopyranosyl azide were transformed in excellent yield into the corresponding glycobiosyl [122] and glucopyranosyl [123, 124] 1,2,3-triazoles by reaction with various acetylene dicarboxylic acid esters (methyl, ethyl, *tert*-butyl). Compound **124**, obtained from protected cellobiosyl azide **123** (Scheme 31), was used as glycosyl donor in the syn-

Scheme 31

thesis of an anthracycline-type antibiotic [122]. A lactosaminyl triazole, obtained from protected lactosamine azide by 1,3-dipolar cycloaddition with di-*tert*-butyl acetylenedicarboxylate, has been converted into a glycosyl fluoride [125].

Al-Masoudi and Al-Soud reported [126, 127] the preparation of pseudo-nucleosides containing 1,2,3-triazole and 1,2,3-triazolo[4,5-*e*][1,3]-diazepin-4-one moieties as potential agents against breast, lung or central nervous system tumors. These compounds were prepared starting from per-*O*-acetylated β-D-glucopyranosyl azide **87**, which was subjected to a thermal 1,3-dipolar cyclization with dimethyl acetylenedicarboxylate to give the corresponding glucosyl triazole **125** (Scheme 32). 4,5-Dicarboxamide derivatives **126–128** were obtained upon treatment of **125** with the corresponding aliphatic amines. Treatment of **125** with piperazine, morpholine and thiomorpholine in the presence of NaOMe afforded derivatives **129–131** bearing a carboxamide moiety on C-4 and a carboxylate group on C-5 of the triazole ring.

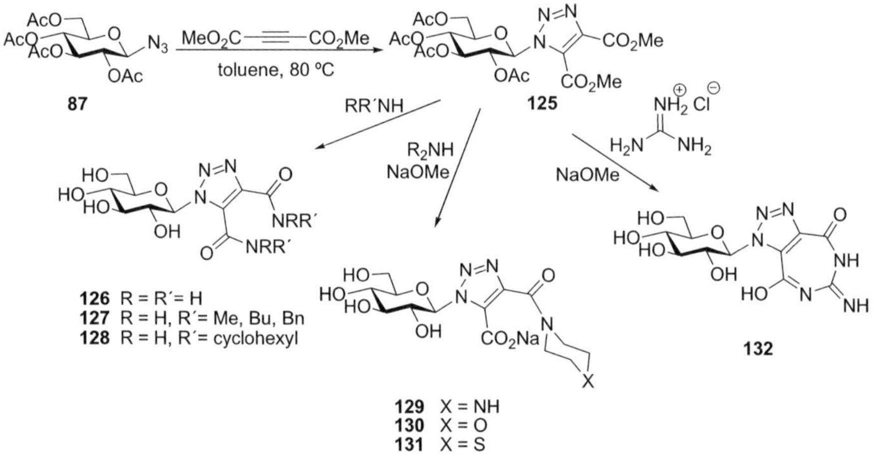

Scheme 32

Finally, when the reaction on **125** was carried out using guanidine 1,2,3-triazolo[4,5-*e*]-1,3-diazepin-4-one **132** was obtained. These pseudo-nucleosides were tested against 60 human tumor cell lines, but showed poor or no anticancer activity [127].

Phospha [128], such as **135** and **136**, and carba [129, 130] sugar nucleosides have been synthesized by 1,3-dipolar cycloaddition reaction of azido pseudo-sugar derivatives with alkynes (Scheme 33). The strong electron-withdrawing phosphoryl group of **134** exerted no effect over regioselectivity, but steric effects played an important role in the selectivity [128]. The car-

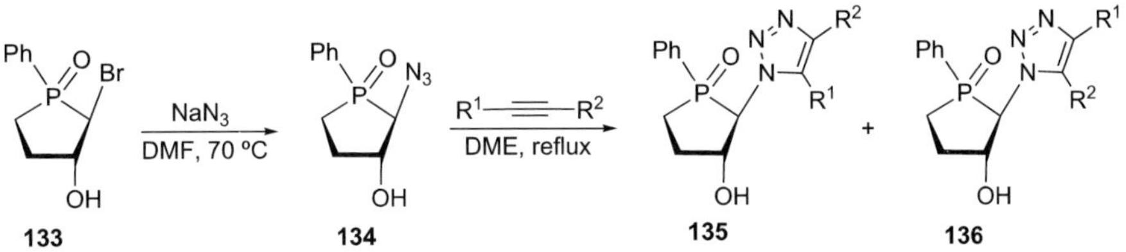

Scheme 33

bocyclic analogues of anti-HCV ribavirin displayed moderate activity against HIV-1 [129].

4.4.5
Cu(I)-Mediated 1,3-Dipolar Reactions

Cu(I) salts make the reaction between azides and alkynes faster (up to 10^7 times) and regiospecific [131, 132] (only the 1,4-regioisomer is obtained). The reaction usually proceeds at room temperature and is compatible with a plethora of functional groups present in the substrates. Many different sources of Cu(I) species have been used, such as Cu(I) salts [132], Cu(I) complexes soluble in organic media [133], and Cu(I) derivatives [134] stabilized with polytriazolylamines. Furthermore, Cu(I) species can be also be obtained by in-situ reduction of Cu(II) derivatives [131]. A standard procedure involves the use of a mixture of Cu(II) sulfate and sodium ascorbate as a source of Cu(I).

Reaction by a stepwise mechanism has been postulated [116, 131, 135], instead of a concerted one, the first step being the formation of an acetylene-Cu(I) π complex that reduces the pK_a value of the terminal acetylene. This can be deprotonated even in water [135], allowing the formation of the copper acetylide (Scheme 34). Next, attack of the azido derivative provokes the formation of a Cu adduct that undergoes intramolecular cyclization to give

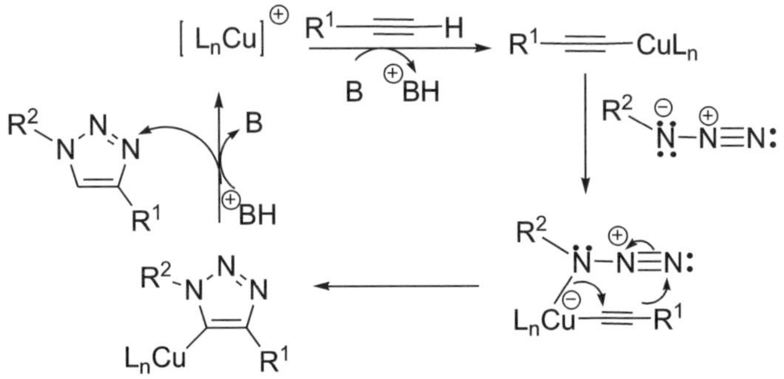

Scheme 34

a 1,2,3-triazole-Cu derivative; final protonation of the triazole ring regenerates the Cu(I) catalyst.

Due to its extraordinary efficiency, the Cu(I)-catalysed Huisgen cycloaddition has become the prototype of *click chemistry*, a concept introduced by Sharpless in 2001 [136] to refer to a group of reactions that obey a series of criteria, such as being modular, stereoespecific and high yielding, using readily available starting materials and benign conditions, with easy product isolation and having a large thermodynamic driving force. Usually reactions involving carbon–heteroatom bond formation are included in the click chemistry concept, such as epoxide and aziridine opening, formation of ureas and amides, additions to double carbon–carbon bonds such as epoxidation and dihydroxylation, and 1,3 dipolar and Diels–Alder cycloadditions [136].

Click chemistry has proved to be a versatile strategy and has undergone a spectacular development in the last few years, as a broad spectrum of applications has been found. Thus, for example, this concept has been applied to the preparation of polymers [137–140], dendrimers [141–144], glycoclusters [145] and pharmacophores [118]. The 1,2,3-triazole moiety was found to be a mimic of the amide moiety, characterized by an enhanced stability towards hydrolysis and oxidations [118].

Click chemistry has also been applied in carbohydrate chemistry. Thus, Huisgen cyclization involving a sugar azide or alkyne has been extensively used for the preparation of a wide range of carbohydrates bearing a triazole moiety in different positions. In this volume, Santoyo González and Hernández-Mateo present a contribution concerning azide–alkyne 1,3-dipolar cycloadditions, mostly devoted to non-anomeric positions.

Glycosyl triazoles have been prepared from glycosyl azides by different authors [146, 147] using Cu(I)-catalysed Huisgen cycloaddition. The same reaction has been employed for the preparation of polyvalent glycoconjugate clusters (Scheme 35). For example, treatment of O-protected glucopyranosyl

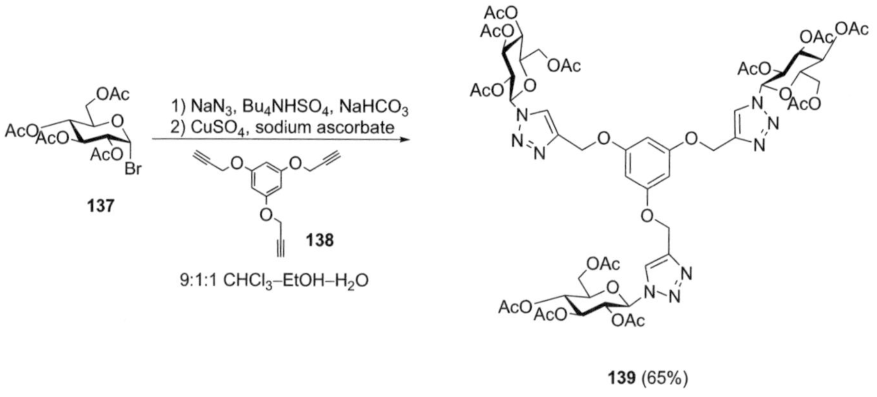

Scheme 35

bromide **137** with trivalent compound **138**, in the presence of $CuSO_4$ and sodium ascorbate, gave dendritic derivative **139** in moderate yield [146].

Another selective and effective preparation of 1,2,3-triazole-linked glyco-dendrimers was accomplished by Joosten et al. [148]. It was observed that for these compounds the standard conditions for Huisgen-catalysed reactions led to slow reactions and that only partial coupling between the polyvalent alkynes and the sugar azides took place. However, when trying to optimize the procedure, microwave-assisted cyclizations proved to occur faster and the reaction went to completion when an excess of the azido derivative was used (Scheme 36). For instance, using per-*O*-acetylated β-D-galactopyranosyl azide and a tetravalent second generation core molecule, dendrimer **140** was prepared in almost quantitative yield. This reaction was also successfully applied [148] to glycobiosyl azides and more branched core molecules.

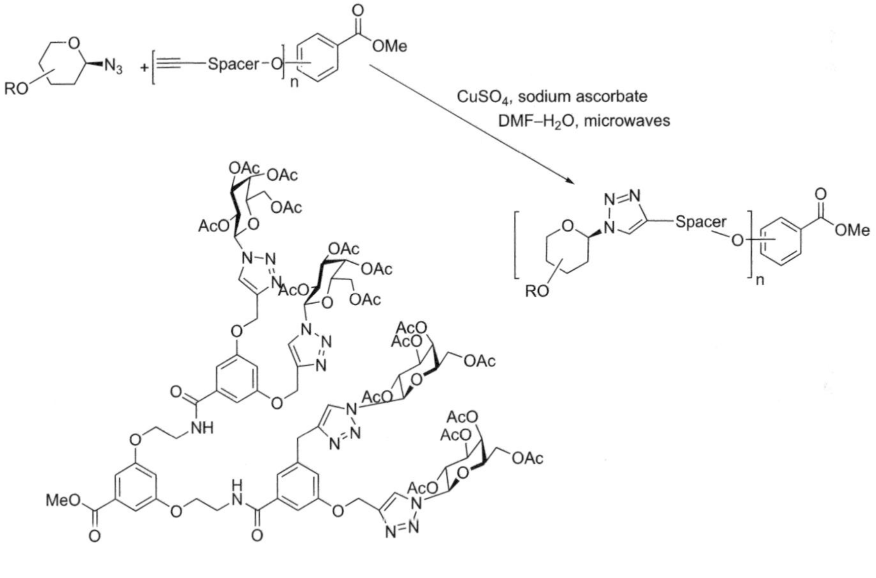

140 (98%)

Scheme 36

Dörner and Westermann [149] reported the preparation of the pseudo-disaccharide **143** with a triazole tether (Scheme 37) as a prototype for the synthesis of carbohydrate-containing macrocycles, through a Huisgen-catalysed reaction and ring-closing methathesis as the key steps. Standard conditions were used, $Cu(OAc)_2$ and sodium ascorbate as the Cu(I) source and a mixture of water and *t*-BuOH as solvent, for the conversion of protected glucopyranosyl azide **141** into pseudo-disaccharide **143** by using dialkyne **142** in a 77% yield.

Scheme 37

Temelkoff et al. accomplished [150] the preparation of neoglycotrimers in which the sugar units are connected through a linkage composed of an amide moiety and a 1,2,3-triazole ring (Scheme 38). The authors claim that interest in final compounds, such as **147**, relies on structural similarity to carbopeptoids, compounds with great potential in medicinal chemistry [150].

Scheme 38

Synthesis is carried out by direct coupling of glycosyl azide **87** and alkyne **144**, prepared from the corresponding glucuronic acid [151]. The corresponding pseudo-disaccharide **145** was converted into glycopyranosyl azide **146** by hydrobromolysis followed by displacement of the transient glycosyl bromide with sodium azide.

The second iteration under the standard cycloaddition conditions to obtain **147** failed and only traces of the desired compound were obtained. A plausible explanation for this could be the very low solubility of **146** in the aqueous medium. The use of an organic solvent, together with $Cu(PPh_3)_3Br$ as catalyst [133] provided **147** in a 54% yield (Scheme 38).

A much more complex oligosaccharidic structure was achieved by Marmuse and co-workers [152], who carried out the assembly of starch fragments using click chemistry as the key step (Schemes 39 and 40). They reported the preparation of amylopectine analogues such as **154**, a pseudo-starch fragment comprised of two linear maltoheptaose chains connected through a maltose unit bearing two triazole rings. This strategy, as compared to the classi-

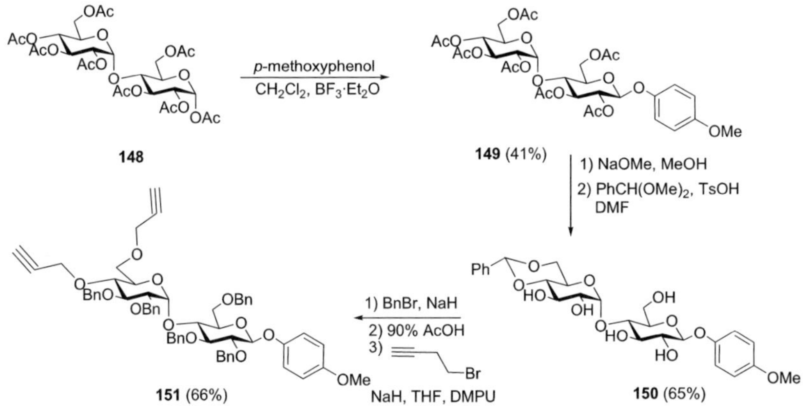

Scheme 39

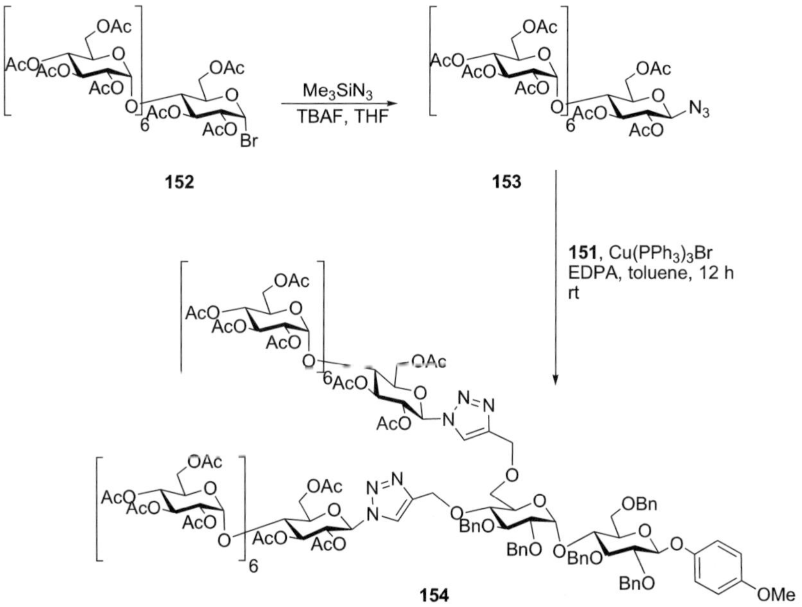

Scheme 40

cal oligosaccharide synthesis, lacks the problems associated with anomeric selectivity.

Thus, per-O-acetylated maltose **148** was glycosylated with p-methoxyphenol using $BF_3 \cdot Et_2O$ as catalyst (Scheme 39); β-glycoside **149** was crystallized from the anomeric mixture in a 41% yield. Deacetylation in basic medium, followed by benzylidene protection of the second saccharidic unit afforded derivative **150**. Dipropargylated maltose derivative **151** was obtained by benzylation of the free hydroxyl groups of **150** and introduction of the propargylic units after removal of the benzylidene protective group [152].

Maltoheptaosyl bromide **152** [153] was converted (Scheme 40) into glycosyl azide **153** under mild conditions using trimethylsilyl azide in the presence of tetrabutylammonium fluoride (TBAF). Final cycloaddition reaction between propargylic derivative **151** and glycosyl azide **153** in the presence of $Cu(Ph_3P)_3Br$ as catalyst [133], at room temperature and prolonged reaction times, afforded **154** in a 27% yield.

Bodine et al. accomplished [154, 155] the preparation of cyclic pseudo-oligosaccharides bearing triazole rings that resemble the structure of cyclodextrins, maltooligosaccharides that have become of great interest in drug transportation [156] and supramolecular chemistry [157].

For instance, the preparation of **160** was carried out [155] starting from partially protected mannopyranoside **155** [158], which was converted into the propargylic derivative **156** by alkylation with propargyl bromide followed by removal of the aglycon in oxidative conditions (Scheme 41). Coupling of **155** with **156** was afforded by activating the reductive derivative with Ph_2SO and Tf_2O; the aglycon was removed as indicated for **155**. Another iteration of this process for **157** gave trisaccharide **158**. Activation of **158** and treatment with Me_3SiN_3 afforded α-glycosyl azide **159**, which bears alkyne and

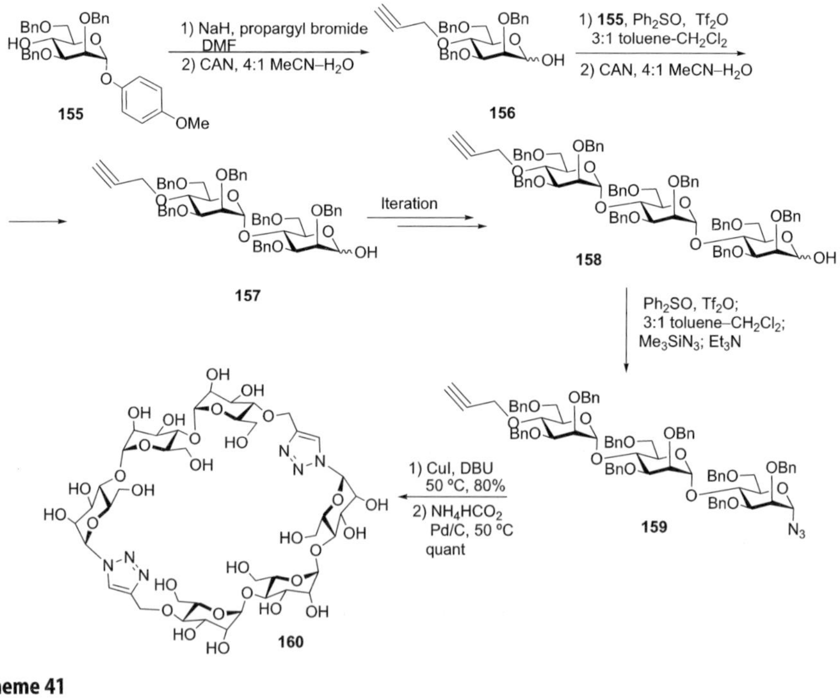

Scheme 41

azido moieties in the same molecule. Finally, cyclization gave the corresponding cyclic pseudo-oligosaccharide **160** with two triazole rings tethers. It is remarkable that by using standard Cu(I)-catalysed Huisgen reaction, decomposition took place. However, combination of CuI and DBU afforded the desired cyclodimerization, which was favoured under these conditions as compared to potential linear oligomerization.

The complexation properties of **160** were studied [155] with 8-anilino-napthalene-1-sulfonate as substrate by detecting changes in the fluorescence spectra. As compared to β-cyclodextrin (cyclomaltoheptaose), it turned out that compound **160** retained the binding properties of the natural cyclodextrin; this feature enhances the importance of these artificial analogues.

As an attempt at preparing biologically active compounds, Ermolat'ev et al. reported [159] the preparation of pseudo-*N*-nucleosides **164**, by assembling a 2(1*H*)-pyrazinone scaffold to a saccharidic residue through a 1,2,3-triazole tether, as indicated in Scheme 42. 2(1*H*)-Pyrazinone derivative **161** was converted into acetilenic derivative **162** through a Pd(0)-catalysed microwave-assisted Sonogashira reaction. Triazole derivative **163** was obtained by a Huisgen-Cu(I) catalysed reaction, assisted by microwaves and a polytriazolylamine as a stabilizing agent for the Cu(I) salt [134]. Final dechlorination of the aromatic ring was afforded by standard Pd-catalysed hydrogenolysis to give compounds **164** in moderate yields.

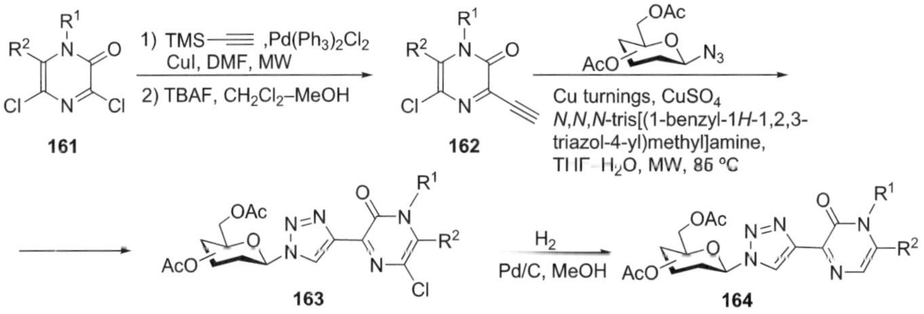

Scheme 42

Acknowledgements The authors thank the Ministerio de Educación y Ciencia for a grant (CTQ2005-01830/BQU) and the Junta de Andalucía (FQM134) for financial support.

References

1. Kaneshiro CM, Michael K (2006) Angew Chem Int Ed 45:1077
2. Elizabeth C, Maljaars P, Halkes KM, de Oude WL, van der Poel S, Pijnenburg NJM, Kamerling JP (2005) J Carbohydr Chem 24:353
3. Ishiwata A, Takatani M, Nakahara Y, Ito Y (2002) Synlett, p 634

4. Taylor CM (1998) Tetrahedron 54:11317
5. Mustafina SR, Baltina LA Jr, Kondratenko RM, Baltina LA, Galin FZ, Tolstikov GA (2006) Chem Nat Compd 42:67
6. Brun MA, Disney MD, Seeberger PH (2006) ChemBioChem 7:421
7. Hiratake J, Sakata K (2003) Methods Enzymol 363:421
8. El Ashry ESH, Rashed N, Shobier AHS (2000) Pharmazie 55:403
9. Knauer S, Kranke B, Krause L, Kunz H (2004) Curr Org Chem 8:1739
10. Isbell HS, Frush HL (1958) J Org Chem 23:1309
11. Isbell HS, Frush HL (1980) Methods Carbohydr Chem 8:255
12. Frush HL, Isbell HS (1951) J Res Natl Bur Standards 47:239
13. Likhosherstov LM, Novikova OS, Derevitskaya VA, Kochetkov NK (1986) Carbohydr Res 146:C1
14. Kallin E, Loenn H, Norberg T, Elofsson M (1989) J Carbohydr Chem 8:597
15. Likhosherstov LM, Novikova OS, Shibaev VN (2002) Dokl Chem 383:89
16. Likhosherstov LM, Novikova OS, Shibaev VN (2003) Dokl Chem 389:73
17. Likhosherstov LM, Novikova OS, Zheltova AO, Shibaev VN (2004) Russ Chem Bull 53:709
18. Hackenberger CPR, O'Reilly MK, Imperiali B (2005) J Org Chem 70:3574
19. Danishefsky SJ, Bilodeau MT (1996) Angew Chem Int Ed 35:1380
20. Dorsey AD, Barbarow JE, Trauner D (2003) Org Lett 5:3237
21. Nicolaou KC, Snyder SA, Nalbandian AZ, Longbottom D (2004) J Am Chem Soc 126:6234
22. Györgydeák Z, Thiem J (2006) Adv Carbohydr Chem Biochem 60:103
23. Hayashi M, Kawabata H (2003) Recent Dev Carbohydr Res 1:195
24. Györgydeák Z, Szilagyi L, Paulsen H (1993) J Carbohydr Chem 12:139
25. Bräse S, Gil C, Knepper K, Zimmermann V (2005) Angew Chem Int Ed 44:5188
26. Temelkoff DP, Smith CR, Kibler DA, McKee S, Duncan SJ, Zeller M, Hunsen M, Norris P (2006) Carbohydr Res 341:1645
27. Bianchi A, Bernardi A (2006) J Org Chem 71:4565
28. Root YY, Bailor MS, Norris P (2004) Synth Commun 34:2499
29. Kovács L, Osz E, Györgydeák Z (2002) Carbohydr Res 337:1171
30. Bianchi A, Ferrario D, Bernardi A (2006) Carbohydr Res 341:1438
31. Lang S, Murphy JA (2006) Chem Soc Rev 35:146
32. Gómez Sánchez A, Gómez Guillén M, Cert Ventulá A, Scheidegger U (1968) An Quím 64:579
33. Gómez-Sánchez A, Borrachero Moya P, Bellanato J (1984) Carbohydr Res 135:101
34. Babiano Caballero R, Fuentes Mota J, Galbis Pérez JA (1986) Carbohydr Res 154:280
35. Fuentes J, Olano D, Gasch C, Pradera MA (2000) Tetrahedron: Asymmetry 11:2471
36. Benito JM, Ortiz Mellet C, Sadalapure K, Lindhorst TK, Defaye J, García Fernández JM (1999) Carbohydr Res 320:37
37. Fuentes J, Moreda W, Ortiz C, Robina I, Welsh C (1992) Tetrahedron 48:6413
38. Fuentes Mota J, García Fernández JM, Ortiz Mellet C, Pradera Adrián MA, Babiano Caballero R (1989) Carbohydr Res 188:35
39. López Ó, Maza S, Maya I, Fuentes J, Fernández-Bolaños JG (2005) Tetrahedron 61:9058
40. Fuentes J, Molina JL, Olano D, Pradera MA (1996) Tetrahedron: Asymmetry 7:203
41. Pradera MA, Olano D, Fuentes J (1995) Tetrahedron Lett 36:8653
42. McNaught AD (1996) Pure Appl Chem 68:1919
43. Fuentes J, Olano D, Pradera MA (1997) Tetrahedron: Asymmetry 8:3443

44. Fuentes J, Sayago FJ, Illangua JM, Gasch C, Angulo M, Pradera AM (2004) Tetrahedron: Asymmetry 15:603
45. Lee RE, Smith MD, Nash RJ, Griffiths RC, McNeil M, Grewal RK, Yan W, Besra GS, Brennan PJ, Fleet GWJ (1997) Tetrahedron Lett 38:6733
46. Siciliano T, De Leo M, Bader A, De Tommasi N, Vrieling K, Braca A, Morelli I (2005) Phytochemistry 66:1593
47. Singh B, Sahu PM, Jain SC, Singh S (2002) Pharm Biol 40:581
48. Ji L-L, Zhao X-G, Chen L, Zhang M, Wang Z-T (2002) Toxicon 40:1685
49. Lee ST, Schoch TK, Stegelmeier BL, Gardner DR, Than KA, Molyneux RJ (2001) J Agric Food Chem 49:4144
50. Pradera MA, Sayago FJ, Illangua JM, Angulo M, Gasch C, Fuentes J (2004) Tetrahedron: Asymmetry 15:2003
51. Pradera MA, Sayago FJ, Illangua JM, Gasch C, Fuentes J (2003) Tetrahedron Lett 44:6605
52. van Well RM, Kaerkkaeinen TS, Kartha KPR, Field RA (2006) Carbohydr Res 341:1391
53. Mitsudo K, Matsuda W, Miyahara S, Tanaka H (2006) Tetrahedron Lett 47:5147
54. Deng S, Gangadharmath U, Chang C-WT (2006) J Org Chem 71:5179
55. Adamo R, Kovac P (2006) Eur J Org Chem, p 2803
56. Witczak ZJ, Culhane JM (2005) Appl Microbiol Biotechnol 69:237
57. MacDougall JM, Zhang X-D, Polgar WE, Khroyan TV, Toll L, Cashman JR (2004) J Med Chem 47:5809
58. Fuentes J, Illangua JM, Sayago FJ, Angulo M, Gasch C, Pradera MA (2004) Tetrahedron: Asymmetry 15:3783
59. Fuentes J, Gasch C, Olano D, Pradera MA, Repetto G, Sayago FJ (2002) Tetrahedron: Asymmetry 13:1743
60. Fuentes J, Olano D, Pradera MA (1999) Tetrahedron Lett 40:4063
61. Li H, Blériot Y, Chantereau C, Mallet J-M, Sollogoub M, Zhang Y, Rodríguez-García E, Vogel P, Jiménez-Barbero J, Sinaÿ P (2004) Org Biomol Chem 2:1492
62. Andersen SM, Ekhart C, Lundt I, Stütz AE (2000) Carbohydr Res 326:22
63. Le Merrer Y, Poitout L, Depezay J-C, Dosbaa I, Geoffroy S, Foglietti MJ (1997) Bioorg Med Chem 5:519
64. Ávalos M, Babiano R, Cintas P, Hursthouse MB, Jimenez JL, Lerma E, Light ME, Palacios JC (2006) Tetrahedron Lett 47:1989
65. Tennant G (1979) In: Barton D, Ollis WD (eds) Comprehensive organic chemistry, vol 2. Pergamon, Oxford, p 525
66. Zech G, Kunz H (2003) Angew Chem Int Ed 42:787
67. Weymann M, Pfrengle W, Schollmeyer D, Kunz H (1997) Synthesis, p 1151
68. Weymann M, Schultz-Kukula M, Kunz H (1998) Tetrahedron Lett 39:7835
69. Jarrahpour AA, Shekarriz M, Taslimi A (2004) Molecules 9:29
70. Khalil NSAM (2005) Nucleosides Nucleotides Nucleic Acids 24:1277
71. Palomo C, Aizpurua JM, Ganboa I, Oiarbide M (2004) Curr Med Chem 11:1837
72. Venturini A, González J (2006) Mini Rev Org Chem 3:185
73. Hernando JIM, Laso NM, Anaya J, Gero SD (1997) Synlett, p 281
74. Anaya J, Gero SD, Grande M, Hernando JIM, Laso NM (1999) Bioorg Med Chem 7:837
75. Thiering ST, Sowa CE, Thiem J (2001) J Chem Soc Perkin Trans 1:801
76. Sowa CE, Kopf J, Thiem J (1995) Chem Commun, p 211
77. Böttcher C, Burger K (2003) Tetrahedron Lett 44:4223
78. Böttcher C, Spengler J, Henning L, Albericio F, Burger K (2005) Monatsh Chem 136:577

79. Böttcher C, Burger K (2002) Tetrahedron Lett 43:9711
80. Kovács J, Pintér I, Messmer A, Tóth G (1985) Carbohydr Res 141:57
81. Ichikawa Y, Matsukawa Y, Isobe M (2006) J Am Chem Soc 128:3934
82. Kovács P, Pintér I, Tóth G, Györgydeák Z, Köll P (1993) Carbohydr Res 239:95
83. Zsoldos-Mády V, Sohár P, Kovács J, Pintér I, Szakács Z (2005) J Carbohydr Chem 24:19
84. Ichikawa Y, Matsukawa Y, Isobe M (2004) Synlett, p 1019
85. Pintér I, Kovács J, Tóth G (1995) Carbohydr Res 273:99
86. López O, Maya I, Fuentes J, Fernández-Bolaños JG (2004) Tetrahedron 60:61
87. Maya I, López O, Fernández-Bolaños JG, Robina I, Fuentes J (2001) Tetrahedron Lett 42:5413
88. Friant-Michel P, Marsura A, Kovacs J, Pinter I, Rivail J-L (1997) THEOCHEM 395–396:61
89. Damkaci F, DeShong P (2003) J Am Chem Soc 125:4408
90. Kovács J, Pintér I, Kajtár-Peredy M, Argay G, Kálmán A, Descotes G, Praly JP (1999) Carbohydr Res 316:112
91. Praly JP, Péquery F, Di Stèfano C, Descotes G (1996) Synthesis, p 577
92. Kovács J, Pintér I, Kajtár-Peredy M, Praly JP, Descotes G (1995) Carbohydr Res 279:C1
93. Yashiro A, Nishida Y, Ohno M, Eguchi S, Kobayashi K (1998) Tetrahedron Lett 39:9031
94. Prato M, Li QC, Wudl F, Lucchini V (1993) J Am Chem Soc 115:1148
95. Praly JP, Péquery F, Hetzer G, Steng M (1999) J Carbohydr Chem 18:833
96. Di Stèfano C, Descotes G, Praly JP (1994) Tetrahedron Lett 93:93
97. Pitt N, Duane RM, O'Brien A, Bradley H, Wilson SJ, O'Boyle KM, Murphy PV (2004) Carbohydr Res 339:1873
98. O'Neil IA, Thompson S, Murray CL, Kalindjian SB (1998) Tetrahedron Lett 39:7787
99. Carmona AT, Fialova P, Kren V, Ettrich R, Martínková L, Moreno-Vargas AJ, González C, Robina I (2006) Eur J Org Chem, p 1876
100. Katrusiak A, Skierska A, Katrusiak A (2005) J Mol Struct 751:65
101. Boddy IK, Briggs GG, Harrison RP, Jones TH, O'Mahony MJ, Marlow ID, Roberts BG, Willis RJ, Bardsley R, Reid J (1996) Pestic Sci 48:189
102. Pati HN, Wicks M, Holt HL Jr, LeBlanc R, Weisbruch P, Forrest L, Lee M (2005) Heterocycl Commun 11:117
103. Chen X-M, Li Z-J, Ren Z-X, Huang Z-T (1999) Carbohydr Res 315:262
104. Safonova TS, Nemeryuk MP, Likhovidova MM, Sedov AL, Grineva NA, Keremov MA, Solov'eva NP, Anisimova OS, Sokolova AS (2003) Pharm Chem J 37:298
105. Brik A, Muldoon J, Lin YC, Elder JH, Goodsell DS, Olson AJ, Fokin VV, Sharpless KB (2003) ChemBioChem 4:1246
106. Lin HN, Walsh CT (2004) J Am Chem Soc 126:13998
107. Stimac A, Leban I, Kobe J (1999) Synlett, p 1069
108. Krivopalov VP, Shkurko OP (2005) Russ Chem Rev 74:339
109. Jeselnik M, Jaksa S, Kobe J (2004) Croat Chem Acta 77:153
110. Yang Q, Li Z-J, Chen X-M, Huang Z-T (2002) Heteroat Chem 13:242
111. Huang Z-T, Li Z-J, Chen X-M, Ren Z-X, Wang M-X, Liu B, Wang L-B, Wang H-T (2001) Chinese Patent ZL 96107062.5
112. Huang Z-T, Wang M-X (1992) J Org Chem 57:184
113. Velázquez S, Álvarez R, Pérez C, Gago F, De Clercq E, Balzarini J, Camarasa M-J (1998) Antiviral Chem Chemother 9:481
114. Olgen S, Chu CK (2001) Naturforsch B: Chem Sci 56:804

biological and pharmacological properties [19–21]. These heterocycles have also been isolated as hydrolysis products of naturally occurring glucosinolates having a hydroxyalkyl side chain [22–26]. In glycochemistry some nucleosides and related structures have also been prepared via 1,3-oxazolidine-2-thiones by reaction of aldoses and ketoses with thiocyanic acid [27–30], or with benzyl isothiocyanate [31]. Furthermore, the 1,3-oxazolidine-2-thione motif has been used extensively as a chiral auxiliary in organic synthesis [32–34].

1,3-Oxazolidine- and 1,3-oxazinane-2-thiones can be generated when an isothiocyanate and a hydroxyl group are simultaneously present in a sugar molecule. Facility and viability of this process are aspects that depend on the ring size as well as on the relative arrangement of both functional groups, and thus, on the inherent strain of the cyclic structure [6]. Usually only β- and γ-located hydroxyl groups with respect to the isothiocyanate moiety can undergo cyclization, although *trans*-diaxial arranged isothiocyanate and hydroxyl groups of a conformationally rigid pyranose ring do not lead to an oxazolidine-2-thione [35].

In this context, García Fernández and co-workers have described [36] the reactivity of methyl 6-deoxy-6-isothiocyanato-α-D-glucopyranoside **1** towards vicinal hydroxyl groups in the preparation of bicyclic sugar-derived thiocarbamates. The authors found that no reaction takes place even on heating **1** in DMF for prolonged time. Nevertheless, it underwent a fast intramolecular cyclization with the hydroxyl group on C-4 upon heating in the presence of a catalytic amount of Et$_3$N (Scheme 1) to afford the corresponding 1,3-oxazinane-2-thione **3**.

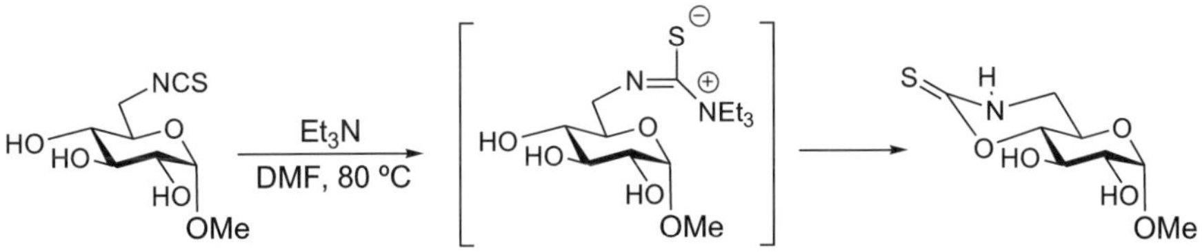

Scheme 1

It was postulated [36, 37] that the formation of **3** proceeds via a zwitterionic complex between the isothiocyanate moiety and triethylamine, and subsequent nucleophilic displacement by the γ-located hydroxyl group of the glucopyranoside.

The same behavior [38] was found for 6,6′-dideoxy-6,-6′-diisothiocyanato-α,α′-trehalose **5** (Scheme 2), prepared by isothiocyanation of **4**. Reactivity of fully-*O*-protected 6-deoxy-6-isothiocyanato derivatives with ammonia in several solvents to provide either thioureas or cyclic thiocarbamates has also been reported [39].

Scheme 2

The same group also reported [37] the synthesis of cyclic thiocarbamates by spontaneous annelation of an isothiocyanate motif and a hydroxyl group of a sugar derivative. Thus, treatment of 6-amino-6-deoxy-D-glucose, -D-galactose and -D-mannose **7–9** with thiophosgene in acetone-water containing $CaCO_3$ led to the pseudo-*C*-nucleosides **13–15** bearing an oxazolidine-2-thione motif in a 82–88% yield (Scheme 3), presumably through transient isothiocyanates **10–12** although they were not detected even at low temperature. It is remarkable that cyclization occurs with total regioselectivity, as the only observed thiocarbamates were those with a furanose structure, and no pyranose-derived thiocarbamates were obtained.

Scheme 3

Partial deprotection of di-*O*-isopropylidene-3-isothiocyanato furanose **16** with aqueous TFA (trifluoroacetic acid) at low temperature led to transient 3-deoxy-3-isothiocyanato derivative **17** that finally evolved to 1,3-oxazinane-2-thione **18** [40]. The outcome of the removal of the 1,2-di-*O*-isopropylidene group was strongly dependent on experimental conditions (Scheme 4). Thus, treatment of **18** with 90% aqueous TFA at 40 °C for 8 h resulted in **19** by hydrolysis of the protective group while the oxazinane-2-thione ring was kept. However, by reducing the rate of TFA and increasing the temperature up to 50 °C, opening of the thiocarbamate took place to afford 3-deoxy-3-isothiocyanato-D-glucopyranose **20**. More prolonged reaction times led to trihydroxyalkyl tetrahydro and dihydro 1,3-oxazine-2-thiones **21** (major) and **22** (minor), obtained by nucleophilic attack of the γ-located OH group of the hydrated form of the aldehyde group of the free sugar. This reaction is simi-

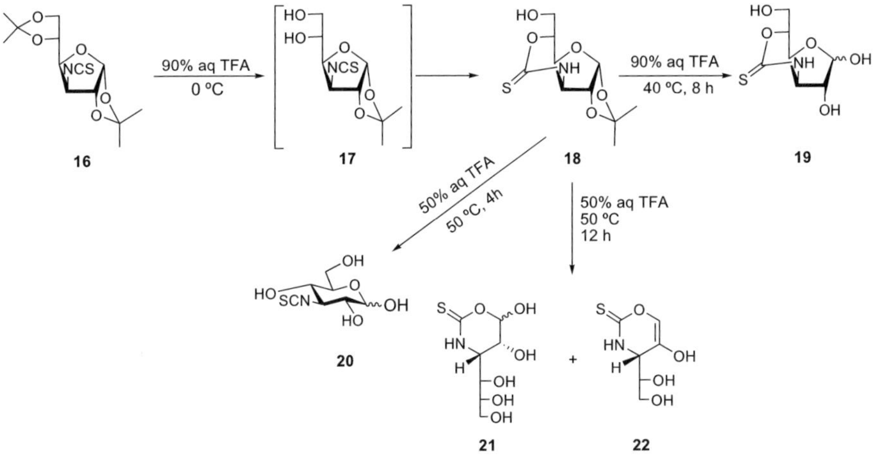

Scheme 4

lar to that observed in β-oxo isothiocyanates in the preparation of mesoionic 1,3,4-oxadiazole-2-thiones [41].

The synthesis of bicyclic N-thiocarbonyl azasugars as glycomimetics structurally related to potent glycosidase inhibitors such as natural nojirimycin, 1-deoxynojirimycin, or castanospermine [42] has also been reported [43, 44]. Thus, thiocarbamate **24** was obtained by spontaneous cyclization of a transient isothiocyanate upon treatment of **23** with carbon disulfide (Scheme 5). Removal of the protective group in acidic conditions led to the corresponding reducing furanose derivative, as an anomeric mixture. Final neutralization with a basic Amberlite resin allowed equilibrium to be shifted to the bicyclic 2-oxaindolizidine **26**, a castanospermine analogue, obtained by nucleophilic attack of the nitrogen to the masked aldehyde group [44].

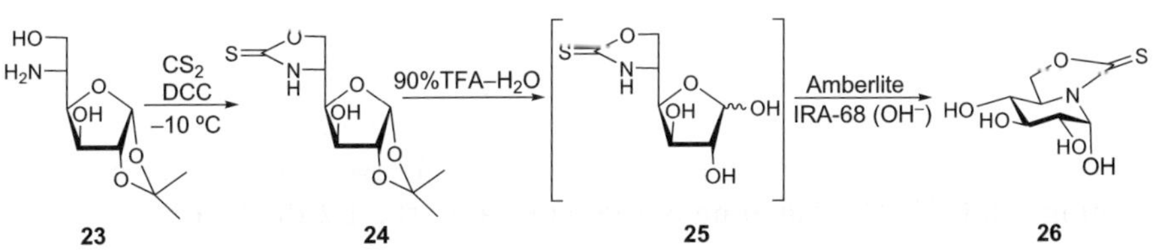

Scheme 5

It is noteworthy that compound **26** exists only with the pseudoanomeric hydroxyl group in the axial position; the explanation for this strong anomeric effect might be the favorable interaction between the π-type lone pair orbital of the nitrogen atom in the ground state of the thiocarbamate moiety and the σ^* antibonding orbital of the vicinal axial $C-O$ bond. Compound **26** was tested [44] against several glucosidases and showed a potent and selective activity ($K_i = 40\,\mu M$) against α-glucosidase from yeast, and a poor activity against rice α-glucosidase, amylase, and β-glucosidase.

Following the same strategy, García Fernández and co-workers [44] prepared some other castanospermine analogues, but replacing the oxazolidine-2-thione moiety by a imidazolidine scaffold, the synthetic pathway being depicted in Scheme 6. Staudinger reduction with triphenylphosphine and aqueous ammonia of the 5,6-diazido derivative **27**, followed by isothiocyanation with carbon disulfide of the corresponding diamino **28**, and acidic removal of the isopropylidene group in **29** afforded final imidazolidine derivative **30**, again as the single α-pseudoanomer. This compound turned out to be an extremely weak inhibitor of the tested glucosidases.

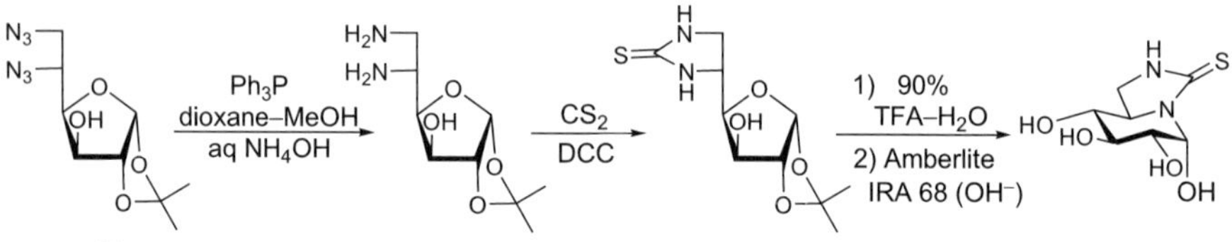

Scheme 6

Analogous imidazolidine-2-thiones of L-*ido* [44], D-*manno*, D-*allo*, D-*galacto*, and L-*gulo*-configurations [45] were also prepared, but showed either moderate inhibition or total absence of activity.

The synthesis of methyl 6,7-dideoxy-7-isothiocyanato-α-D-heptapyranosides of D-*gluco*, D-*manno*-, and D-*galacto* configurations was accomplished [46] by homologation reaction and proved that these compounds were stable even upon heating in the presence of triethylamine, and thus, no seven- or higher-membered thiocarbamate derivatives were obtained. This behavior was different to that observed for D-*gluco*-heptapyranose derivative **33** [46], whose preparation was carried out as shown in Scheme 7. Reduction of heptafurano nitrile **31** using borane-dimethyl sulfide and isothiocyanation with carbon disulfide, followed by acidic removal of the acetal moiety afforded isothiocyanate **33** also stable in the absence of base. Heating at 80 °C with triethylamine in DMF led to 1,3-oxazinane-2-thione **35** as a mixture of anomers. The same compound could be obtained by first base-induced intramolecular annelation of **32**, followed by acidic deprotection of **34**.

Fernández-Bolaños and co-workers, in the search of fully *O*-unprotected sugar-containing isothiocyanates and their derivatization, found the first examples of isothiocyanates being in equilibrium with cyclic thiocarbamates by spontaneous annelation with a vicinal hydroxyl group [47]. Preparation of glucosamine-derived isothiocyanate **37** was accomplished (Scheme 8) starting from benzamide **36**, by basic hydrolysis, followed by buffering the medium with carbon dioxide and treatment with thiophosgene. NMR study

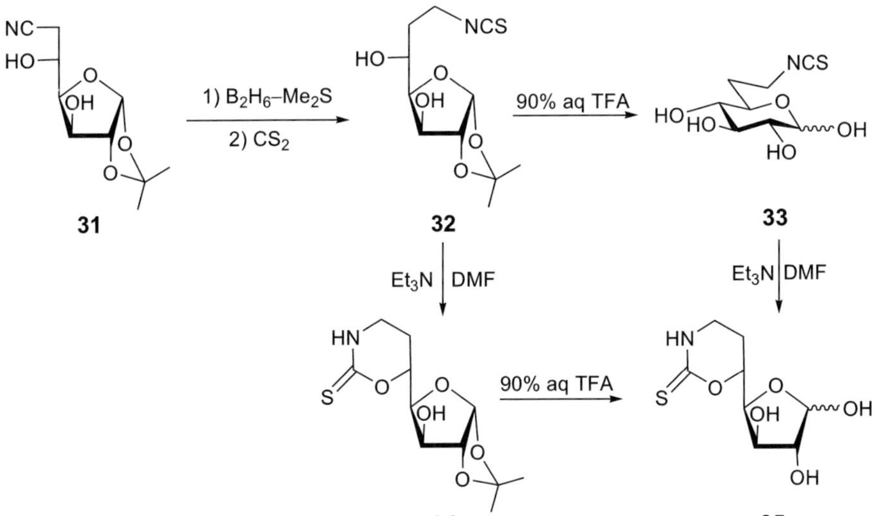

Scheme 7

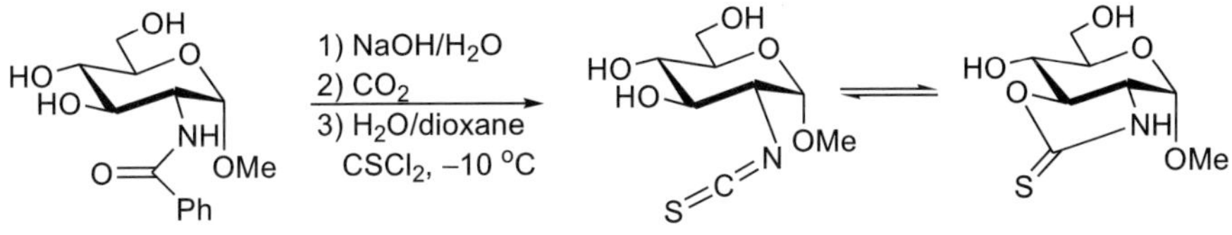

Scheme 8

revealed [47] a solvent and temperature-dependent equilibrium (Table 1) between isothiocyanate **37** and bicyclic *trans*-fused thiocarbamate **38**. The strain caused by the *trans* arrangement of the two rings allows thiocarbamate opening and, thus, the equilibrium with **37**.

Table 1 Data on isothiocyanate **37**/thiocarbamate **38** equilibrium, measured by ^{1}H-RMN

Solvent[a]	37/38 Molar ratio	Temperature, °C (in DMSO-d_6)	37/38 Molar ratio
$(CD_3)_2SO$	14 : 86	24	14 : 86
D_2O	26 : 74	60	26 : 74
CD_3OD	45 : 55	70	29 : 71
–	–	99	45 : 55
–	–	103	50 : 50

[a] 24 °C

The equilibrium mixture of these two compounds can be directly used for the preparation of sugar-derived thioureas by coupling with alkyl and aryl amines [47]. This was the first example in the literature of a thiocarbamate acting as a latent isothiocyanate for the synthesis of thioureas.

As depicted in Table 1, equilibrium is shifted towards thiocarbamate **37** at lower temperatures, as a result of entropy decrease in annelation. Furthermore, the higher proportion of **38** in DMSO-d_6 might be due to stronger hydrogen bonding of the NH moiety, as compared to D_2O or CD_3OD.

It has recently been shown that β-D-glycopyranosyl isothiocyanates behave similarly [10]. Thus, treatment of β-D-gluco- or β-D-galactopyranosylamines **39** and **40** with thiophosgene in a buffered aqueous medium afforded the expected isothiocyanates **41** and **42**, respectively (Scheme 9). This was the first isolation and characterization of fully O-unprotected glycosyl isothiocyanates. Previously, unprotected glycopyranosyl isothiocyanates were used for glycosidase inhibition and protein labeling in glucose transportation through membranes [48, 49]; nevertheless, decomposition was described [50] as occurring in physiological conditions and no more chemistry involving these intermediates was reported up to a recent study [51].

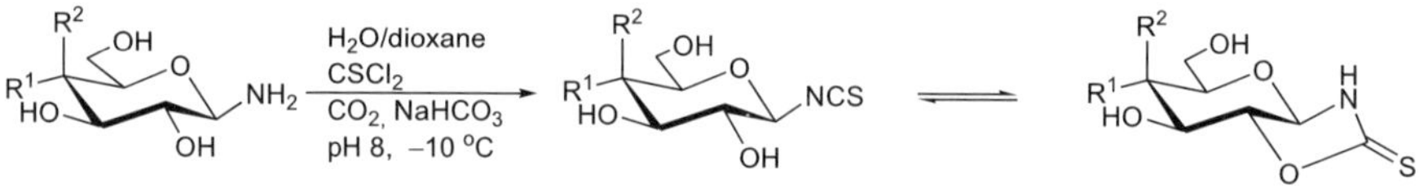

39 R^1= OH, R^2= H
40 R^1= H, R^2= OH

41 R^1= OH, R^2= H
42 R^1= H, R^2= OH

43 R^1= OH, R^2= H
44 R^1= H, R^2= OH

Scheme 9

Analogously to that indicated above for **37**, glycosyl isothiocyanates **41** and **42** were shown to be in a solvent-dependent equilibrium with bicyclic *trans*-fused thiocarbamates **43** and **44** by spontaneous annelation with a hydroxyl group on C-2, showing a similar preference for thiocarbamates when DMSO-d_6 was used as the solvent [10]. This mixture could again be used for the one-pot preparation of O-unprotected glycopyranosyl thioureas, demonstrating that this procedure is mild enough for the preparation of glycoconjugates prior to isothiocyanate decomposition [10].

The mixture of glucopyranosyl isothiocyanate **41** and thiocarbamate **43** was recently tested [52] as a potential inhibitor of glucose transporter (EII^{Glc}) of the phototransferase system of *Escherichia coli*, leading to its rapid inactivation.

Nevertheless, the isothiocyanation reaction of a glycosylamine or an amino sugar with the amino moiety and the vicinal hydroxyl group in a *cis*-arrangement led to a stable *cis*-fused thiocarbamate. For example, when D-glucosamine was treated [53] with thiophosgene in a buffered aqueous medium, *cis*-arranged glucopyranoso[2,1-*d*]oxazolidine-2-thione **47** was the only isolated compound, obtained by spontaneous annelation of the transient isothiocyanate **46** with the hydroxyl group of the α-anomer (Scheme 10). The other two possible thiocarbamates are in a *trans*-fused arrangement, and if formed, might be in equilibrium with the isothiocyanate, and this equilibrium shifted to more stable derivative **47**.

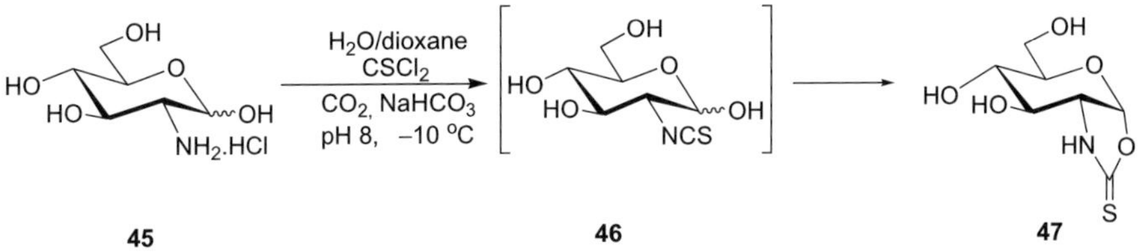

Scheme 10

Analogously, isothiocyanation of D-mannosamine hydrochloride **48** led [53] to *cis*-fused thiocarbamate **50** as a mixture of anomers through non-detected isothiocyanate **49** (Scheme 11). It is worth noting that cyclization with anomeric hydroxyl groups of **49** were not observed.

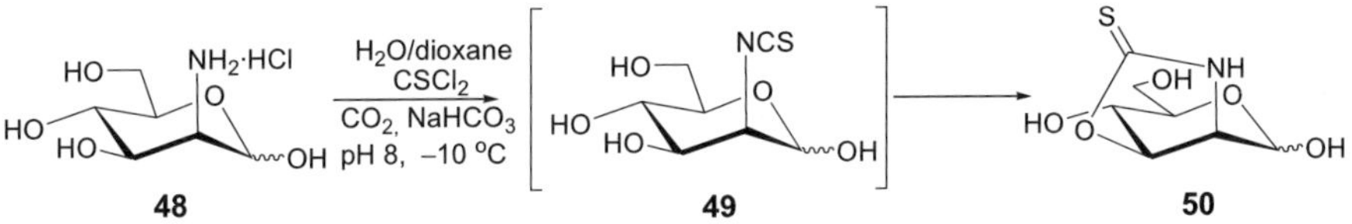

Scheme 11

In a similar way, isothiocyanation of β-D-mannopyranosylamine **51** only gave [10] *cis*-fused thiocarbamate **52** that did not react with aromatic amines (Scheme 12), proving the inherent stability of this kind of *cis*-arranged bicyclic systems, contrary to what was observed in *trans*-fused thiocarbamates [10, 47]. The remarkable difference in reactivity between *cis*- and *trans*-fused sugar thiocarbamates might be explained considering that in *cis*-arrangement, flattening [54] of the six-membered ring is much more favorable than in the *trans*-counterparts.

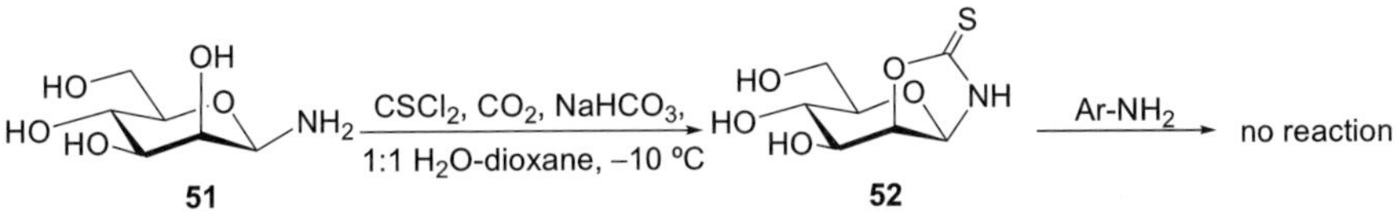

Scheme 12

An alternative procedure for the preparation of *cis*-1,2-fused oxazolidine-2-thiones in carbohydrates was reported [55] by Beaupere and co-workers starting from 1,2-*O*-sulfinyl-α-D-glycopyranose or glycofuranose. Treatment of the α-1,2-*O*-sulfinyl derivative with sodium thiocyanate led to a β-thiocyanate that, under the reaction conditions was isomerized to a transient α-isothiocyanate via an oxocarbenium ion. The isothiocyanate underwent spontaneous annelation with a vicinal hydroxyl group to afford the corresponding glycopyrano and glycofurano *cis*-fused thiocarbamates (Scheme 13).

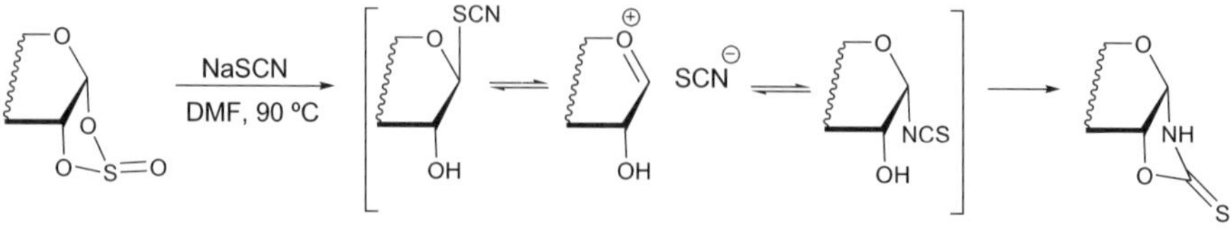

Scheme 13

3
Synthesis of 2-Amino-2-Oxazolines (Cyclic Isoureas)

2-Amino-2-oxazolines or cyclic isoureas have frequently been targeted [56–59] in the last few years in organic synthesis, as some biologically active compounds bear this motif in their structure. They exhibit octapamine [60] and adrenoreceptor agonist [61] activity, or histamine [62] and adenosine [63] receptor antagonist activity. They also might be useful in hypertension treatment [64, 65], as anti-obesity drugs [66], or as feromone biosynthesis inhibitors [67], i.e., as potential pest control agents [68].

In the carbohydrate context, there are some naturally occurring bicyclic isoureas that are potent and selective glycosidase inhibitors. Among them, it is worth mentioning trehazolin **53** [69, 70] and allosamidin **54** [71], potent in vitro trehalase and quitinase inhibitors, respectively (Fig. 1). They have been considered as candidates for pest control, but their highly polar structure has precluded the practical use of trehazolin and allosamidin as in vivo pesticidal agents [68].

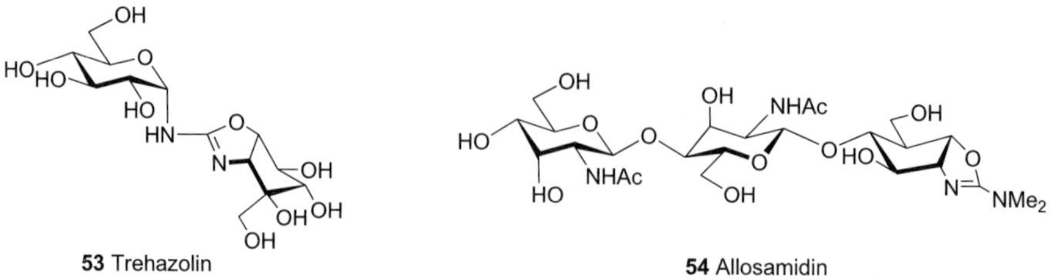

53 Trehazolin   **54** Allosamidin

Fig. 1 Trehazolin **53**, allosamidin **54**

Isolation and discovery of these biologically active compounds has encouraged researchers to carry out not only their total synthesis but also to accomplish the synthesis of the aglycon [72–74] or modifications both in the cyclitol moiety and in the carbohydrate residue [75–82]. For instance, there are examples where the aminocyclopentitol motif has been replaced by either a six-membered polyhydroxylated carbocycle [83] or by a tetrahydropyrane moiety [84].

Uchida and Ogawa attempted [85] the preparation of glycosylamino oxazolines without any substituents in the heterocyclic moiety as trehazolin analogues. Thus, coupling of 2,3,4,6-tetra-*O*-benzyl-α-D-galactopyranosyl isothiocyanate **55** [86] with 2-aminoethanol afforded per-*O*-protected galactosylthiourea **56** (Scheme 14). Cyclodesulfurization with yellow HgO in Et$_2$O gave 2-galactopyranosylamino-oxazoline **57** through a transient carbodiimide. Attempts to carry out the deprotection of **57** under Birch conditions did not allow the isolation of the expected isourea **58**; instead, 1,2-fused glycopyranoso and glycofuranoso-oxazolines **59** and **60** were isolated in a roughly 1 : 1 ratio.

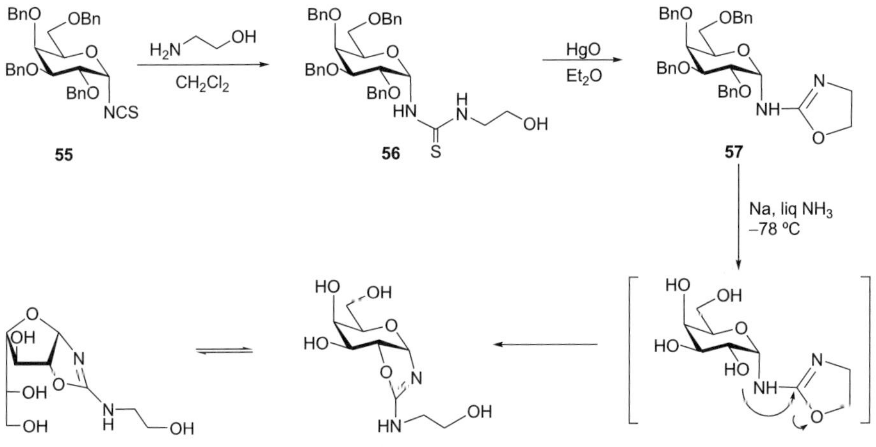

Scheme 14

The same behavior was observed for the glucose epimer [85]. On the contrary, starting from benzylated β-D-gluco and β-D-galactopyranosyl isothiocyanates, or from acetylated α-D-mannopyranosyl isothiocyanate **61** (Scheme 15), due to the *trans* disposition of substituents on C1 and C-2 no rearrangement took place after deprotection, and the corresponding 2-glycopyranosylamino-oxazolines were obtained [85]. The α-mannopyranosyl-oxazoline **63** turned out to be a moderate inhibitor of α-mannosidase ($K_i = 98\,\mu$M).

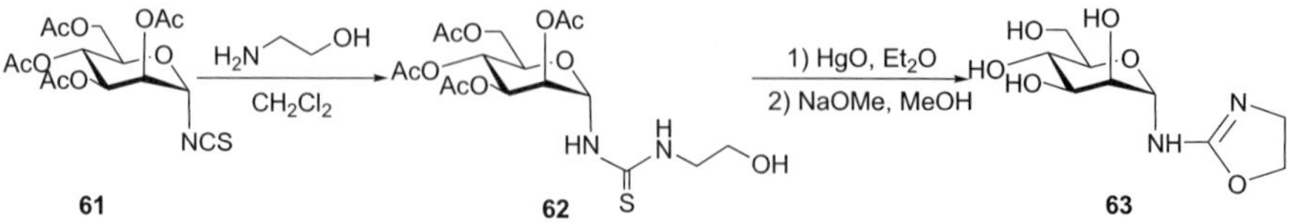

Scheme 15

In this context, Fernández-Bolaños and co-workers [87] have developed an efficient one-pot three-step procedure for the preparation of 2-(alkylamino, dialkylamino, arylamino)dihydroglucopyranoso[1,2-*d*]oxazoles starting from β-D-glucopyranosylamine (Scheme 16). Treatment of β-D-glucopyranosyl-amine **39** with thiophosgene in a buffered 1 : 1 water–dioxane medium, followed by addition of primary and secondary amines led to the corresponding O-unprotected glucopyranosyl thioureas **64**, through the above-described isothiocyanate **41**. In situ treatment with yellow HgO afforded the corresponding *trans*-fused bicyclic isoureas **65** in a 55–84% overall yield for the three steps.

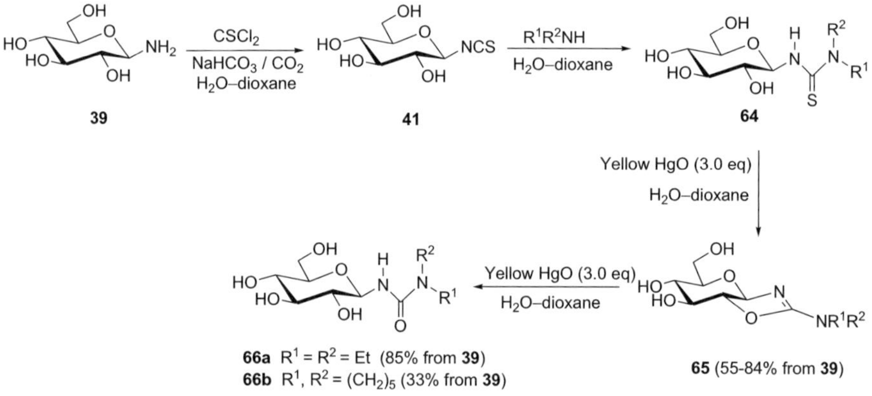

Scheme 16

The one-pot three-step procedure could be extended one more step; thus, more prolonged reaction times for the cyclodesulfurization step of *N,N*-disubstituted thioureas (diethyl- and piperidine-derived thioureas) afforded the corresponding β-D-glucopyranosyl ureas [10] in moderate to good yields for the four-step procedure (Scheme 16). It is noteworthy that this reaction did not work for *N,N'*-disubstituted thioureas and that reaction time differed significantly for both compounds, the time being only two hours for *N,N*-diethylthiourea, and for piperidine derivative even after three months the reaction was not complete. A mechanism involving reversible formation of the *trans*-fused bicyclic isoureas through a carbodiimidium ion was devised to explain the different behavior of the tested isoureas **65**.

The one-pot methodology for the cyclodesulfuration of sugar thioureas strongly contrasts with procedures reported in the literature so far, as those involved the use of *O*-protected sugar thioureas, together with freshly prepared mercury (II) oxide [88–91] or 2-chloro-3-ethylbenzoxazolium tetrafluoroborate [77, 84], both employed in anhydrous solvents. Thus, the preparation of a 1-aza-2-(β-D-glucopyranosyl)imino-3-oxabicyclo[3.3.0]octane by treatment of a 2,5-dihydroxymethyl-*N*-(2,3,4,6-tetra-*O*-acetyl-β-D-glucopyranosyl)thiocarbamoyl pyrrolidine with mercury (II) oxide in acetonitrile has been recently described [92].

Isourea **70**, that shows a great structural resemblance to trehazolin **53**, was obtained [87] in an 89% yield by treatment of thiourea **68** with yellow HgO in aqueous dioxane (Scheme 17). The same isourea was previously prepared [84] from protected thiourea **67** by desulfurization using Mukaiyama's reagent [93] in anhydrous acetonitrile to give after debenzylation of **69** a non-resolved mixture of the expected isourea **70** (17%), together with the *N,N'*-bis(β-D-glucopyranosyl)urea (32%).

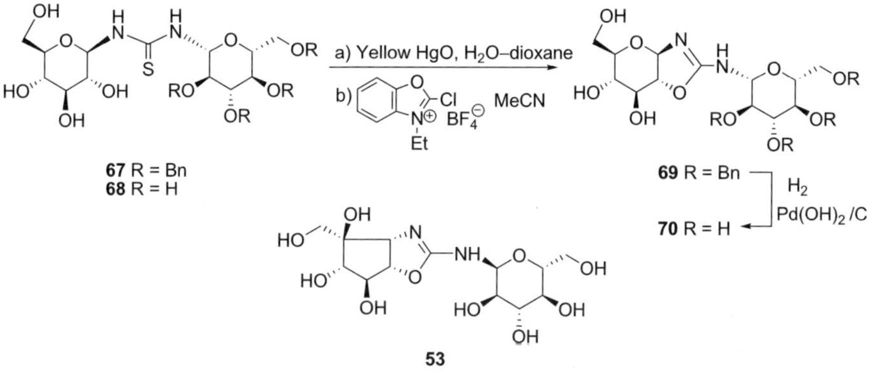

Scheme 17

4
Synthesis of 2-Amino-2-Thiazolines (Cyclic Isothioureas)

Santoyo-Gonzalez and co-workers [94, 95] explored the versatility of 2-deoxy-2-iodoglycosyl isothiocyanates in the synthesis of glycopyranoso-fused thiazolines. 2-Deoxy-2-iodoglycosyl isothiocyanates, such as **71**, are easily prepared by reaction of glycals with silica-supported potassium thiocyanate and iodine in chloroform [95]. The simultaneous presence of two active electrophilic groups in the molecule, make them exceptional building blocks for the synthesis of glycopyranoso-fused thiazoles through thiocarbamates, dithiocarbamates, or thioureas, which are formed by reaction with alcohols, thiols, or amines, respectively (Scheme 18). The subsequent nucleophilic displacement of the *trans*-vicinal iodine by the sulfur atom of the thiocarbonyl group in the presence of a base led to bicyclic structures such as **73**, **75**, and **76** in good to high yields.

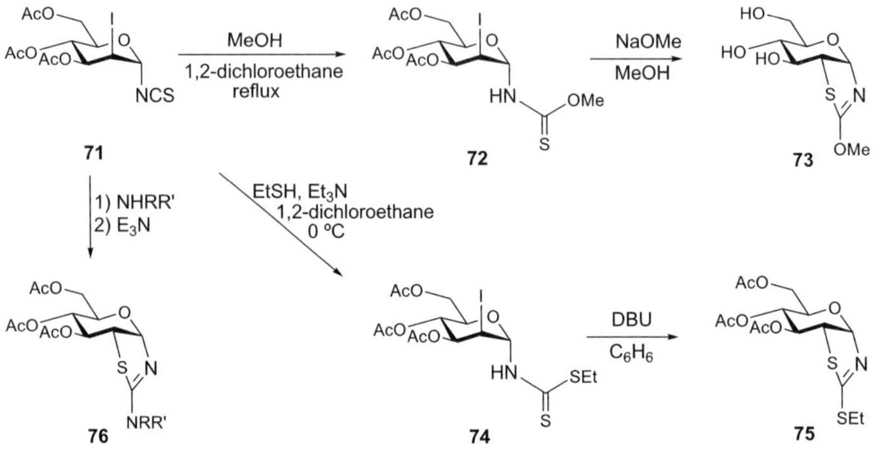

Scheme 18

Application of this methodology allowed the preparation of several multivalent neoglycoconjugates bearing some glycopyranoso[1,2-*d*]thiazoline motifs when reaction was carried out between the 2-iodo-glycopyranosyl isothiocyanate and a symmetric polyamine [94]. As an example, the preparation of compounds **77** and **78** is remarkable, derived from **71** and a calix[4]arene or a triamine, respectively (Fig. 2).

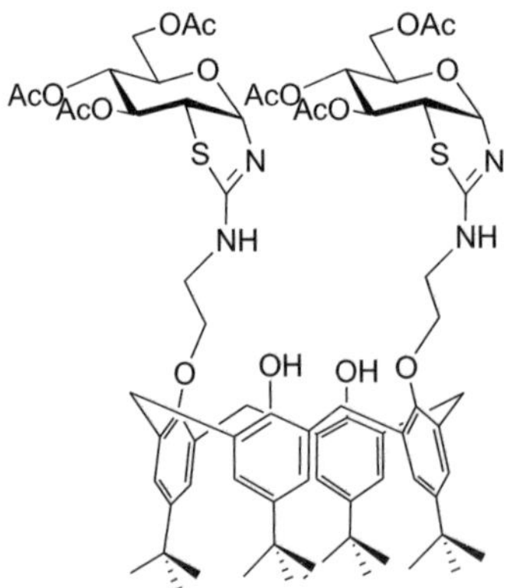

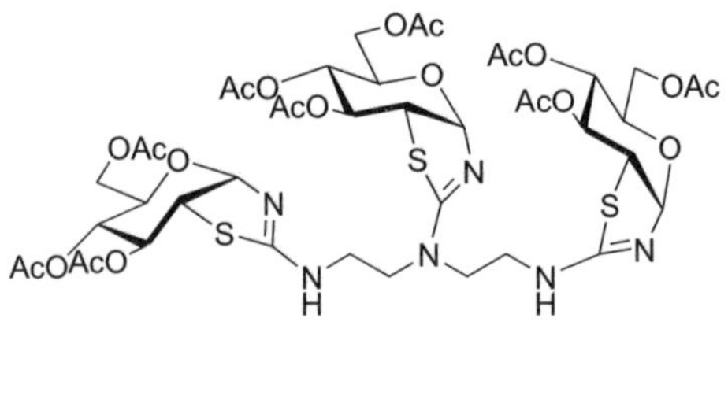

Fig. 2 Calixarene **77**, triamine derivative **78**

Chiara and co-workers have recently accomplished [96] the synthesis of an isosteric derivative of trehazolin; thus, they carried out the replacement of the endocyclic oxygen atom of the aglycon by sulfur, to afford a thiazoline ring (Scheme 19). Partially protected aminocyclopentitol **80** was obtained by the same authors from D-mannose, using as a key step a stereoselective reductive carbocyclization reaction using SmI_2 [80, 97]. Coupling of **80** with O-protected α-D-glucopyranosyl isothiocyanate **79** afforded the expected thiourea **81**. Treatment of the latter with triflic anhydride gave a transient triflate on the secondary free hydroxyl group, that was displaced by nucleophilic attack of the sulfur atom. Final deprotection afforded thiotrehazolin **83**. This compound turned out to be a nanomolar, tight-binding inhibitor of porcine trehalase ($K_i = 30.4$ nM), only roughly 15 times less active than natural trehazolin ($K_i = 2.1$ nM).

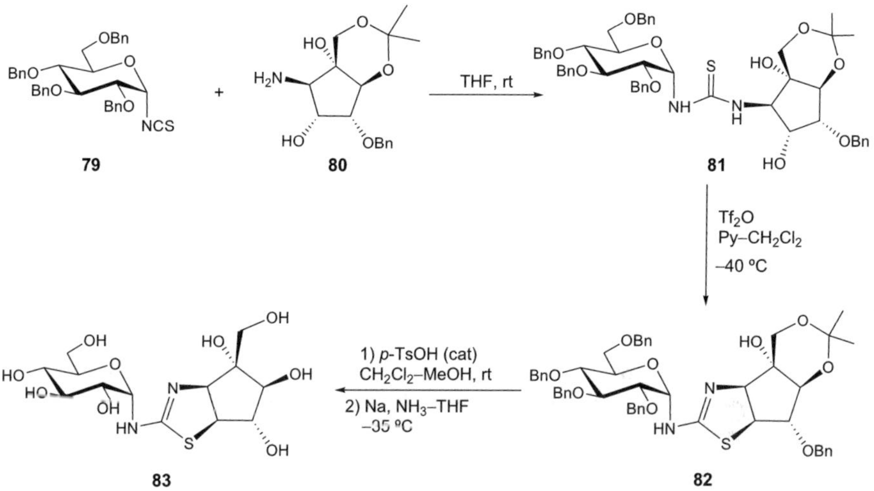

Scheme 19

5
Synthesis of Nucleosides and Glycosylamino-Heterocyles

Nucleosides are compounds exhibiting pharmacological activity of great interest [98, 99]; thus, some of these heterocyclic compounds have been used as anti-tumoral [100, 101] or antiviral agents, including AIDS [102, 103] and hepatitis B [104, 105] treatments.

There are numerous families of nucleosides that have an active role in cancer chemotherapy and infectious diseases. Among them, compounds like L-nucleosides [106, 107], carbocyclic nucleosides [108, 109], 4′-thionucleosides [110, 111], *C*-nucleosides [112, 113], or acyclic derivatives [114–116] have been synthesized and tested for biological activities.

The well-known reaction of alkyl or aryl isothiocyanates with 2-amino-2-deoxy-aldoses is an excellent synthetic pathway for the preparation of imidazole-derived *C*-nucleosides and acyclic *C*-nucleosides [117–120]. The reaction carried out using sugar-derived isothiocyanates gives access to compounds that are simultaneously *C*-nucleosides and *N*-nucleosides. Thus, Fuentes and co-workers described [121] the preparation of 1-*N*-glycosyl-4-(D-*arabino*-tetritolyl)imidazolidine-2-thiones by reaction between 2-amino-2-deoxy-D-glucosamine and *O*-acetylated glycosyl isothiocyanates derived from mono, di, and trisaccharides, and with *O*-acetylated-2-deoxy-2-isothiocyanato-β-D-glucose.

In a similar way, the synthesis of pseudo-nucleosides **84** and **86–88** was accomplished [122] using the synthetic route depicted in Scheme 20. Treatment of D-glucosamine with methyl 2-deoxy-2-isothiocyanato-α-D-glucopyranoside **37** in equilibrium with its corresponding thiocarbamate **38** (see above) afforded, through a transient thiourea, 5-hydroxy-imidazolidine-2-thione **84** as a diastereomeric mixture of the 5*R* and 5*S* isomers in a solvent-dependent equilibrium (5*R* : 5*S* = 93 : 7 in D_2O and 98 : 2 in DMSO-d_6). Cyclodehydration of the tetritolyl chain was afforded by heating in aqueous acetic acid to give bicyclic glucofuranoso-imidazolidine-2-thione **85**, that underwent isomerization into imidazoline-2-thione **86** upon treatment with methanolic TFA at rt.

Subsequent treatment of **86** with aqueous TFA allowed [122] a cyclodehydration in the polyhydroxylated chain to afford the α and β-D-erithrofuranosyl-*C*-nucleosides **87** in a roughly 1 : 4 α/β ratio (Scheme 20). Furthermore, prolonged treatment of imidazolidine-2-thione **86** with TFA in refluxing methanol gave access to *S*-methylimidazol derivatives **88** as a mixture of anomers, isolated in a 1 : 7 α/β ratio.

A different strategy for preparing nucleosides is based on the reaction of glycosyl isothiocyanates with stabilized carbanions, due to the strong electrophilic character of the isothiocyanato moiety [123]. An example of this reaction is depicted in Scheme 21. Reaction of per-*O*-acetylated β-D-glucopyranosyl isothiocyanate **89** with diethyl acetonyl (phenacyl) malonate in the presence of sodium hydride afforded the corresponding thioamides **90** and **93**. In the case of compound **90**, the thioamide moiety was found to be in an equilibrium with the corresponding pyrrolidine structure **91**. This equilibrium was shifted towards the latter, roughly 1 : 10 at rt.

Cyclodehydration of the thioamide-derived structures was induced by treatment with an acetic anhydride and phosphoric acid mixture. In the case of the mixture of **90** and **91**, *N*-glucopyranosyl tetrahydrothioxopyridine **92** was obtained upon acidic treatment [123]. Under the same conditions, compound **93** evolved to *N*-glucopyranosyl thioxopyrroline **94**.

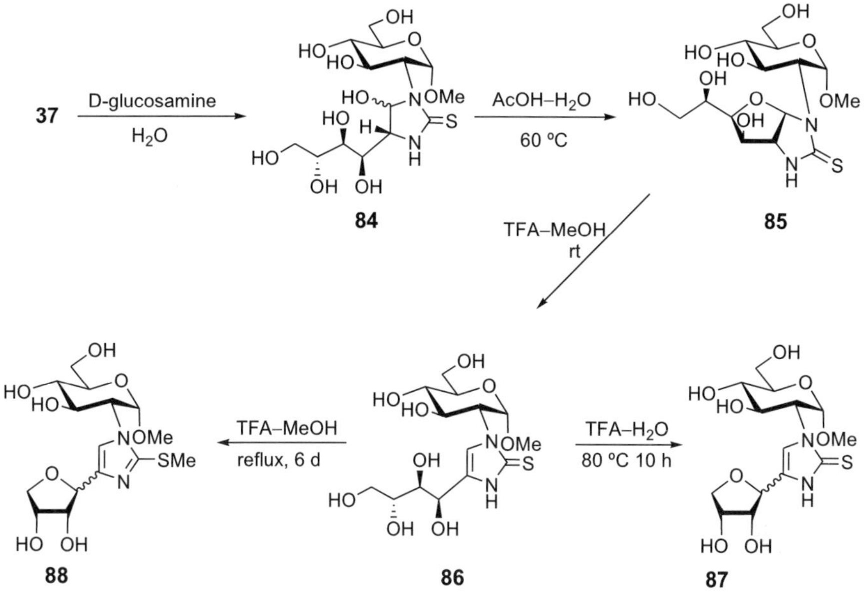

Scheme 20

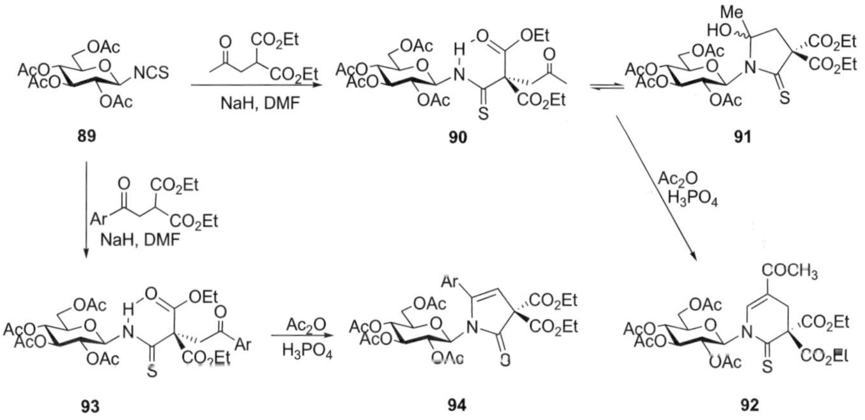

Scheme 21

Furthermore, treatment of isothiocyanate **89** with the carbanion generated from malonodinitrile, followed by addition of phenacyl bromide afforded [123] fully O-protected β-D-glucopyranosylamino thiophene derivative **96** via non-isolated thiolate **95** (Scheme 22).

There are some other examples in literature concerning cyclization of thiocarbamoyl derivatives to give pseudo-nucleosides [7]. For instance, Marino and co-workers [124] prepared 3-β-D-galactofuranosyl-4-oxo-imidazolidine-2-thione **99** by cyclization of thiourea **98**, that was obtained [125] by

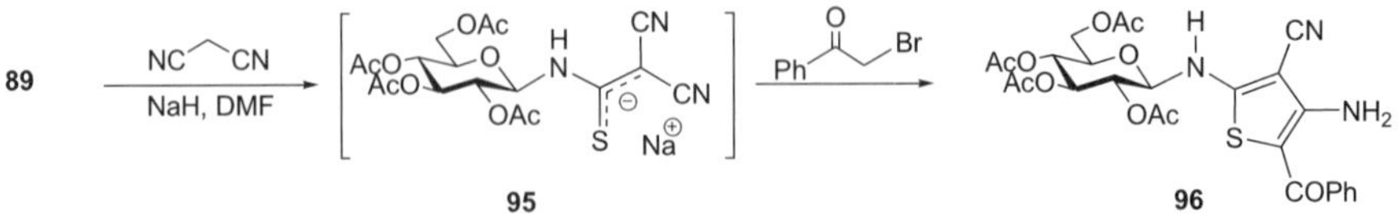

Scheme 22

coupling of galactofuranosyl isothiocyanate **97** [126] with glycine ethyl ester hydrochloride (Scheme 23). In a similar way, condensation of isothiocyanate **97** with aminoacetaldehyde dimethylacetal gave the corresponding thiourea **100**, whose masked aldehyde was deprotected in acidic conditions, and underwent cyclization with the glycosyl nitrogen to furnish 4-methoxyimidazolidine-2-thione **101** and imidazoline-2-thione **102**, after removal of the ester groups. The authors claimed a tautomeric equilibrium imidazoline-2-thione/mercaptoimidazole shifted to the latter form, as deduced from NMR data [124]; however, the chemical shift found for C-2 (162.0 ppm) in the ^{13}C NMR spectrum of **102** is in agreement with the data reported for the imidazoline-2-thione structure [122, 127], a more stable structure than its tautomer mercaptoimidazole [128]. Compound **101** was found to be a moderate inhibitor of *exo* β-D-galactofuranosidase (IC$_{50}$ 100 μM).

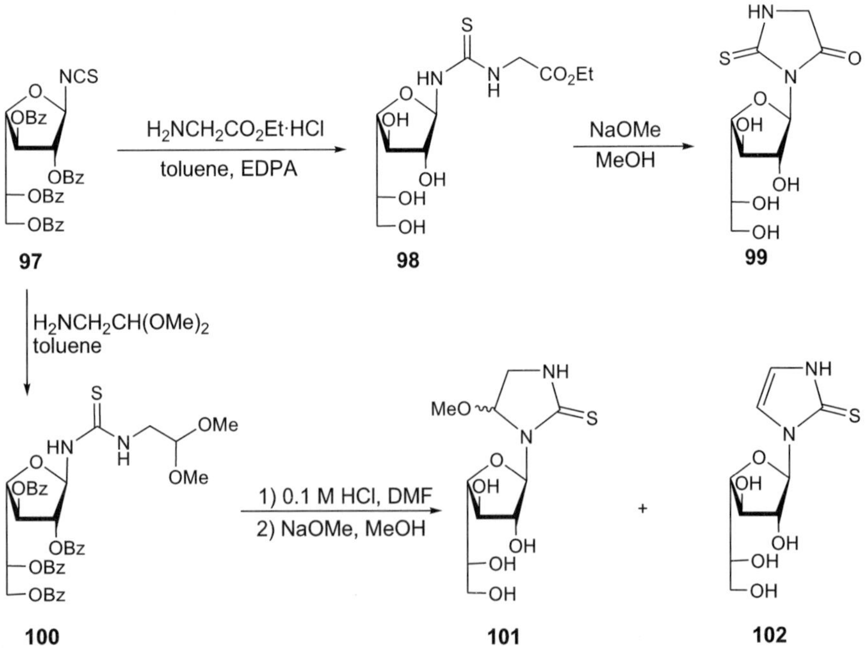

Scheme 23

Saleh reported [129] the preparation of nucleosides of dihydropyrimido[5,4-*b*]indol-4-one **107-109** using glycopyranosyl isothiocyanates as starting materials (Scheme 24). Condensation of *O*-acetylated sugar isothiocyanates with 3-aminoindole derivative **104** afforded the corresponding thioureas **105**, which underwent cyclization in mild conditions promoted by Lewis acid $ZnCl_2$ to give the pyrimidoindol scaffold **106**. Deprotection, *S*-alkylation, and oxidation on MCPBA (*m*-chloroperbenzoic acid) yielded the sulfonyl compounds **109**.

Scheme 24

In a similar way, *N*-glycopyranosylamino thiazoles were prepared by reaction of glycosyl isothiocyanates with 3-indolylaminomethyl-ketone hydrochloride followed by cyclodehydration with acetic anhydride [130].

Another different approach for accessing nucleosides is based on cycloaddition reactions involving the isothiocyanate moiety, exploiting the high versatility of this functional group. Deniaud and co-workers have described [131, 132] a [4 + 2] cycloaddition reaction involving sugar isothiocyanates and diazadienium iodide **114** in the synthesis of pyrimidine nucleoside analogues.

Glycopyranosyl isothiocyanates **103** were obtained by the method described by Camarasa and co-workers [133], involving treatment of protected glycopyranosyl bromides with potassium thiocyanate under phase-transfer conditions (Scheme 25). Diazadienium iodide **114** was prepared by sulfhydration of commercially available 3-dimethylaminoacrylonitrile **112**, followed by methylation on the sulfur atom of thioamide **113** (Scheme 25).

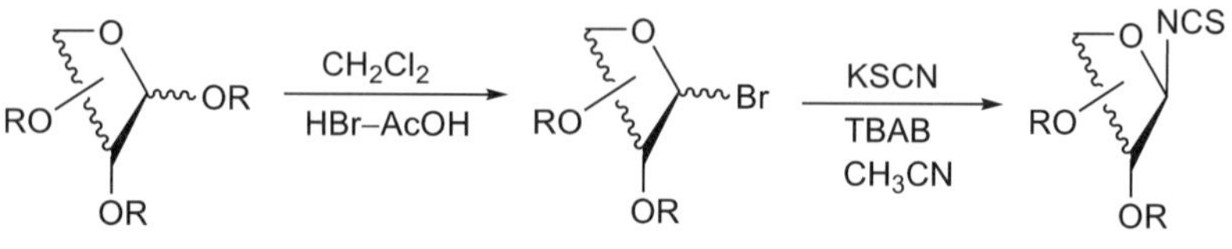

sugar= per-*O*-acetylated glucopyranosyl, galactopyranosyl, cellobiosyl, lactosyl
per-*O*-benzoylated ribofuranosyl, xylofuranosyl

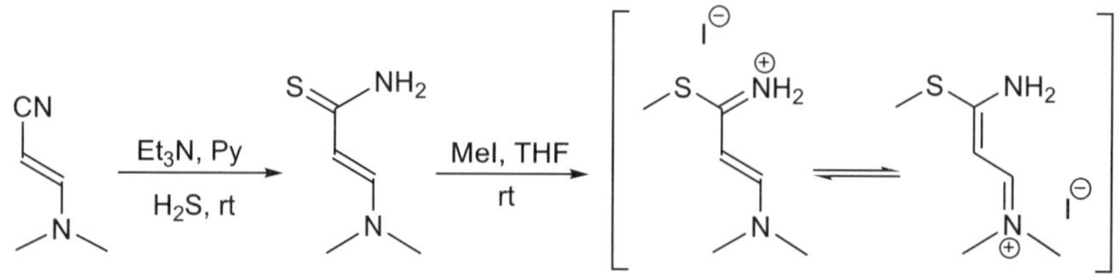

Scheme 25

Treatment of the glycosyl isothiocyanates **103** (2 equiv) with **114** afforded the corresponding dihydropyrimidine-2-thiones **116** in good yields (72–84%) in a regiocontrolled manner. The dimethylamine, generated in the deamination of the non isolated intermediate **115**, reacted with a second equivalent of isothiocyanate to provide glycosyl thioureas **117** as a side-product (Scheme 26). The methylsulfanyl group can be easily substituted by hydroxyl, amino, or benzylamino groups [132].

Jochims' [134] and Hassan's [135] groups reported the efficient use of 1-aza-2-azoniaallene salts **119** and *O*-acetylated β-D-glucopyranosyl isothiocyanate **89** for the preparation of glucopyranosylimino-1,3,4-thiadiazoles **121** and **122**. These salts, prepared from chloroalkylazo derivatives **118** and antimony (V) chloride, can undergo 1,3-dipolar cycloadditions with multiple bonds-containing compounds, such as alkynes, alkenes, isocyanates, isothio-

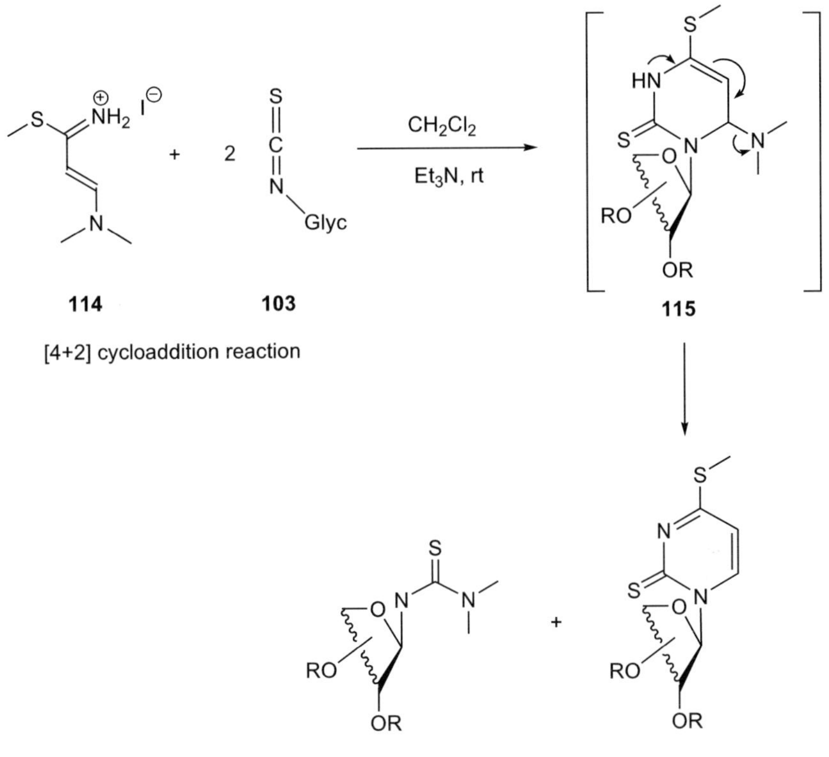

[4+2] cycloaddition reaction

CH_2Cl_2

Et_3N, rt

114   **103**   **115**

117   **116**

Scheme 26

cyanates, carbodiimides, or nitriles (Scheme 27) [134–138] to afford a plethora of heterocyclic derivatives. For instance, treatment [134] of 2,3,4,6-tetra-O-acetyl-β-D-glucopyranosyl isothiocyanate **89** with the azo compound **123** furnished glycosylimino-thiadiazole derivative **124** in a 63% yield.

Al-Masoudi et al. applied the same methodology to the synthesis of thiosugar nucleosides [139, 140]. Thus, acetylated 5-thio-β-D-glucopyranosyl isothiocyanate **126**, prepared from 5-thioglycosyl bromide **125** and trimethyl-silyl isothiocyanate, was transformed into iminothiadiazoles **127** and **129** using azo compounds **123** and **128**, respectively (Scheme 28). Reaction of isothiocyanate **126** with aminoacetone or chloroethylamine also afforded a N-(5-thio-β-D-glucopyranosyl)-imidazoline-2-thione nucleoside **130** and a 2-(5-thio-glucopyranosylamino)-2-thiazolidine derivative **131** [140], whose glucosyl counterparts having oxygen in the ring had been previously described [141–143].

Desulfurization of thiocarbonyl compounds has also proved to be an alternative key step in furnishing glycosyl isothiocyanate-mediated pseudonucleosides. Thus, Gama and co-workers extended [144] their studies of desul-

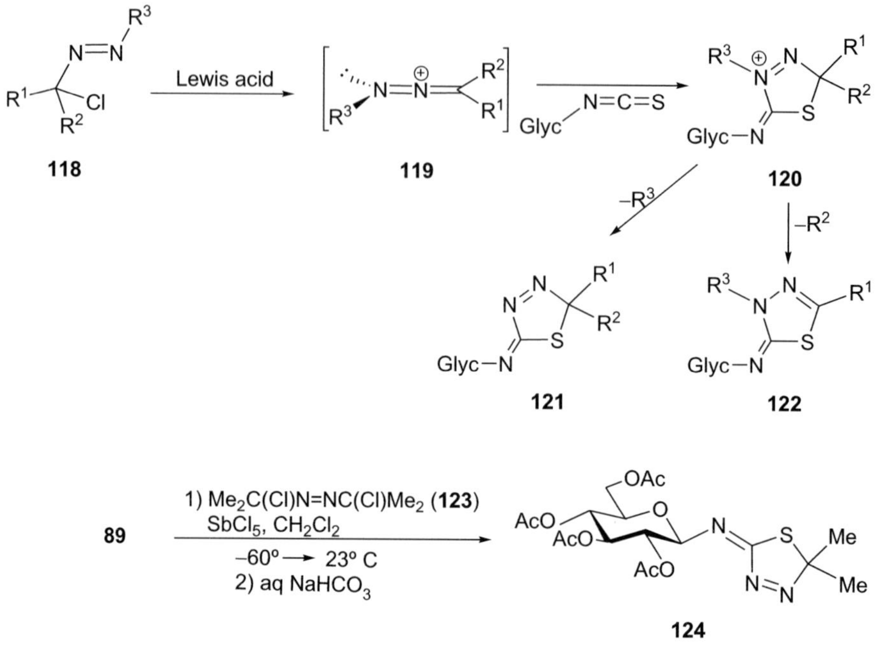

Scheme 27

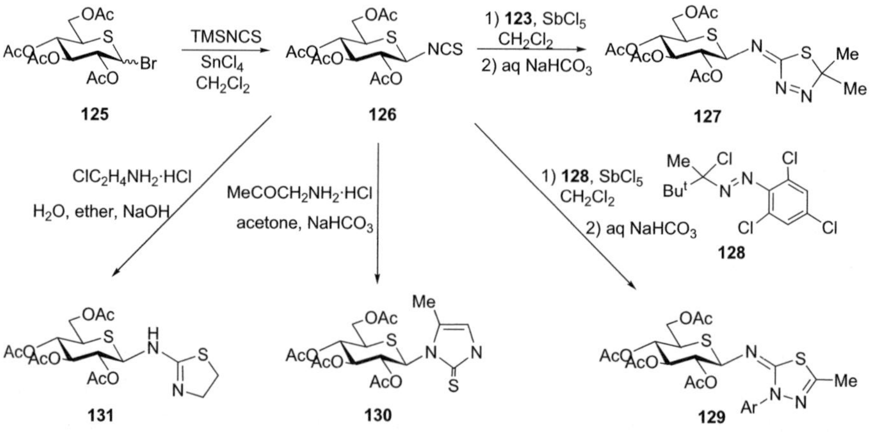

Scheme 28

furization reactions of glycosyl thioureas with silver cyanate to accomplish the preparation of glycosyl-5-azauracil derivatives **135**. Coupling of known protected glycopyranosyl isothiocyanates with secondary amines afforded the corresponding thioureas **132**, that upon treatment with silver cyanate gave the corresponding heterocyclic derivatives **135** in good yields (Scheme 29). It was postulated [144] that a 1 : 1 thiourea-silver cyanate complex **133** must

Scheme 29

be involved in the reaction. This complex undergoes a nucleophilic attack by a second cyanate ion, with release of silver sulfide. Final cyclization of **134** between the glycosyl nitrogen atom and the second cyanate fragment gives access to the 5-azauracil scaffold **135**. In a similar way, cyclodesulfurization of *N*-aryl-*N'*-glycosyl thioureas with dimethylcyanamide and silver triflate led to 2-glycosylamino-quinazolines in good yields [145].

The same group prepared *N*-(β-D-glucopyranosyl)oxazolidine-2,4-diones [146] and *N*-(β-D-glucopyranosyl)imidazolidine-2,4-diones [147] by desulfurization-condensation of glucosyl isothiocyanate with α-hydroxyacids and *N*-substituted α-aminoacids, respectively, in the presence of silver trifluoroacetate and triethylamine. 4-Glucosyl-1,2,4-oxadiazolidine-5-thiones were also prepared by reaction of protected glucosyl isothiocyanates with an oxaziridine ring [148].

The synthesis of dithioxopyrimidine nucleosides **138** from acetylated β-D-glucopyranosyl isothiocyanate **89** has been reported by Fuentes and co-workers [149]. *N*-Glucosylthioamides **137**, prepared by reaction of **89** with *N*-aryl-enaminoesters or enaminones **136** in basic medium, were transformed into glucosyl dithioxopyrimidines nucleosides **138** by treatment with thiophosgene (Scheme 30). The concomitant glycosylamides formation

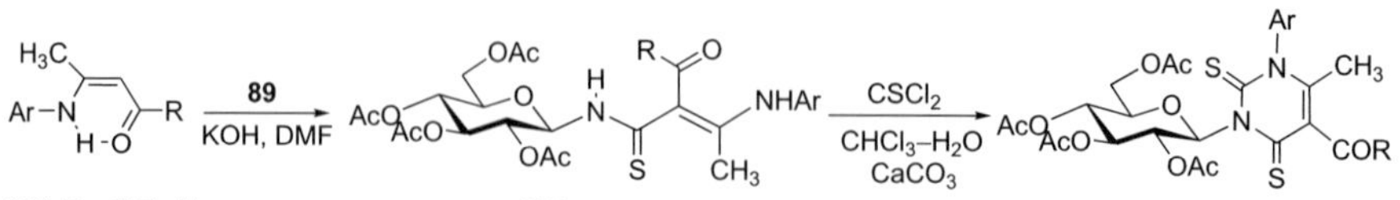

Scheme 30

due to basic hydrolysis of **137** was responsible for the low yield (18–24%) observed.

The contribution made by Sim and co-workers [150] to the synthesis of 2-glycosylamino-dihydropyrimidinones **141** as a novel type of guanidinoglycosides through the formation of asymmetrically substituted carbodiimides **139**, should also be highlighted. The authors synthesized the carbodiimides from β-D-glycosyl isothiocyanates and different azides, following the tandem Staudinger–aza-Wittig reaction via iminophosphoranes [151]. Treatment of β-glycosyl carbodiimides **139** with methyl (S)-3-amino-3-phenylpropanoate afforded the transient open-chain guanidine **140**, that evolved to pyrimidone **141** in a selective fashion, as **142**, involving cycli-

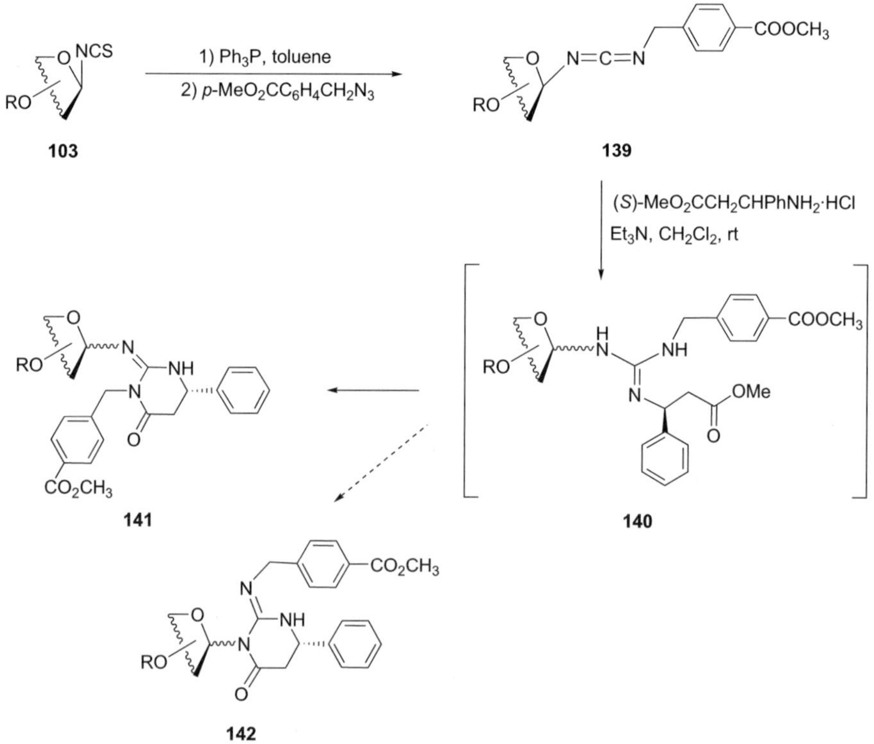

Scheme 31

zation through the glycosyl nitrogen, was not observed (Scheme 31). Compounds **141** underwent anomerization during the guanylation step. This novel family of compounds, after deprotection, was screened for some biological activity as antibiotics against MRSA (methicillin resistant *Staphylococcus aureus*) and *E. Coli*. They proved to be inactive against these organisms and they lack cytotoxicity.

6
Synthesis of Spironucleosides

Anomeric spironucleosides is a type of sugar derivative with spiranic structure, in which the anomeric carbon belongs simultaneously to the sugar ring and to a heterocyclic base [152, 153]. The area of synthetic anomeric spironucleosides [152–167] has gained considerable importance since the discovery in 1991 of hydantocidin [168, 169], a natural spironucleoside (Fig. 3) with potent herbicidal [170] and plant growth inhibitory activities, and without toxicity to microorganisms and animals [171].

The reaction of sugar isothiocyanates with aminosugars described above has also been studied with 1-amino (alkyl and arylamino)-1-deoxy-D-fructose **143** (Scheme 32). Thus, treatment of different *O*-acylated sugar isothiocyanates with D-fructosamines **143** gave *N*-glycosyl-5-hydroxy-5-tetritolyl-imidazolidine-2-thiones **144** as pairs of diastereomers in almost quantitative yield [152]. The reaction times were shorter (15–20 min) for fructosamine or its *N*-methyl derivatives than in the cases of *N*-aryl derivatives (8–24 h). Acid-catalyzed dehydration of **144** with ethanolic TFA at rt led to **147** in 65–77% yield. These compounds can be considered simultaneously as *N*-nucleosides and acyclic *C*-nucleosides of imidazoline-2-thiones.

The dehydration of β-D-ribofuranosyl derivatives **144** in milder conditions, using ethanolic Dowex 50W-X8 resin, gave a resoluble mixture of **147** and the spironucleosides **146**. The spiro compounds were formed as a mixture of two epimers in C-5 by cyclodehydration between the hydroxyl group on C-5 of the imidazolidine ring and the OH-3 of the polyhydroxyalkyl chain, through the transient stabilized cation **145**. From the reaction mixture the major stereoisomers of 5*S* configuration were isolable (40–16%), whereas the minor one only could be isolated (7%) for R = *p*-MeC$_6$H$_4$ [152].

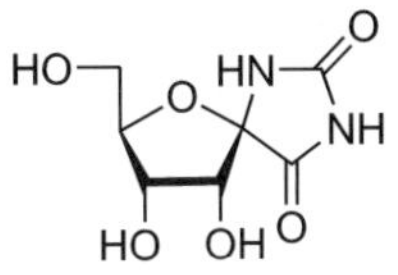

(+)-Hydantocidin

Fig. 3 (+)-Hydantocidin

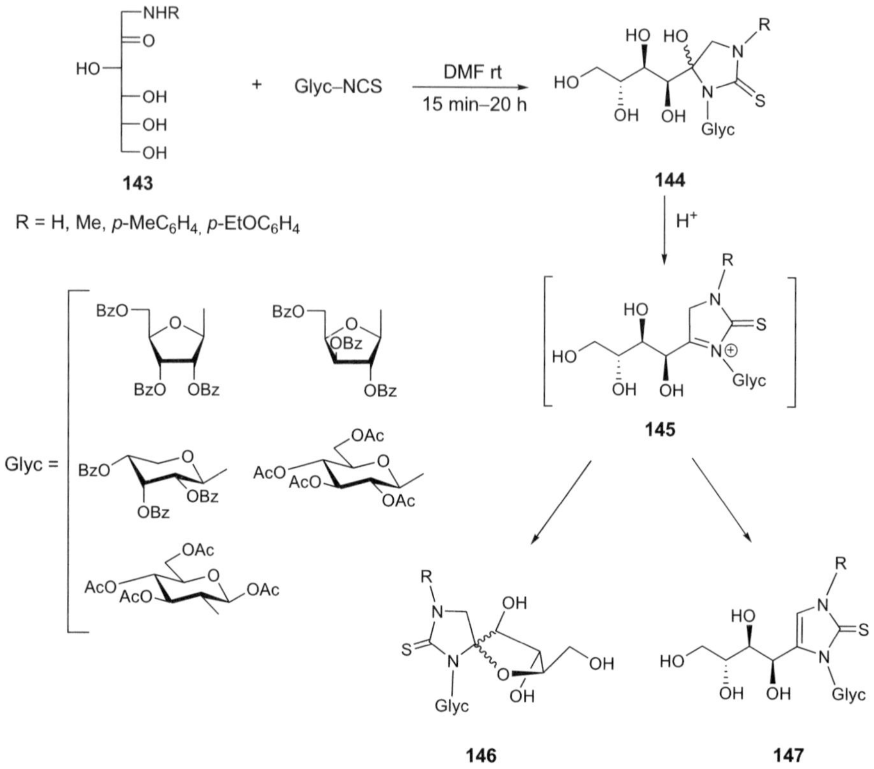

143

R = H, Me, *p*-MeC$_6$H$_4$, *p*-EtOC$_6$H$_4$

Glyc =

Scheme 32

A stereocontrolled synthesis of furanoid and pyranoid spironucleosides of oxazolidine2-thiones (**150** and **151**), and oxazinane-2-thiones (**154** and **155**) via isothiocyanato sugar derivatives (Scheme 33) has been reported by Fuentes and co-workers [156]. The α-(**150**) and β-(**151**) spirooxazolidines were prepared from psicofuranose spiroacetal **148** by opening of the acetal ring with trimethylsilyl azide and trimethylsilyl triflate, followed by hydrogenation, desilylation, isothiocyanation with thiocarbonyl diimidazole, and spontaneous cyclization of the intermediate isothiocyanates **149**. In a similar way, opening of the spiroacetal **148** with trimethylsilyl cyanides and trimethylsilyl triflate, followed by reduction with lithium aluminum hydride and treatment with thiophosgene led to isothiocyanates **152** and **153**, which could be isolated. Base-catalyzed intramolecular cycloaddition of the isothiocyanates and the free hydroxyl groups gave spirooxazinanes **154** and **155** in high yields.

Following a similar sequence of reactions, spironucleosides **159** and **160** were obtained [156] from fructopyranose spiroacetal **156** via non-isolated isothiocyanates **157** and **158**, respectively (Scheme 34).

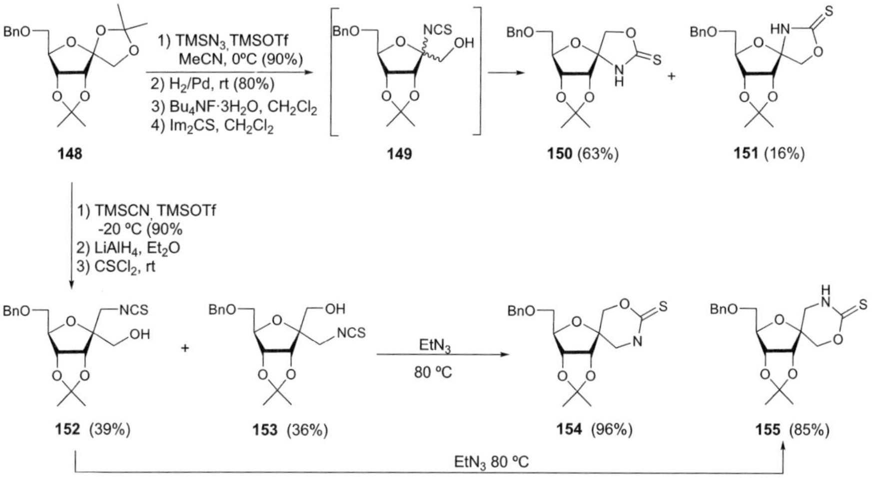

Scheme 33

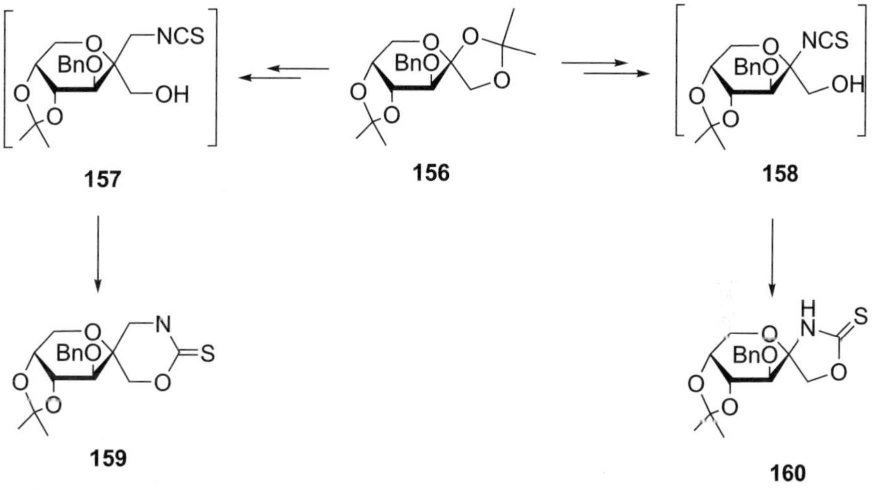

Scheme 34

In the area of spironucleoside chemistry, a considerable effort has been devoted to the synthesis of 2-thiohydantocidin, its 5-epimer [154, 155, 166, 167, 172] and other thiohydantoin spironucleosides [154, 160, 162, 173, 174]. Thus, Fuentes and co-workers have prepared recently [154, 155] 2-thiohydantoin spironucleosides **164α** and **164β** (R = H, alkyl, aryl, glycosyl) from glycosylaminoesters **161** or from glycosyl isothiocyanates **162** via the transient thioureas **163**, as depicted in Scheme 35. Aminoesters **161** and isothiocyanates **162** were obtained from psicofuranose spiroacetal **148** as well as from fructopyranose spiroacetal **156**.

164α

161α

161β

163α

162α

163β

162β

164β   R = H, alkyl, aryl, glycosyl

Scheme 35

In a similar way, reaction of methyl 2-aminohept-2-ulofuranosate **165** [175] with acylated β-D-glucopyranosyl or β-D-ribofuranosyl isothiocyanates resulted in spirohydantoins **166** or **167** (Scheme 36), in almost quantitative yield [154]. It is remarkable that, although both anomers of **165** are in equilibrium under the reaction conditions, only spironucleosides of β anomeric configuration were formed. Compounds **166** and **167** can be considered simultaneously N-nucleosides and spironucleosides.

165 (α/β ratio 1:5)   **166** (94%)   **167** (94%)

Scheme 36

Shiozaki [166, 167] prepared 2-thiohydantocydin **171** and its epimer **170** from 2,3-isopropylidene-D-ribono-1,4-lactone **168** in eight steps via the α-chloro ester **169**, in a 16.5 and 9.2% overall yield, respectively (Scheme 37). These compounds were also prepared by Sano and co-workers through aza-Wittig reaction of an α-azidocarboxamide derivative with tri-*n*-butylphosphine and carbon disulfide [172].

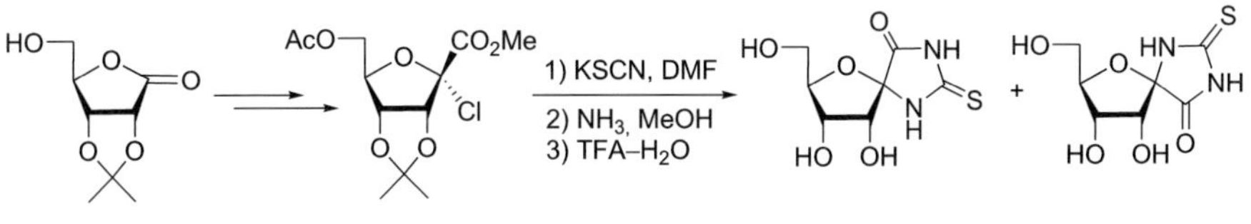

Scheme 37

Somsák and co-workers have described the synthesis of 2-thiohydantocidin analogues with pyranose ring structure. Thus, spiro compound **175** was prepared (Scheme 38) by displacement of the bromide in **172** with thiocyanate ion [176]. The ring-closure reaction proceeded with inversion of the anomeric configuration through a transient isothiocyanate derivative **173**. Mechanisms involving either radical SCN or a classical S_N2 route with complete inversion have been proposed [162]. Analogues of **175** with D-*gluco*, D-*arabino*, and D-*ribo* configurations have also been prepared following the same procedure [162, 176]. The D-*gluco* analogue has been shown to be a potent inhibitor of muscle and liver glycogen phosphorylases [162] and of human salivary α-amylase [177].

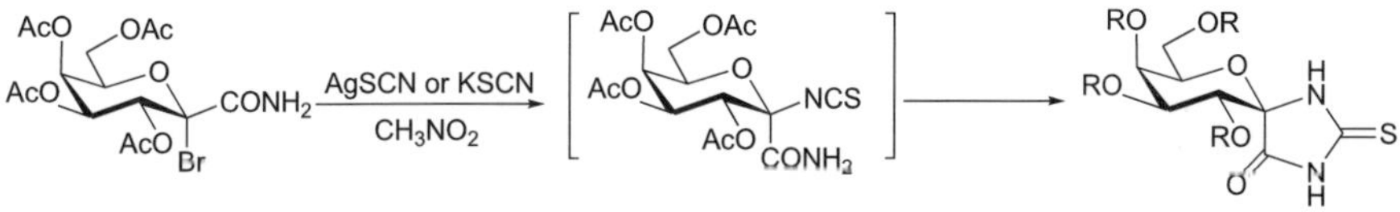

Scheme 38

Acknowledgements The authors thank the Ministerio de Educación y Ciencia for a grant (CTQ2005-01830/BQU) and the Junta de Andalucía (FQM134) for financial support.

References

1. Morris ME, Telang U (2006) Isothiocyanates and Cancer Prevention. In: Awad AB, Bradford PG (eds) Nutrition and Cancer Prevention. CRC Press, Boca Raton, p 435
2. Bianchini F, Vainio H (2004) Drug Metab Rev 36:655
3. Braverman S, Cherkinsky M, Birsa ML (2005) Sci Synth 18:65
4. Mukerjee AK, Ashare R (1991) Chem Rev 91:1

5. Brandsma L (2001) Eur J Org Chem, p 4569
6. García Fernández J, Ortiz Mellet C (2000) Adv Carbohydr Chem Biochem 55:35
7. Witczak ZJ (1986) Adv Carbohydr Chem Biochem 44:91
8. García Fernández JM, Ortiz Mellet C (1996) Sulfur Rep 19:61
9. Goodman I (1958) Adv Carbohydr Chem 13:215
10. López Ó, Maya I, Fuentes J, Fernández-Bolaños JG (2004) Tetrahedron 60:61
11. Fuentes J, Angulo M, Pradera MA (2002) J Org Chem 67:2577
12. Isac-García J, Calvo-Flores FG, Hernández-Mateo F, Santoyo-González F (2001) Eur J Org Chem, p 383
13. Taylor CM (1998) Tetrahedron 54:11317
14. Shi H-F, Cao L-H (2005) Youji Huaxue 25:1066
15. Machón Z, Mielczarek I, Wieczorek J, Mordarski M (1987) Arch Immunol Ther Exp 35:609
16. Benito JM, Gómez-García M, Ortiz Mellet C, Baussanne I, Defaye J, García Fernández JM (2004) J Am Chem Soc 126:10355
17. Patel A, Lindhorst TK (2002) Eur J Org Chem, p 79
18. Saitz-Barrie C, Torres-Pinedo A, Santoyo-González F (1999) Synlett, p 1891
19. Chopade RS, Bahekar RH, Khedekar PB, Bhusari KP, Rao ARR (2002) Arch Pharm 335:381
20. Vanheusden V, Munier-Lehmann H, Froeyen M, Busson R, Rozenski J, Herdewijn P, van Calenbergh S (2004) J Med Chem 47:6187
21. Dallemagne P, Pham Khanh L, Alsaidi A, Varlet I, Collot V, Paillet M, Bureau R, Rault S (2003) Bioorg Med Chem 11:1161
22. Hashem FA, Wahba HE (2000) Phytother Res 14:284
23. Leoni O, Bernardi R, Gueyrard D, Rollin P, Palmieri S (1999) Tetrahedron: Asymmetry 10:4775
24. Mekonnen Y, Draeger B (2003) Planta Med 69:380
25. Seo B, Yun J, Lee ST, Kim M, Hwang K, Kim J, Min KR, Kim Y, Moon D (1999) Planta Med 65:683
26. Brertelsen F, Giseel-Nielsen G, Kjaer A, Skrydstrup T (1988) Phytochemistry 27:3743
27. Tatibouët A, Lawrence S, Rollin P, Holman GD (2004) Synlett, p 1945
28. Gosselin G, Bergogne MC, Imbach JL (1990) Nucleos Nucleot 9:81
29. Grouiller A, Mackenzie G, Najib B, Shaw G, Ewing D (1988) J Chem Soc Chem Commun, p 671
30. Leconte N, Silva S, Tatibouët A, Rauter AP, Rollin P (2006) Synlett, p 301
31. Kruse HP, Heydenreich M, Engst W, Schilde U, Kroll J (2005) Carbohydr Res 340:203
32. Velázquez F, Olivo HF (2002) Curr Org Chem 6:303
33. Crimmins MT, She J (2004) Synlett, p 1371
34. Ortiz A, Quintero L, Hernández H, Maldonado S, Mendoza G, Bernes S (2003) Tetrahedron Lett 44:1129
35. Elbert T, Cerny M (1985) Collect Czech Chem Commun 50:2000
36. García Fernández JM, Ortiz Mellet C, Fuentes J (1992) Tetrahedron Lett 33:3931
37. García Fernández JM, Ortiz Mellet C, Fuentes J (1993) J Org Chem 58:5192
38. García Fernández JM, Ortiz Mellet C, Jiménez Blanco JL, Fuentes Mota J, Gadelle A, Coste-Sarguet A, Defaye J (1995) Carbohydr Res 268:57
39. García Fernández JM, Ortiz Mellet C, Díaz Pérez VM, Jiménez Blanco JL, Fuentes J (1996) Tetrahedron 52:12947
40. García Fernández JM, Ortiz Mellet C, Jiménez Blanco JL, Fuentes J (1994) J Org Chem 59:5565
41. Grashey R, Keramaris N, Baumann M (1970) Tetrahedron Lett 11:5087

42. Lillelund VH, Jensen HH, Liang X, Bols M (2002) Chem Rev 102:515
43. Jiménez Blanco JL, Díaz Pérez VM, Ortiz Mellet C, Fuentes J, García Fernández JM, Díaz Arribas JC, Cañada FJ (1997) Chem Commun 1969
44. Díaz Pérez VM, García Moreno MI, Ortiz Mellet C, Fuentes J, Díaz Arribas JC, Cañada FJ, García Fernández JM (2000) J Org Chem 65:136
45. Díaz Pérez P, García-Moreno MI, Ortiz Mellet C, García Fernández JM (2005) Eur J Org Chem, p 2903
46. Benito JM, Ortiz Mellet C, García Fernández JM (2000) Carbohydr Res 323:218
47. Fernández-Bolaños JG, Zafra E, Robina I, Fuentes J (1999) Carbohydr Lett 3:239
48. Mullins RE, Langdon RG (1980) Biochem 19:1199
49. Mullins RE, Langdon RG (1980) Biochem J 19:1205
50. Rees WD, Gliemann J, Holman GD (1987) Biochem J 241:857
51. Maya I, López Ó, Férnandez-Bolaños JG, Robina I, Fuentes J (2001) Tetrahedron Lett 42:5413
52. García Alles LF, Navdaeva V, Haenni S, Erni B (2002) Eur J Biochem 269:4969
53. Fernández-Bolaños JG, López Ó, Maya I (2003) Synthesis of O-unprotected sugar isothiocyanates and their transformation into thioureas and cyclic isoureas. In: Pandalai S (ed) Recent Developments in Carbohydrate Research, Transworld Research Network, Trivandrum, India, p 81
54. Eliel EL, Wilen SH (1994) Stereochemistry of Organic Chemistry. Wiley, New York, p 771
55. Beaupere D, El Meslouti A, Lelievre P, Uzan R (1995) Tetrahedron Lett 36:5347
56. García-Moreno MI, Díaz-Pérez P, Ortiz Mellet C, García Fernández JM (2003) J Org Chem 68:8890
57. Ogawa S, Mori M, Takeuchi G, Doi F, Watanabe M, Sakata Y (2002) Bioorg Med Chem 12:2811
58. García-Moreno MI, Díaz-Pérez P, Ortiz Mellet C, García Fernández JM (2002) Chem Commun, p 848
59. Cordi AA, Berque-Bestel I, Persigand T, Lacoste J-M, Newman-Tancredi A, Audinot V, Millan MJ (2001) J Med Chem 44:787
60. Hirashima A, Morimoto M, Kuwano E, Eto M (2003) Bioorg Med Chem 11:3753
61. Wong WC, Sun W, Cui W, Chen X, Forray C, Vaysse PJ-P, Branchek TA, Gluchowski C (2000) J Med Chem 43:1699
62. Bosc JJ, Jarry C, Martínez B, Molimard M (2001) J Pharm Pharmacol 53:923
63. Massip S, Guillon J, Bertarelli D, Bosc J-J, Leger J-M, Lacher S, Bontemps C, Dupont T, Mueller CE, Jarry C (2006) Bioorg Med Chem 14:2697
64. Bruban V, Feldman J, Greney H, Dontenwill M, Schann S, Jarry C, Payard M, Boutin J, Scalbert E, Pfieffer B, Renard P, Vanhoutte P, Bousquet P (2001) Br J Pharmacol 133:261
65. Venteclef N, Guillard R, Issandou M (2005) Biochem Pharmacol 69:1041
66. Tao R, Fray A, Aspley S, Brammer R, Heal D, Auerbach S (2002) Eur J Pharmacol 445:69
67. Hirashima A, Eiraku T, Watanabe Y, Kuwano E, Taniguchi E, Eto M (2001) Pest Manag Sci 57:713
68. López Ó, Fernández-Bolaños JG, Gil MV (2005) Green Chem 7:431
69. Ando O, Satake H, Itoi K, Sato A, Nakajima M, Takahashi S, Haruyama H, Ohkuma Y, Kinoshita T, Enokita R (1991) J Antibiot 44:1165
70. Nakayama T, Amachi T, Murao S, Sakai T, Shin T, Kenny PT, Iwashita T, Zagorski M, Komura H, Nomoto KJ (1991) Chem Soc Chem Commun, p 919
71. Sakuda S, Isogai A, Matsumoto S, Suzuki A (1987) J Antibiot 40:296
72. Feng X, Duesler EN, Mariano PS (2005) J Org Chem 70:5618

73. Akiyama M, Awamur T, Kimura K, Hosomi Y, Kobayashi A, Tsuji K, Kuboki A, Ohira S (2004) Tetrahedron Lett 45:7133
74. Chiara JL, García A (2005) Synlett, p 2607
75. Berecibar A, Granjean C, Siriwardena A (1999) Chem Rev 99:779
76. Uchida C, Ogawa S (1999) Recent Res Dev Pure Appl Chem 3:161
77. Kobayashi Y (1999) Carbohydr Res 315:3
78. Crimmins MT, Tabet EA (2001) J Org Chem 66:4012
79. Seepersaud M, Al-Abed Y (2001) Tetrahedron Lett 42:1471
80. Storch de Gracia I, Bobo S, Martín-Ortega MD, Chiara JL (1999) Org Lett 1:1705
81. Sakuda S, Sugiyama Y, Zhou Z-Y, Takao H, Ikeda H, Kakinuma K, Yamada Y, Nagasawa H (2001) J Org Chem 66:3356
82. Clark M, Goering BK, Lee J, Ganem B (2000) J Org Chem 65:4058
83. Miyazaki H, Kobayashi Y, Shiozaki M, Ando O, Nakajima M, Hanzawa H, Haruyama H (1995) J Org Chem 60:6103
84. Shiozaki M, Mochizuki T, Hanzawa H, Haruyama H (1996) Carbohydr Res 288:99
85. Uchida C, Ogawa S (1996) Bioorg Med Chem 4:275
86. Uchida C, Kitahashi H, Yamagishi T, Iwaisaki Y, Ogawa S (1994) J Chem Soc Perkin Trans 1:2775
87. López Ó, Maya I, Ulgar V, Robina I, Fuentes J, Fernández-Bolanos JG (2002) Tetrahedron Lett 43:4313
88. Boiron A, Zillig P, Faber D, Giese B (1998) J Org Chem 63:5877
89. Uchida C, Kimura H, Ogawa S (1997) Bioorg Med Chem 5:921
90. Ledford BE, Carreira EM (1995) J Am Chem Soc 117:11811
91. Knapp S, Purandare A, Rupitz K, Withers SG (1994) J Am Chem Soc 116:7461
92. García-Moreno IM, Rodríguez-Lucena D, Ortiz Mellet C, García Fernández JM (2004) J Org Chem 69:3578
93. Mukaiyama T (1979) Angew Chem Int Ed Engl 18:707
94. Isac-García J, Hernández-Mateo F, Calvo-Flores FG, Santoyo-González F (2004) J Org Chem 69:202
95. Santoyo-González F, García-Calvo-Flores F, Isac-García J, Hernández-Mateo F, García-Mendoza P, Robles-Díaz R (1994) Tetrahedron 50:2877
96. Chiara JL, Storch de Gracia I, García A, Bastida A, Bobo S, Martín-Ortega MD (2005) ChemBiochem 6:186
97. Chiara JL, Marco-Contelles J, Khiar N, Gallego P, Destabel C, Bernabe M (1995) J Org Chem 60:6010
98. Chu CK (2002) Recent Advances in Nucleosides: Chemistry and Chemotherapy. Elsevier Science BV, Amsterdam
99. (2003) Nucleosides. In: Ferrier RJ (ed) Carbohydrate Chemistry. Monosaccharides, Disaccharides and Specific Oligosaccharides. The Royal Society of Chemistry, London, p 248
100. Lech-Maranda E, Korycka A, Robak T (2006) Mini Rev Med Chem 6:575
101. Robak T, Korycka A, Kasznicki M, Wrzesien-Kus A, Smolewski P (2005) Curr Cancer Drug Targets 5:421
102. Vivet-Boudou V, Didierjean J, Isel C, Marquet R (2006) Cell Mol Life Sci 63:163
103. De Clercq E (2002) Biochim Biophys Acta 1587:258
104. Núñez M, Soriano V (2005) Lancet Infect Dis 5:374
105. Fung SK, Lok ASF (2004) Nat Clin Pract Gastroenterol Hepat 1:90
106. Seela F, Lin W, Kazimierczuk Z, Rosemeyer H, Glaccon V, Peng X, He Y, Ming X, Andrzejewska M, Gorska A, Zhang X, Eickmeier H, La Colla P (2005) Nucleos Nucleot Nucl 24:859

107. Wang J, Jin Y, Rapp KL, Bennett M, Schinazi RF, Chu CK (2005) J Med Chem 48:3736
108. Ludek OR, Balzarini J, Meier C (2006) Eur J Org Chem, p 932
109. Cho JH, Bernard DL, Sidwell RW, Kern ER, Chu CK (2006) J Med Chem 49:1140
110. Gunaga P, Moon HR, Choi WJ, Shin DH, Park JG, Jeong LS (2004) Curr Med Chem 11:2585
111. Fernández-Bolaños JG, Al-Masoudi NAL, Maya I (2001) Adv Carbohydr Chem Biochem 57:21
112. Sheban MAE (1998) Adv Heterocycl Chem 70:163
113. Sun Z, Ahmed S, McLaughlin L (2006) J Org Chem 71:2922
114. Abbasi MM, El-Kousy SM, El-Moghazy YE, El-Kafrawy S (2005) Int J Chem 15:77
115. De Clercq E, Holy A (2005) Nat Rev Drug Discov 4:928
116. Oh C-H, Baek T-R, Hong JH (2005) Nucleos Nucleot Nucl 24:153
117. Fernández-Bolaños Guzmán J, García Rodríguez S, Fernández-Bolaños J, Diánez MJ, López-Castro A (1991) Carbohydr Res 210:125
118. Ávalos M, Babiano R, Cintas P, Jiménez JL, Palacios JC, Silvero G, Valencia C (1999) Tetrahedron 55:4377
119. Ávalos M, Babiano R, Cintas P, Hursthouse MB, Jiménez JL, Light ME, Palacios JC, Silvero G (2005) Tetrahedron 61:7945
120. Fernández-Bolaños J, Fuentes Mota J, Fernández-Bolaños Guzmán J (1988) Carbohydr Res 173:17
121. Fuentes J, Molina JL, Olano D, Pradera MA (1996) Tetrahedron: Asymmetry 7:203
122. Fernández-Bolaños JG, Zafra E, López Ó, Robina I, Fuentes J (1999) Tetrahedron: Asymmetry 10:3011
123. Fuentes J, Molina JL, Pradera MA (1998) Tetrahedron: Asymmetry 9:2517
124. Marino C, Herczegh P, de Lederkremer RM (2001) Carbohydr Res 333:123
125. Marino C, Varela O, de Lederkremer RM (1997) Tetrahedron 53:16009
126. Marino C, Varela O, de Lederkremer RM (1997) Carbohydr Res 304:257
127. Ávalos M, Babiano R, Cintas P, Hursthouse MB, Jiménez JL, Light ME, Palacios JC, Silvero G (2005) Tetrahedron 61:7931
128. Zhu H-J, Ren Y, Ren J, Chu S-Y (2005) Theochem 730:199
129. Saleh MA (2002) Sulfur Lett 25:235
130. Saleh MA (2002) Nucleos Nucleot Nucl 21:401
131. Pearson MSM, Robin A, Bourgougnon N, Meslin JC, Deniaud D (2003) J Org Chem 68:8583
132. Robin A, Julienne K, Meslin J-C, Deniaud D (2006) Eur J Org Chem, p 634
133. Camarasa MJ, Fernández-Resa P, García López MT, de las Heras FG, Méndez-Castrillón PP, San Félix A (1984) Synthesis, p 509
134. Al-Masoudi N, Hassan NA, Al-Soud YA, Schmidt P, Gaafar AE-DM, Weng M, Marino S, Schoch A, Amer A, Jochims JC (1998) J Chem Soc Perkin Trans 1:947
135. El-Gazzar ABA, Hegab MI, Hassan NA (2002) Sulfur Lett 25:161
136. Wei M-J, Fang D-C, Liu R-Z (2004) Eur J Org Chem, p 4070
137. Al-Masoudi NA, Al-Soud YA (2004) J Carbohydr Chem 23:111
138. Wei M-J, Fang D-C, Liu R-Z (2002) J Org Chem 67:7432
139. Al-Soud YA, Al-Masoudi NA (2003) Heteroatom Chem 14:298
140. Al-Masoudi NA, Al-Soud YA, Al-Masoudi WA (2004) Nucleos Nucleot Nucl 23:1739
141. Fuentes Mota J, Ortiz Mellet MC, Pradera Adrián M (1984) An Quím Ser C 80:48
142. Ogura H, Takeda K, Takayanagi H (1989) Heterocycles 29:1171
143. Ávalos M, Babiano R, Cintas P, Chavero MM, Higes FJ, Jiménez JL, Palacios JC, Silvero G (2000) J Org Chem 65:8882
144. Gama Y, Shibuya I, Shimizu M, Goto M (2001) J Carbohydr Chem 20:459

145. Gama Y, Shibuya I, Shimizu M (2002) Chem Pharm Bull 50:1517
146. Gama Y, Shibuya I, Shimizu M (2000) J Carbohydr Chem 19:119
147. Gama Y, Shibuya I, Shimizu M (1999) Synth Commun 29:1493
148. Shimizu M, Gama Y, Shibuya I (1998) Heterocycles 48:1935
149. Fuentes J, Angulo M, Molina JL, Pradera MA (1997) J Carbohydr Chem 16:1457
150. Lin P, Lee CL, Sim MM (2001) J Org Chem 66:8243
151. García Fernández JM, Ortiz Mellet C, Díaz Pérez VM, Fuentes J, Kovács J, Pintér I (1997) Carbohydr Res 304:261
152. Gasch C, Pradera MA, Salameh BAB, Molina JL, Fuentes J (2000) Tetrahedron: Assymmetry 11:435
153. Fernández-Bolaños J, Blasco López A, Fuentes Mota J (1990) Carbohydr Res 199:239
154. Fuentes J, Salameh BAB, Pradera MA, Fernández de Cordoba FJ, Gasch C (2006) Tetrahedron 62:97
155. Gasch C, Salameh BAB, Pradera MA, Fuentes J (2001) Tetrahedron Lett 42:8615
156. Gasch C, Pradera MA, Salameh BAB, Molina JL, Fuentes J (2001) Tetrahedron: Asymmetry 12:1267
157. Singha K, Roy A, Dutta PK, Tripathi S, Sahabuddin S, Achari B, Mandal SB (2004) J Org Chem 69:6507
158. Long DD, Smith MD, Martin A, Wheatley JR, Watkin DG, Mueller M, Fleet GWJ (2002) J Chem Soc Perkin Trans 1:1982
159. Freire R, Martín A, Pérez-Martín I, Suárez E (2002) Tetrahedron Lett 43:5113
160. Somsák L, Nagy V, Docsa T, Tóth B, Gergely P (2000) Tetrahedron: Asymmetry 11:405
161. Renard A, Lhomme J, Kotera M (2002) J Org Chem 67:1302
162. Somsák L, Kovács L, Tóth M, Osz E, Szilágyi L, Gyoergydeák Z, Dinya Z, Docsa T, Tóth B, Gergely P (2001) J Med Chem 44:2843
163. Andersch J, Hennig L, Wilde H (2000) Carbohydr Res 329:693
164. Chatgilialoglu C, Gimisis T, Spada GP (1999) Chem Eur J 5:2866
165. Agasimundin YS, Mumper MW, Hosmane RS (1998) Bioorg Med Chem 6:911
166. Shiozaki M (2002) Carbohydr Res 337:2077
167. Shiozaki M (2001) Carbohydr Res 335:147
168. Haruyama H, Takayama T, Kinoshita T, Kondo M, Nakajima M, Haneishi T (1991) J Chem Soc Perkin Trans 1:1637
169. El Ashry ESH, El Nemr A (2005) Synthesis of Naturally Occurring Nitrogen Heterocycles from Carbohydrates. Blackwell Publishing, Oxford, UK
170. Nakajima M, Itoi K, Takamatsu Y, Kinoshita T, Okazaki T, Kawakubo K, Shindou M, Honma T, Toujigamori M, Haneishi T (1991) J Antibiot 44:293
171. Siehl DL, Subramanian MV, Walters EW, Lee S-F, Anderson RJ, Toschi AG (1996) Plant Physiol 110:753
172. Sano H, Mio S, Kitagawa J, Shindou M, Honma T, Sugai S (1995) Tetrahedron 51:12563
173. Somsák L, Nagy V (2000) Tetrahedron: Asymmetry 11:1719
174. Osz E, Somsák L, Szilágyi L, Kovács L, Docsa T, Tóth B, Gergely P (1999) Bioorg Med Chem Lett 9:1385
175. Dondoni A, Schermann MC (1994) J Org Chem 59:6404
176. Osz E, Sós E, Somsák L, Szilágyi L, Dinya Z (1997) Tetrahedron 53:5813
177. Gyemant G, Kandra L, Nagy V, Somsák L (2003) Biochem Biophys Res Commun 312:334

Top Heterocycl Chem (2007) 7: 101–132
DOI 10.1007/7081_2006_046
© Springer-Verlag Berlin Heidelberg
Published online: 28 February 2007

β-Lactams from Carbohydrates

Bartłomiej Furman · Zbigniew Kałuża · Agnieszka Stencel ·
Barbara Grzeszczyk · Marek Chmielewski (✉)

Institute of Organic Chemistry of the Polish Academy of Sciences, Kasprzaka 44/52,
01-224 Warsaw, Poland
chmiel@icho.edu.pl

Abstract Since more then thirty years carbohydrates have gained much attention as chiral starting materials in stereocontrolled target oriented synthesis. Among variety of applications, synthesis of β-lactam antibiotics from carbohydrate precursors played a special role owing to the importance of these class of compounds in modern chemotherapy. In 1994 we published a review article on the synthesis of β-lactams from carbohydrate precursors. The present survey reports literature published after that date. The streaking feature of the syntheses performed during the last decade is domination of the highly stereoselective direct formation of a four-membered β-lactam ring via [2 + 2]cycloaddition of ketenes to imines and of chlorosulfonyl isocyanate to olefins. Particularly attractive substrates are: in former method, imines derived from glyceraldehyde and in latter one, vinyl ethers of acetal protected sugars.

1
Introduction

Since the early 1970s, the tendency to use low molecular weight carbohydrates as convenient enantiomerically pure substrates for stereocontrolled target-oriented organic synthesis became a world trend often seen in the chemical literature [1–5]. Despite spectacular achievements of contemporary organic synthesis in the field of catalytic enantioselective organic reactions [6–16], carbohydrates still remain very attractive, renewable, starting organic materials, simply because they are very cheap and available in many structural and configurational forms [17, 18].

In 1994 we published a review article on the synthesis of β-lactams from carbohydrate precursors [19]. Carbohydrates were used either as a chiral pool or chiral auxiliaries. Transformations of carbohydrates offered full stereocontrol in the formation of the desired structure but showed the potency of modern organic synthesis rather than practical value. Great effort has been made to select the appropriate sugar synthons since simple carbohydrates display some shortcomings such as overfunctionalization with hydroxyl groups, too many stereogenic centers, the lack of double bonds, and a configuration that is not fixed at the anomeric center.

Although the search for new, more active β-lactam antibiotics, or for new methods of their synthesis does not represent contemporary world research trends, a variety of compounds having β-lactam fragments have been found to display very interesting, but not antibacterial activities [20]. This prompted us to update the report on the synthesis of β-lactams from carbohydrates. The present survey comprises literature published after 1994.

Over the last 12 years direct formation of a four-membered β-lactam ring via [2 + 2]cycloaddition of ketenes to imines and of isocyanates to olefins dominated over other methods. Because of the well-defined transition states, both cycloadditions usually offer an excellent stereoselectivity. In the case of cycloaddition of ketenes or their equivalents to imines, the cycloadducts were used not only for the synthesis of β-lactam antibiotics, but also as intermediates for the preparation of other biologically active compounds.

2
Syntheses Based on [2 + 2]Cycloaddition of Sugar-Derived Imines with Ketenes or their Equivalents

The discovery that E-imines obtained from a variety of isopropylidene protected sugar open-chain aldehydes and ketenes, or ketene equivalents afford *cis*-substituted β-lactam adducts usually with a high asymmetric induction and definite relative geometry depending on the absolute configuration of the stereogenic center next to the imine carbon atom (Scheme 1) [21–24], prompted many laboratories to exploit this reaction in a number of syntheses. Imines derived from both easily available enantiomeric forms of 2,3-*O*-isopropylidene-glyceraldehyde are particularly attractive.

The search for high biological activity prompted Hassan and Soliman to synthesize compounds that were a combination of β-lactam with sulfonamide fragments. A series of potential antibacterial and antiviral agents

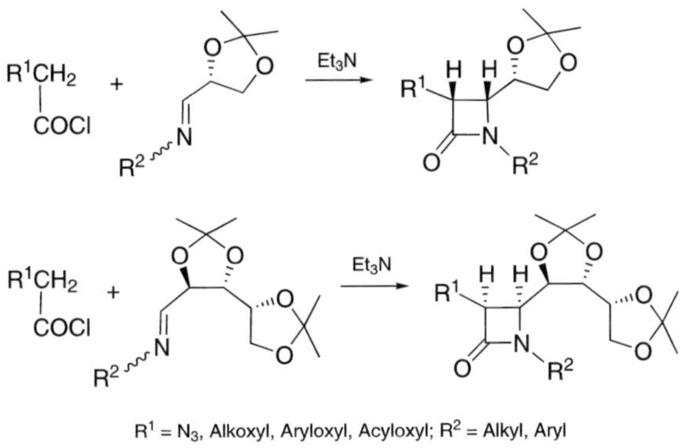

Scheme 1

were prepared by [2 + 2]cycloaddition of imines derived from isopropylidene D-glyceraldehyde and *p*-aminophenyl-sulfonamides *N*-linked to heterocycles with benzyloxyacetyl chloride [25].

An easy deprotection of the isopropylidene residue in **1** and glycolic cleavage of the diol **2** to the aldehyde **3**, or glycolic cleavage followed by the oxidation to the carboxylic function and formation of the ester **4**, provide particularly attractive synthons (Scheme 2). Dirhodium complexes derived from difluoro-azetidinones, obtained in this way, were used as chiral catalysts for enantioselective decomposition of diazocompounds and cyclopropanation, to show, however, a moderate selectivity (Scheme 3) [26].

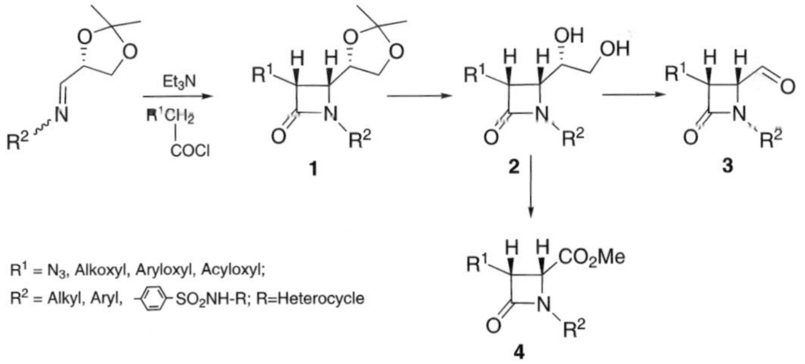

Scheme 2

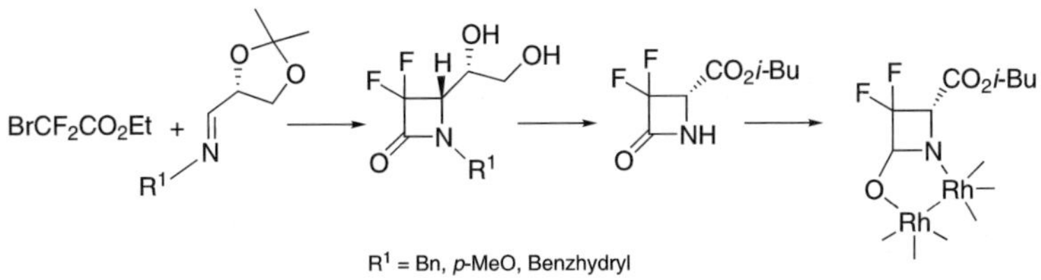

Scheme 3

The Bose group continued their studies on the cycloaddition of ketenes to imines directed at target-oriented organic synthesis [23] and the application of microwave methodology [27, 28]. Recently, highly stereoselective synthesis of α-hydroxy-β-lactams readily available from carbohydrates has been developed [29, 30]. It has been shown [27–30] that the reaction of acetoxyacetyl chloride and *N*-methylmorpholine with Schiff bases having aromatic substituents gives mostly *cis* β-lactams at low energy of microwave irradiation, whereas at higher energy levels more than 90% of the β-lactam formed may be the *trans* isomer. Interestingly, with Schiff bases derived from carbohydrate precursors 5–8 the *cis* β-lactams are formed at all levels of microwave irradiation. Stereocontrolled approaches and microwave-assisted chemical reactions have been utilized for the preparation of the corresponding hydroxyazetidinones and this was followed by their conversion to intermediates for the synthesis of gentosamine (**9**), 6-epilincosamine (**10**), γ-hydroxythreonine (**11**), and polyoxamic acid (**12**) [29, 30].

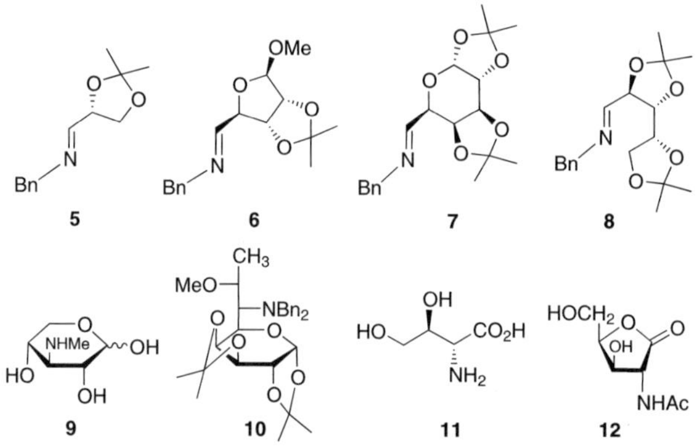

Chart 1

An intramolecular opening of β-lactams with amines as a strategy for the formation of larger heterocyclic compounds was reported by two groups [31, 32]. Banik et al. [31] transformed β-lactam **13**, which was obtained by cycloaddition of a glyceraldehyde-derived imine with *o*-nitrophenoxyacetyl chloride, into oxazine **14** by a sequence of reactions involving reduction of the nitro group in the presence of indium/ammonium chloride, followed by intramolecular transamidation. The same reaction sequence performed in the presence of zinc stopped at the stage of *o*-anisidine derivative **15** (Scheme 4).

Similar transamidation leading to the enlargement of the four-membered ring has been reported by Banfi et al. [32]. In that case, the amino function in **17** was introduced to the end of the 4-allyloxy substituent of the azetidinone **16** by a sequence of reactions involving hydroxylation, introduction of an azide, and its reduction (Scheme 5) [32].

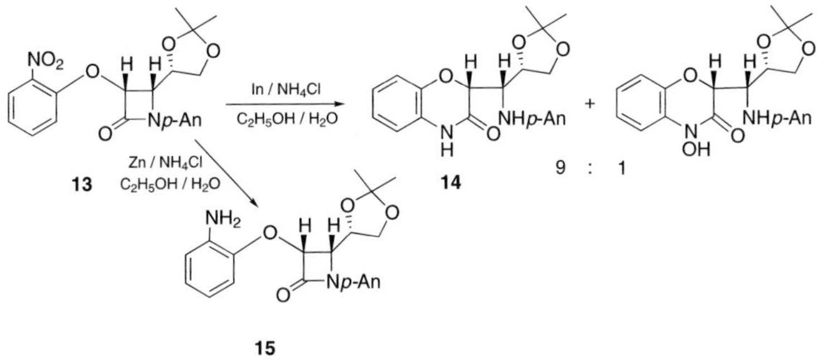

Scheme 4

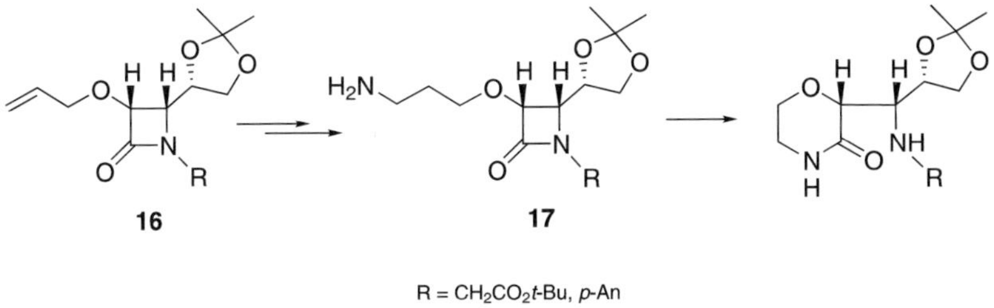

R = CH$_2$CO$_2$t-Bu, p-An

Scheme 5

Azetidinones derived from glyceraldehyde imines and acetoxy-, phenoxy-, or alkoxy-acetyl chlorides have been investigated as intermediates for the synthesis of a variety of β-lactam antibiotics by the Alcaide group [33–35]. Easy hydrolysis of the dioxolane ring in **18** followed by deoxygenation afforded 4-vinyl azetidinone **19**. Alternatively, hydrolysis of **20** and a glycolic cleavage of the resulting diol provided the corresponding optically pure aldehyde **21** which after nucleophilic addition of lithium TMS-acetylide afforded compounds **22**, as diastereomeric mixtures. Both 4-substituted azetidinones **19** and **22** constitute starting materials for a variety of syntheses. Both enyne azetidinones **19** and **22** were used for the synthesis of fused tricyclic β-lactam **23** and diastereomeric tricyclic system **24** by the Pauson–Khand reaction (Scheme 6) [33, 34].

In a similar way, highly functionalized medium-sized rings fused to β-lactams **25–27** were obtained from the aldehyde **21** using radical cyclization of Baylis–Hillman adducts (Scheme 7) [35].

The Wittig–Horner olefination of the aldehyde **28** provided alkenes **29** which were subjected to radical cyclization leading to benzofused tricyclic β-lactams **30**, obtained as single diastereomers (Scheme 8) [36]. A convenient, direct regio- and stereoselective route to optically pure unusually fused or bridged tricyclic β-lactams has been developed by the use of intramolecular nitrone-alkene cycloaddition reactions. For example, the aldehyde **21** can be transformed into nitrone **31** which subsequently was used for a variety

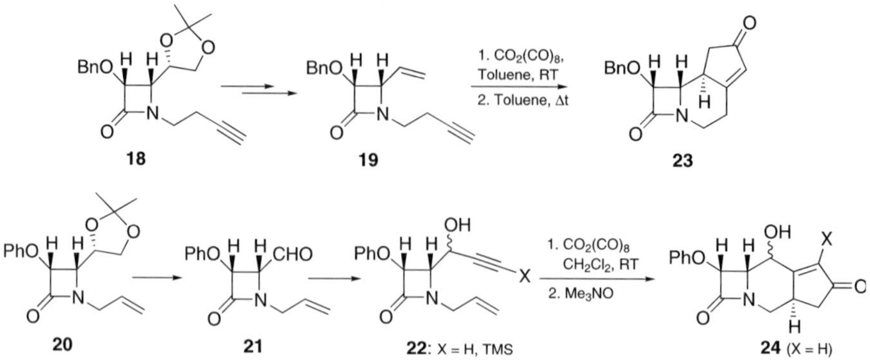

Scheme 6

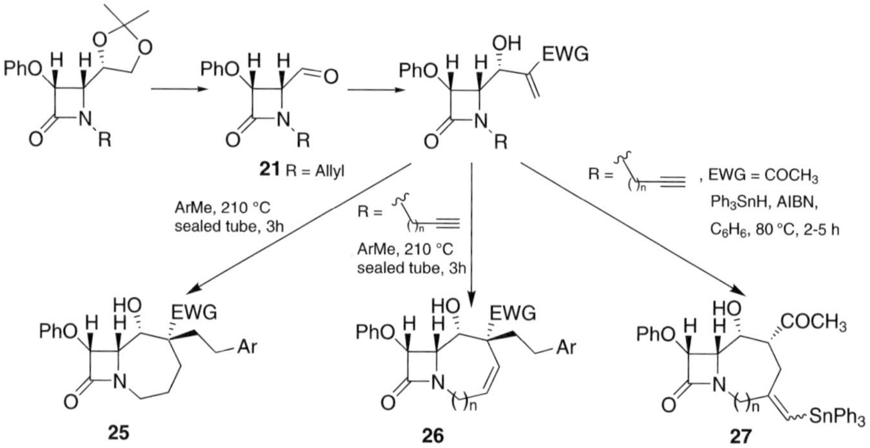

Scheme 7

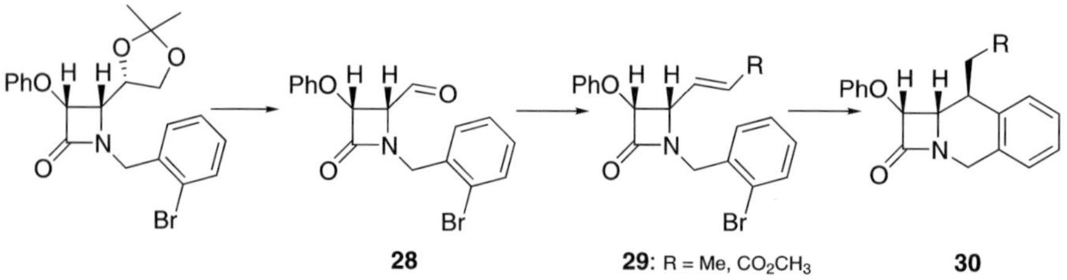

Scheme 8

of cycloaddition reactions providing unusual fused polycyclic β-lactam **32** (Scheme 9) [37].

A thorough investigation of allylation reactions of the adducts **33** and the aldehydes **34** under a variety of conditions was reported. Mesylates of the homoallylic alcohols obtained, having an extra alkene or alkyne at position

Scheme 9

1 or 3 of the lactam ring, were transformed into fused tricyclic β-lactams
35–38 via tandem one-pot elimination-intramolecular Diels–Alder reactions
(Scheme 10) [38].

Scheme 10

The Alcaide group [39, 40] continued investigations leading to unusual fused tricyclic β-lactams **39–42** via aza-cycloadditions/ring closing
metathesis. The diolefinic precursors were obtained, like before, from [2 +
2]cycloadducts of imines derived from glyceraldehydes [39]. A similar strategy has been applied to the synthesis of other unusual β-lactams having
a seven-membered ring fused to the azetidinone **43** and **44** via tin-promoted
radical cyclization [40].

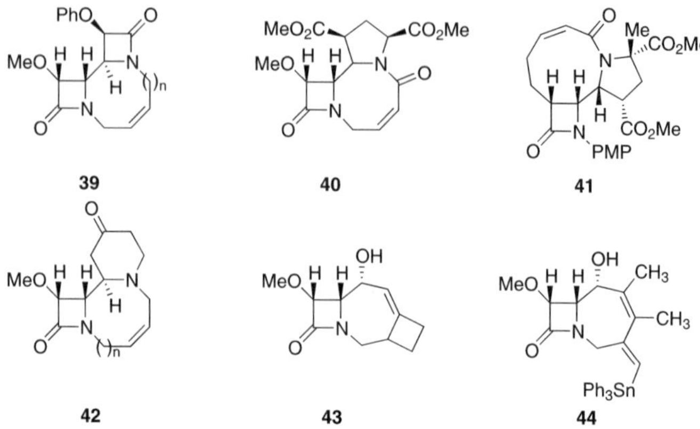

Chart 2

An introduction of an alkyl substituent, bearing a double bond, to the β-lactam nitrogen atom and subsequent transformation of the dioxolane ring stemming from glyceraldehyde into the aldehyde group opened an interesting access to carbacephams. Lewis acid-promoted carbonyl-ene cyclizations of azetidinone-tethered alkenylaldehydes led to a rapid, highly diastereoselective formation of polyfunctionalized compounds having carbacepham skeletons (Scheme 11) [41].

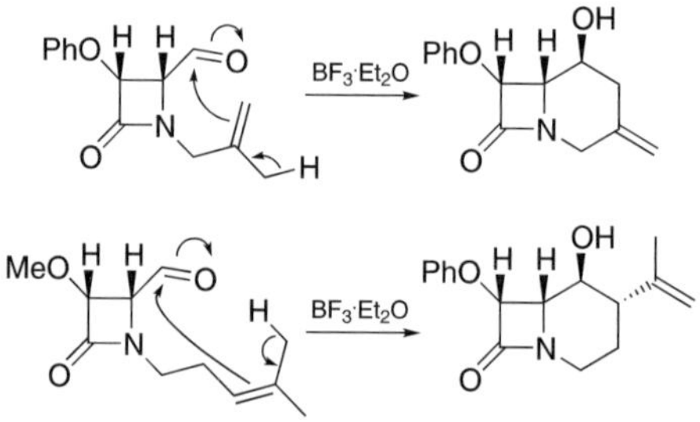

Scheme 11

Azetidinone having a N-dehydroamino acid side chain, structurally related to the active penem and cephem antibiotics, was obtained by the standard phenylselenylation-oxidation-elimination reaction sequence [42]. C-4-substituted aldehydes can also be subjected to a novel N-1–C-4 β-lactam bond cleavage in the presence of 2-(trimethylsilyl)thiazole (TMST) to give enantiopure α-alkoxy-γ-keto amides (Scheme 12) as the major products [43].

Deprotection of the C-3 hydroxy group and subsequent oxidation gave reactive dicarbonyl compounds **45**. Addition of a variety of nucleophilic reagents to the keto group led to C-3 functionalized compounds [44–47]. Using this path, 3-hydroxy-3-allyl-**46** [44], allenyl-**47**, propargyl-**48** [45],

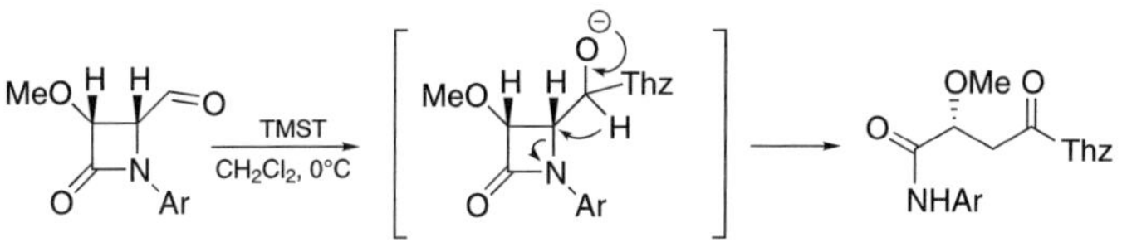

Scheme 12

butadienyl-**49** [46], and methoxycarbonyl-vinyl **50** [47] substituted azetidinones were obtained, and they were used for further transformations leading to 3-functionalized β-lactams (Scheme 13).

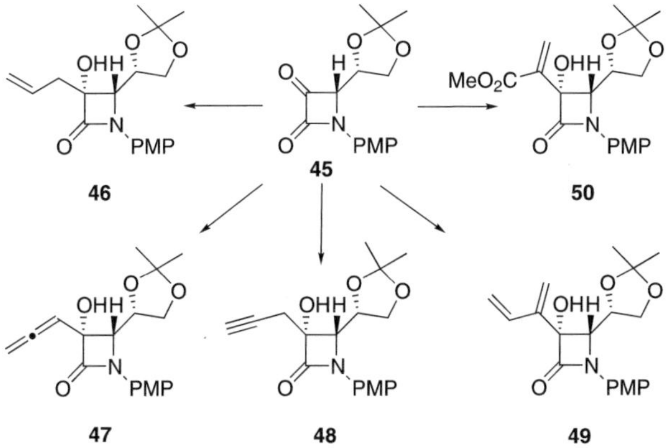

Scheme 13

The dicarbonyl compound **51** was oxidized to the anhydride **52**, which subsequently reacted with primary or secondary amines to form α-amino acids, α-amino amides and dipeptides **53** (Scheme 14) [48]. 3-Hydroxy β-lactams obtained from imines derived from carbohydrates [49, 50] or prepared via the Sharpless AD reaction [51–53] were directly oxidized to anhydrides by treatment with NaOCl and TEMPO. Anhydrides **54–56** were used for the synthesis of compounds related to the family of polyoxins represented by **57** (Scheme 15) [49–53].

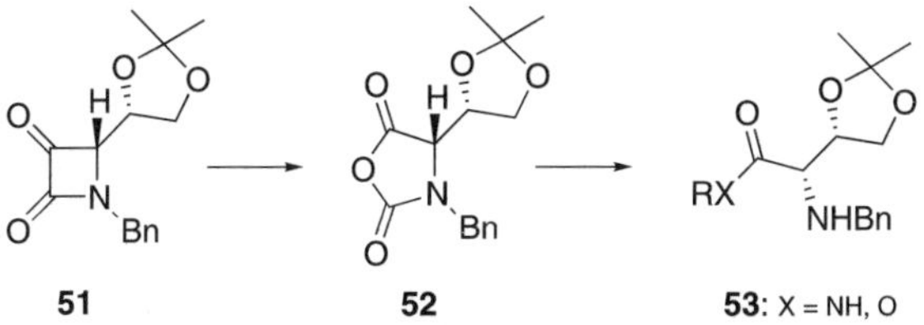

Scheme 14

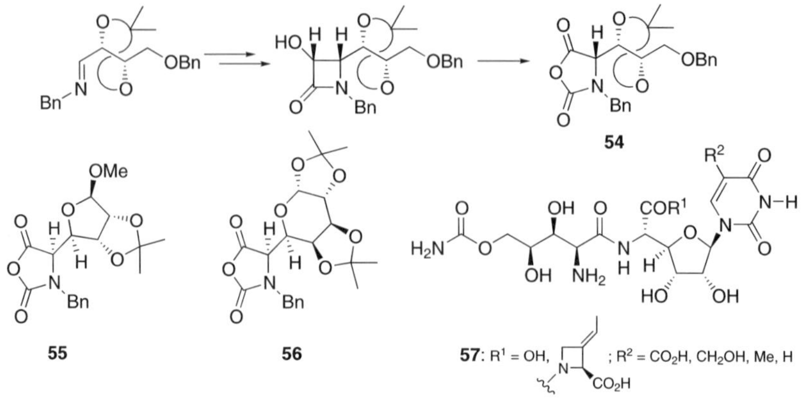

54

55   **56**   **57**: R^1 = OH, ; R^2 = CO$_2$H, CH$_2$OH, Me, H

Scheme 15

Cyclohexylidene-protected glyceraldehyde imines were used by Annunziata et al. [54, 55] for the synthesis of a β-lactam precursor of thrombin and tryptase inhibitors [54], as well as for the inhibitor of serine protease [55].

Glucosamine-derived imines **58** were used for the synthesis of carbapenem and carbacephem antibiotics. [2 + 2]Cycloaddition with methoxyacetyl chloride provided diastereomeric β-lactams with low asymmetric induction [56]. Radical cyclization and oxidation reactions of diastereoisomer **59** led to carbapenems **60** and **61** (Scheme 16) [57]. The same research group continued the transformation of β-lactams derived from D-glucosamine. In particular, a tandem elimination-conjugate addition performed on **62** provided the second ring of the carbacephem, for example **63** (Scheme 17) [58].

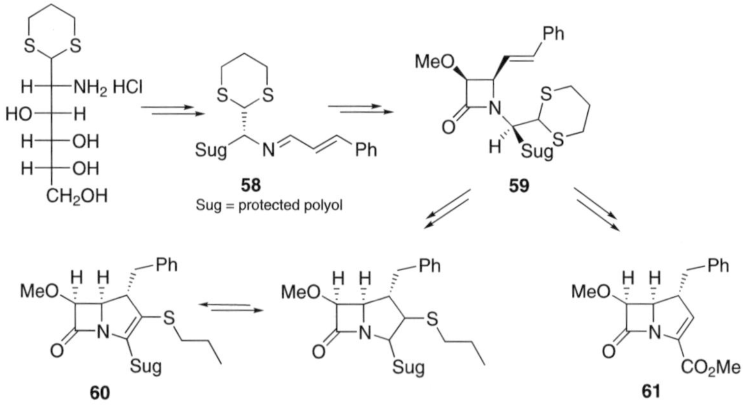

58

Sug = protected polyol

59

60   **61**

Scheme 16

β-Lactams obtained from glucosamine were used also for other transformations leading to a variety of carbacephams. The formation of epoxides

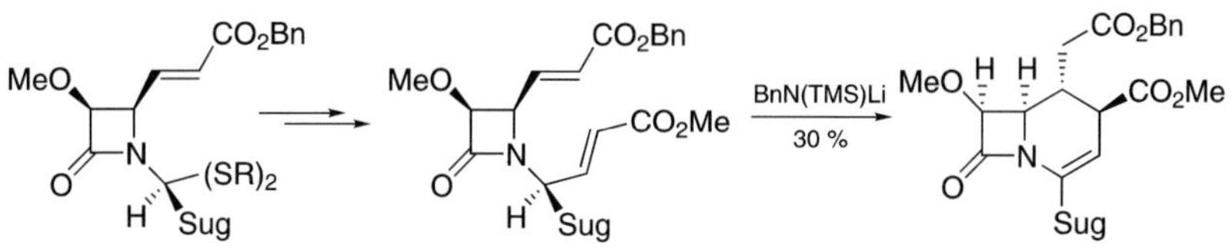

Scheme 17

64–66, their chromatographic separation into pure diastereomers, followed by cyclization in the presence of titanocene monochloride afforded bicyclic and tricyclic β-lactams with high regio- and stereoselectivity [59–61].

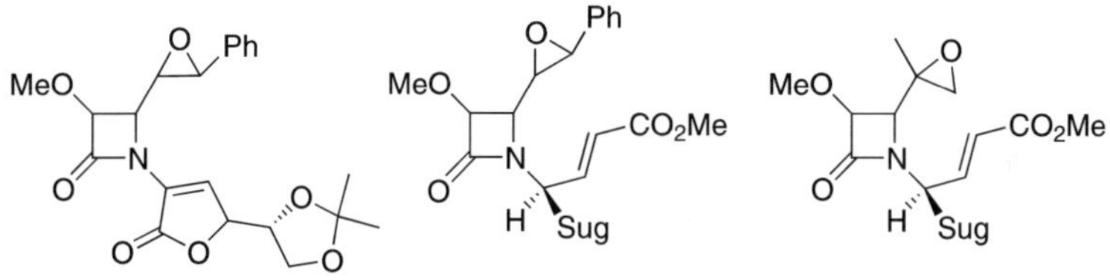

Chart 3

Readily available 3-deoxy-3-iodo-sugar aldehydes derived from D-glucose [62, 63] and D-galactose [29, 30, 49, 50] provide imines attractive for reaction with a variety of ketenes. [2 + 2]Cycloaddition of benzyloxy- or phenoxyacetyl chloride to D-ribo compound 67 proceeded with a low asymmetric induction to afford two *cis* β-lactams 68 and 69 in the ratio of about 1 : 1 [63]; the same observation was also made by another group [64]. On the other hand, D-xylo imines 70 under the same conditions gave only one *cis* diastereomer 71. The intramolecular radical cyclization in 68, 69, and 71 proceeded with high stereoselectivity providing in all cases D-xylo configuration in the sugar part of the bicyclic β-lactam products 72 and 73, respectively (Scheme 18) [63].

The Vasella group [65] reported the synthesis of carboxylic acid 77 and the corresponding phosphonic-acid isoster 78 from D-erythrose derived imines 74–76 (Scheme 19). Despite expectations, none of the synthesized compounds exhibited a significant inhibitory in vitro activity against the sialidases of *Vibrio cholerae*, *Salmonella typhimurium*, *Influenza A*, and *Influenza B* virus.

Oxazolidinone auxiliaries based on D-xylose [66] and D-mannitol [67] residues were converted into corresponding carboxylic acids 79 and 80 which after activation were used for stereoselective Staudinger reaction with diaryl, or aryl-styryl imines. In the first case, an excellent diastereoselectivity was obtained to afford 81 with high preponderance [66]. In the second case,

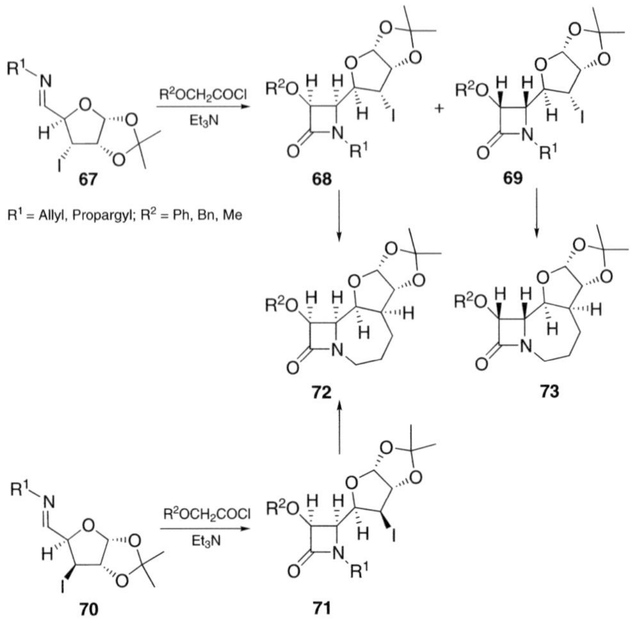

R^1 = Allyl, Propargyl; R^2 = Ph, Bn, Me

Scheme 18

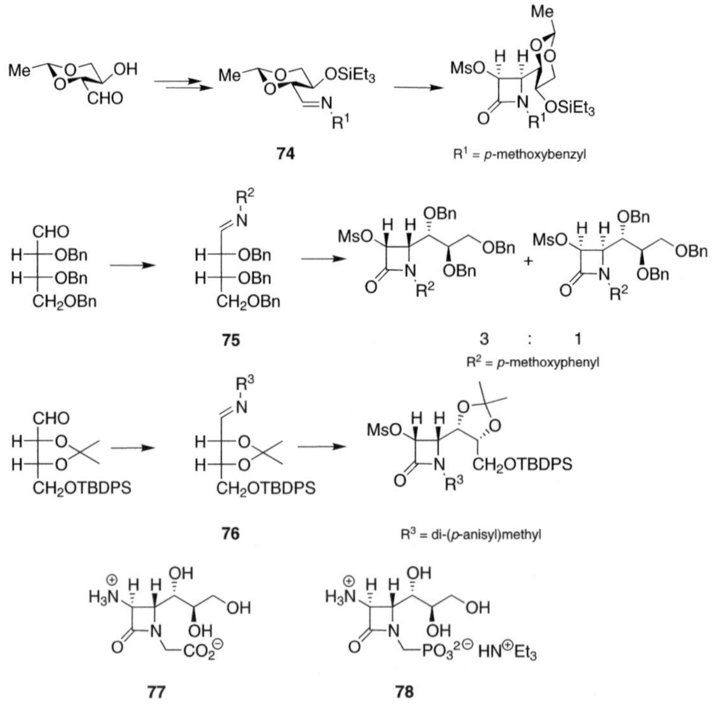

Scheme 19

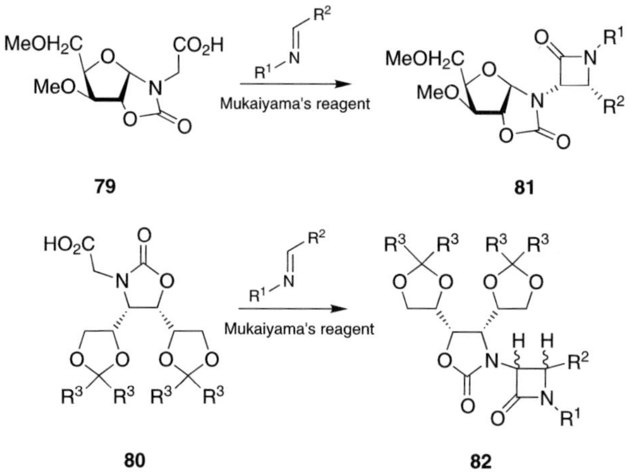

Scheme 20

a mixture of *cis* and *trans* adducts **82** was obtained with high predominance of the *cis* isomer; the cyclohexylidene protection was shown to display a better selectivity than the isopropylidene one (Scheme 20) [67].

3
Synthesis of β-Lactams by Cyclization Methods

Cyclization methods of β-lactam formation shown in Scheme 21 have not been recently widely reported. The most common cyclization of chiral β-amino acids is represented by a few reports only.

Imino aldol reaction of ketene silyl acetals with the chiral imine derived from tartaric acid **83** in the presence of a cation-exchange resin provided the corresponding β-amino esters **84** in a good yield and high diastereoselectivity [68]. The esters **84**, thus obtained, were subjected to the Grignard reagent which promoted β-lactam formation. After a sequence of reactions compound **84** was transformed into the ester **85** [68] which in the past was

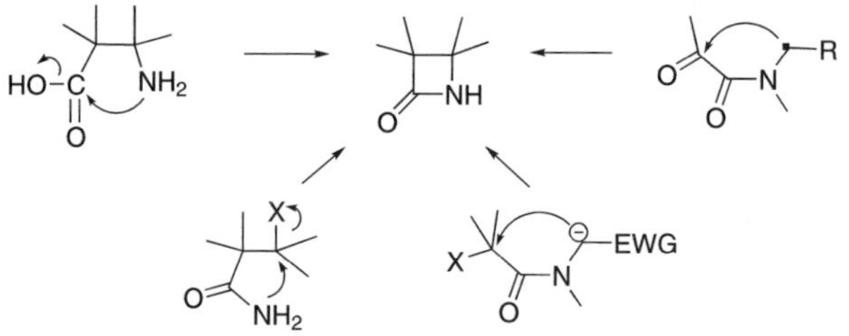

Scheme 21

the key intermediate for 2-isocephem and 2-oxaisocephem antibiotics **86** (Scheme 22) [69].

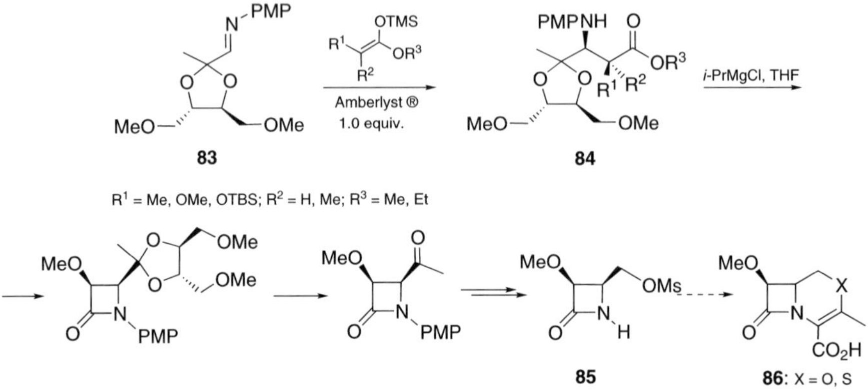

Scheme 22

β-Amino acid **88** obtained from methyl 3-deoxy-3-amino-altropyranoside **87** was treated with methanesulfonyl chloride in the presence of sodium bicarbonate and gave the β-lactam **89**. Compound **89** was subsequently subjected to ring-opening polymerization to afford optically active polyamide **90** (Scheme 23) [70].

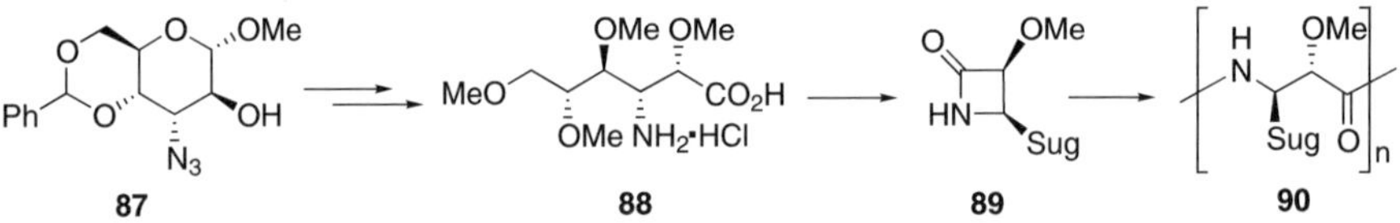

Scheme 23

The Miller group [71] continued investigation on the cyclization of β-hydroxy-O-benzyl hydroxamates in the presence of the Mitsunobu reagent leading to N-benzyloxy-β-lactams. The Ferrier rearrangement of D-glucal **91** followed by formation of hydroxamate and deacetylation provided a substrate **92** suitable for cyclization. In the Mitsunobu reaction conditions **92** was converted into the β-lactam **93**, which was oxidized to the lactone **94** (Scheme 24).

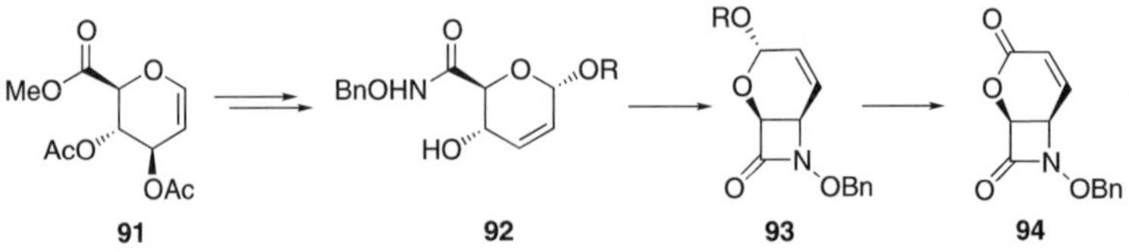

Scheme 24

Compounds having a four-membered *β*-lactam ring fused to the pyranoid ring were synthesized with intention to find a new entry to carbapenems and -cephems. *N*-Tosyl-2-*C*-carbamoyl glycosides with α-L-*arabino*- and *β*-D-*xylo*-configuration provided, under Mitsunobu reaction conditions, the corresponding bicyclic *β*-lactams, even in the presence of a neighboring *trans-vic*-diol fragment. The preference of *β*-lactam formation over *γ*-lactam was noticed (Scheme 25) [72].

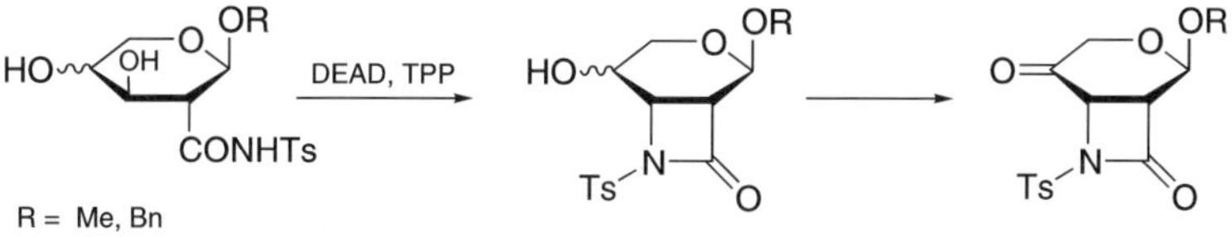

DEAD, TPP

R = Me, Bn

Scheme 25

An interesting stereoselective formation of the *β*-lactam ring was reported by Izquierdo et al. (Scheme 26) [73]. Protected aldehydes **95** and **96** were transformed into the corresponding epoxides **97** and **98** by the method reported earlier by Lopez-Herera et al. [74–76] Cyclization involving carbanions generated from the *N,N*-dibenzylamino group provided a single isomer **99** in the case of the fructopyranose auxiliary (**97**), whereas a mixture of *β*-lactams **100** was obtained in the case of the glyceraldehyde (**96**).

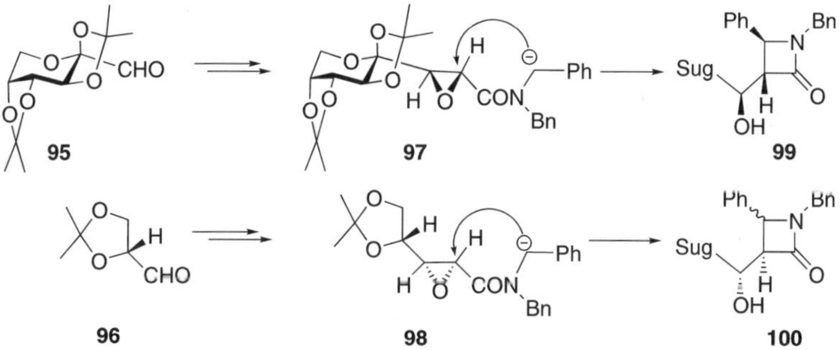

Scheme 26

Tri-*O*-acetyl-D-glucal **101** was used for resolving racemic hydroxy *β*-lactam **102** into two diastereomeric glycosides **103** and **104** in the presence of iodine in THF (Scheme 27). Compounds **103** and **104** were subjected to catalytic transfer hydrogenation under microwave irradiation [77–80].

Carbohydrates or carbohydrate-derived compounds were used to promote asymmetric induction in the crystalline state. In particular, enantioselective photocyclization of α-ketoamide **105** in the clathrate crystalline environment with chiral hosts **106** and **107** derived from tartaric acid provided the *β*-lactam **108** with high enantiomeric purity (Scheme 28) [81].

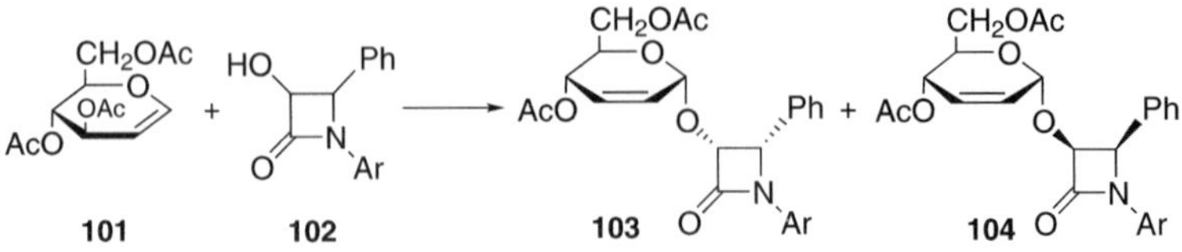

Scheme 27

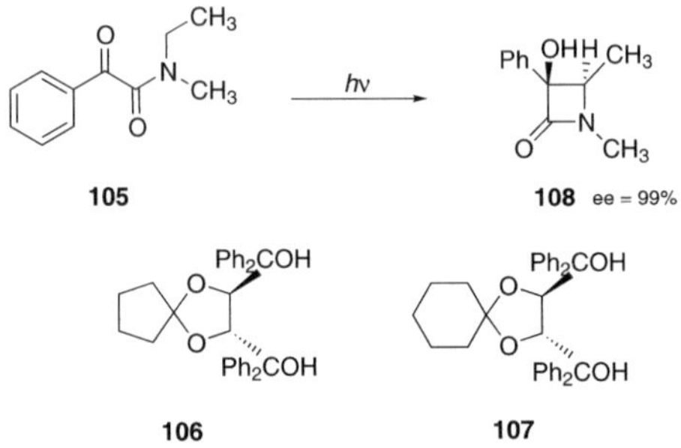

Scheme 28

The other report describes irradiation of achiral and chiral α-ketoamides as crystalline complexes with α-, β-, and γ-cyclodextrins leading to 3-hydroxy-β-lactams [82].

4
Synthesis of β-Lactams via [2 + 2]Cycloaddition of Isocyanates to Chiral Vinyl Ethers

Owing to the interesting antibacterial, antifungal, and β-lactamase inhibitory activity, oxygen analogs of penicillins and cephalosporins have attracted the attention of many research laboratories [82]. Among the variety of possible methods of construction of 4-alkoxy-substituted azetidinones [83], [2 + 2]cycloaddition of isocyanates to vinyl ethers plays a special role because it leads directly to the suitably substituted four-membered ring and offers the possibility of stereocontrol of the configuration of the bridgehead carbon atom of clavams and 5-oxacephams.

We have shown, that adducts of isocyanates and glycals can serve as substrates for the synthesis of clavams [84] and oxacephams [85, 86]. The stereochemical control of the synthetic strategy was established on the observation that the [2 + 2]cycloaddition proceeded exclusively *anti* to the substituent at the C-3 carbon atom of the sugar ring (Scheme 29). Syntheses based on this strategy have employed a glycolic cleavage for the opening of the sugar ring [84–86]. This led to the need of discrimination between carbon atoms which were split in the cleavage step. The opening of the sugar ring by other

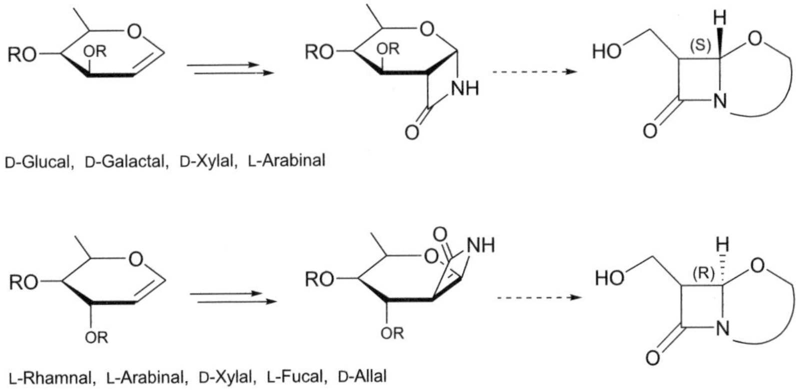
RO
OR
O
NH
HO
H
(S)
N
D-Glucal, D-Galactal, D-Xylal, L-Arabinal
(R)
L-Rhamnal, L-Arabinal, D-Xylal, L-Fucal, D-Allal

Scheme 29

means than the glycolic cleavage has also been investigated (Schemes 30 and 31) [87, 88]. The model studies of the Beckmann rearrangement led to the introduction of an amino function at C-3 of the azetidinone ring, the final product **108**, however, was found not to be stable enough to be used for fur-

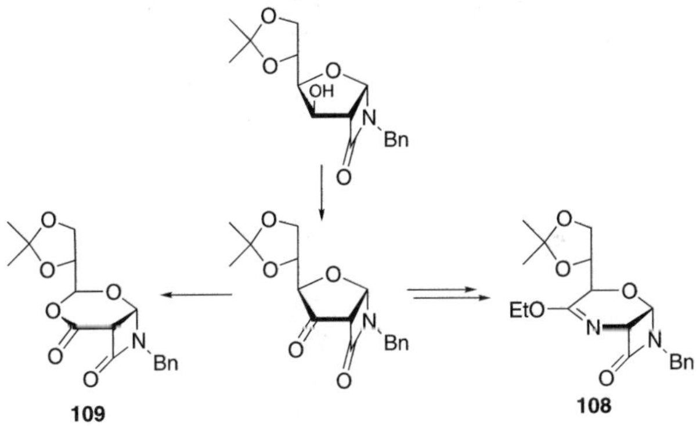
OH
N
Bn
EtO
109
108

Scheme 30

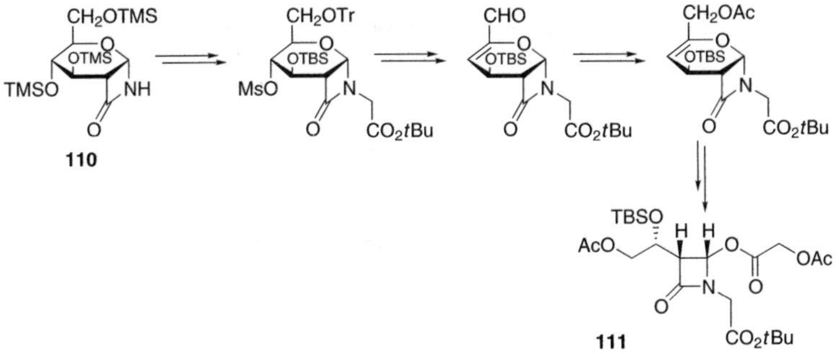
CH2OTMS
OTMS
TMSO
NH
110
CH2OTr
OTBS
MsO
CO2tBu
CHO
CH2OAc
TBSO
AcO
H H
OAc
111

Scheme 31

ther transformations [87]. On the other hand, the Baeyer–Villiger oxidation cleaved the C-3–C-4 bond [87] and consequently provided compound **109** having the same constitution as the one obtained by the glycolic cleavage methodology (Scheme 30) [84–86].

The most attractive opening of the sugar ring in the β-lactam **110**, other than the glycolic cleavage, consisted of the introduction of the double bond to the pyranoid ring via the β-elimination process, this was subsequently subjected to ozonolysis, reduction of the aldehyde group, and acetylation affording 4-acyloxy azetidinone **111** (Scheme 31) [88].

The study of [2 + 2]cycloaddition of chlorosulfonyl isocyanate (CSI) to a variety of 1,2-O-isopropylidene-3-O-vinyl-D-glycofuranoses (Scheme 32) showed that the observed high selectivity of these reactions was sterically controlled and depended on the size of the substituent at C-5 of the sugar [89]. The β-lactams **113** and **114**, obtained from vinyl ethers **112**, were transformed into the corresponding cephams **115–118** by the intramolecular alkylation of the β-lactam nitrogen atom [89, 90]. [2 + 2]Cycloaddition of CSI to (Z) and (E) vinyl ethers proceeds stereospecifically with asymmetric induction in the range reported for simple vinyl ethers to afford corresponding *cis* and *trans* β-lactams [91, 92].

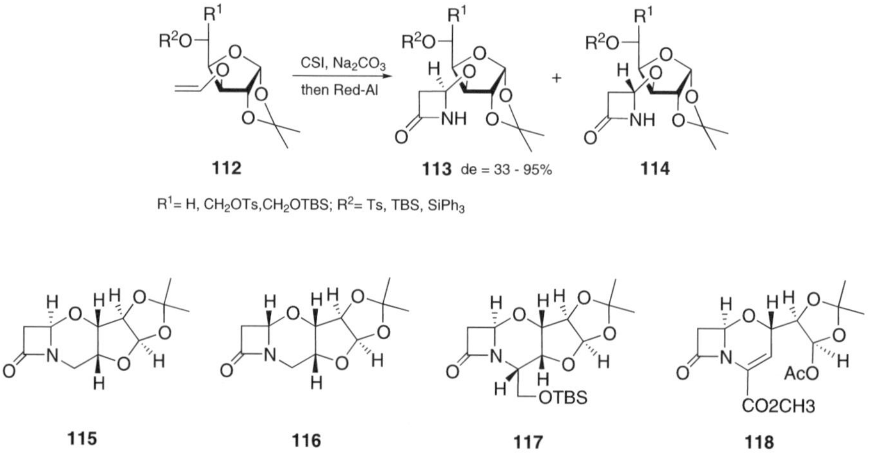

112   **113** de = 33 - 95%   **114**

R^1= H, CH$_2$OTs, CH$_2$OTBS; R^2= Ts, TBS, SiPh$_3$

115   **116**   **117**   **118**

Scheme 32

Bearing in mind the stereospecificity of [2 + 2]cycloaddition, a concerted mechanism of the reaction has been proposed [91–94]. This proposition has been further partially supported by the *ab initio* calculations performed by Ugalde et al. [95, 96]. They concluded that in the gas phase and in some cases in solution the mechanism is concerted.

Cycloadditions involving terminal C-silylated vinyl ethers proceeded stereospecifically except reactions with olefins having a bulky *tert*-butyldimethylsilyl moiety [97]. It was found that a silyl substituent located next to the carbonyl group is easily removed, even under very mild conditions [97].

The [2 + 2]cycloaddition of CSI to 5-*O*-vinyl derivatives of 1,2-*O*-isopropyl-idene-α-D-glycofuranoses **119** shows that in this case the stereoselectivity of the reaction can also be controlled (Scheme 33) [98]. The presence of a small substituent at the C-3 carbon atom on the top of the furanose ring, or even a large substituent localized at the bottom (D-*allo* configuration), gave an excellent asymmetric induction to form preferentially 4-alkoxy-azetidin-2-ones **120** with (*S*) configuration at the C-4 carbon atom [98]. The intramolecular alkylation of the nitrogen atom in **120** led to clavams **121**, structurally related to the natural antifungal compounds **122** [99–104].

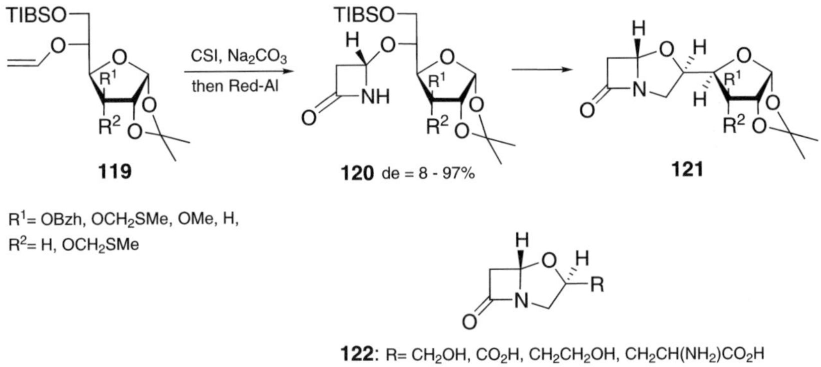

CSI, Na$_2$CO$_3$
then Red-Al

119

120 de = 8 - 97%

121

R^1= OBzh, OCH$_2$SMe, OMe, H,
R^2= H, OCH$_2$SMe

122: R= CH$_2$OH, CO$_2$H, CH$_2$CH$_2$OH, CH$_2$CH(NH$_2$)CO$_2$H

Scheme 33

The methodology based on [2 + 2]cycloaddition of CSI to vinyl ethers was compared with the alternative one based on condensation of the corresponding alcohols with 4-acetoxy-azetidin-2-one (**144**). In all cases the cycloaddition offered much better stereoselectivity than that found in the condensation [105].

Structurally related to **119**, vinyl ether **123** was synthesized from L-tartaric acid [106], and then subjected to the cycloaddition-cyclization reaction sequence providing clavam **125** related to clavulanic acid **124** with a high diastereomeric purity (Scheme 34). Compound **124** exhibited pronounced activity as an inhibitor of β-lactamase enzymes [106].

Taking into account that the sugar fragments of compounds **112**, **119**, and **123** have acid-labile protective groups, instead of simple vinyl ethers, the (*Z*)-propenyl ethers were used. Owing to the stereospecificity of the cycloaddition [91, 93, 94], such a modification should not change the preferred approach of both reactants (Scheme 35) [107].

Selection of sugar auxiliaries **126** and **127** which represent the enantiomeric relation in the part of the molecules where the cycloaddition occurs, demonstrates a stereocontrolled entry to 5-oxacephams having the desired

Scheme 34

Scheme 35

configuration at the bridgehead carbon atom, which is crucial for the biological activity of the final product [108, 109]. The cycloaddition of CSI to **126** and **127** provided only single diastereoisomers **128** and **129** in each case having configuration 4(R) and 4(S), respectively (Scheme 35). The intramolecular alkylation of the nitrogen atom in **128** and **129** afforded the corresponding cephams **130** and **131** [108, 109].

Using the high yielding [2 + 2]cycloaddition [110–112], new β-lactam disaccharides **132** were synthesized as inhibitors of elongating α-D-mannosyl

132: R=H, Bu

Chart 4

phosphate transferase (eMPT), a key enzyme of the *Leishmania* parasite [113].

5
[2 + 2]Cycloaddition of CSI to Alkoxyallenes

[2 + 2]Cycloaddition of CSI to chiral alkoxyallenes derived from sugars has been studied [114–116] to follow Buynak's [117–119] entry to the 3-alkylidene-azetidin-2-ones. The value of this strategy is that the *exo*-double bond not only exists in a number of antibiotics and β-lactamase inhibitors [120–128], but also can be transformed into a variety of substituents which are important for biological activity of the antibiotic [117–119].

The [2 + 2]cycloaddition of CSI to alkoxyallenes **133–137** is sterically controlled and proceeds with a significantly lower asymmetric induction in comparison to that found for the corresponding vinyl ethers (Scheme 36). Introduction of a trimethylsilyl group to the allene fragment **136** and **137** increases both the yield of the reaction and asymmetric induction [114]. The mixture **138** (de = 40%) was transformed into the corresponding cephams **139** which were easily separated into pure components [116]. Both cephams were used as substrates in a wide range of transformations leading to the introduction of isopropyl, hydroxyisopropyl, hydroxy, and *N*-acyl functions all located next to the β-lactam carbonyl group (Scheme 37) [116].

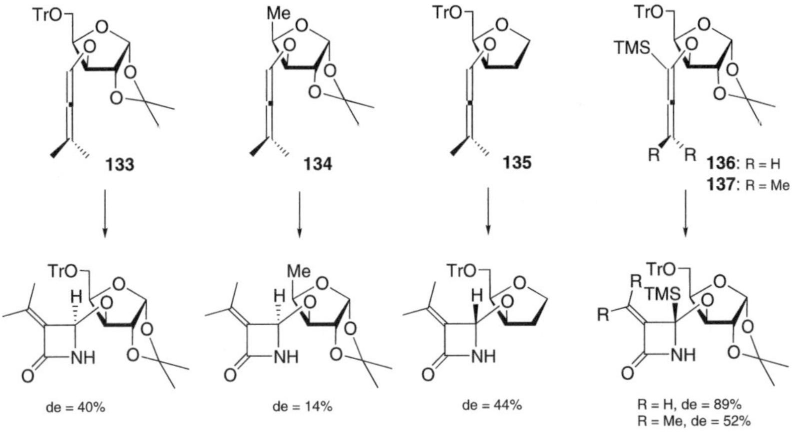

Scheme 36

Alkoxyallene **140** derived from 1,3-benzylidene-L-erythritol treated with chlorosulfonyl isocyanate provided diastereomeric β-lactams **141** with a moderate stereoselectivity, these after the intramolecular alkylation of the nitrogen atom afforded compounds having oxacepham skeletons which were

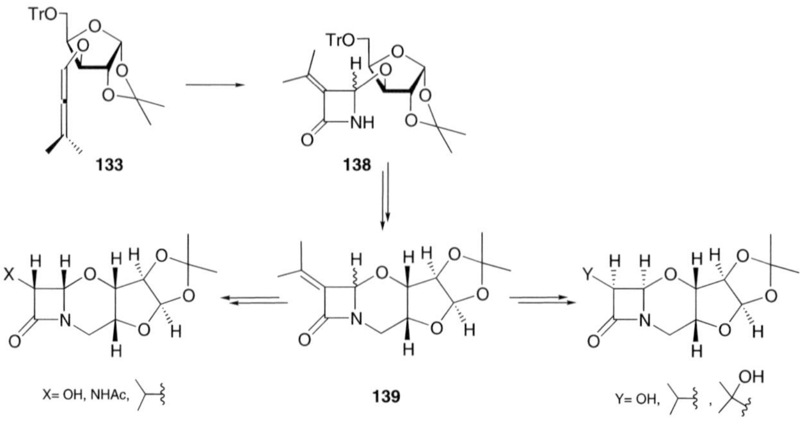

Scheme 37

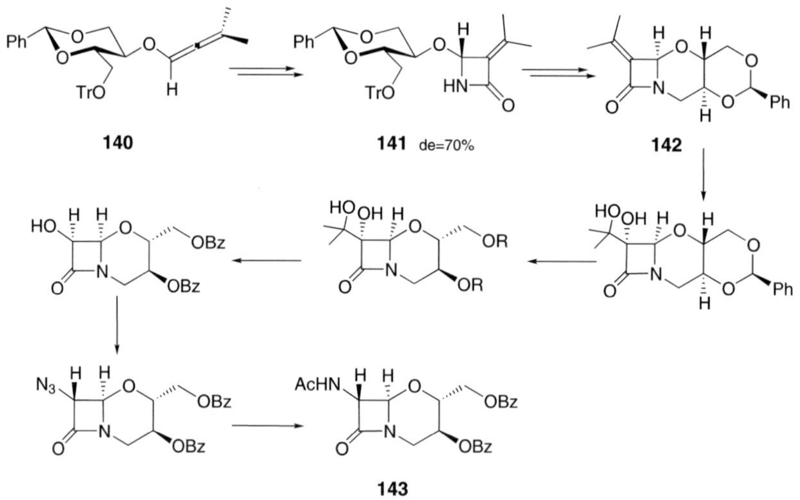

Scheme 38

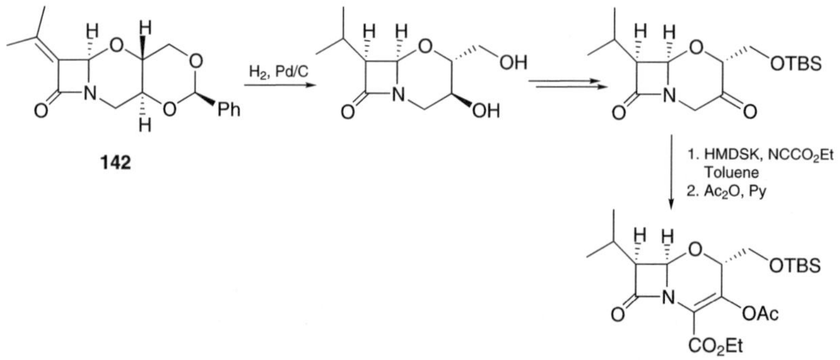

Scheme 39

separated into pure components [114]. The *exo*-isopropylidene group allows one to introduce several substituents to the C-7 carbon atom. For example, the cepham **142** enabled introduction of the amino function (**143**) by the sequence of reactions shown in Scheme 38 [129]. On the other hand, removal of the benzylidene protection in **142** followed by oxidation of 3-OH to the ketone resulted in carboxylation of the C-2 carbon atom (Scheme 39) [129].

6
Synthesis of 5-Oxacephams using 4-Vinyloxy-Azetidin-2-one

Reactions between 4-acetoxy-azetidin-2-one (**144**) and a chiral alcohol proceed with a low asymmetric induction [130–133]. In order to improve stereoselectivity of the nucleophilic substitution at C-4 of azetidin-2-one, the reaction has been performed as an intramolecular process. This required alkylation of the β-lactam nitrogen atom prior to the formation of a bicyclic skeleton via nucleophilic substitution. Because of the low stability in the presence of bases, compound **144** could not be alkylated at the nitrogen atom, therefore 4-vinyloxy-azetidin-2-one (**145**), readily available from divinyl ether and chlorosulfonyl isocyanate is used [134–136]. The vinyloxy group can be subsequently transformed into the formyloxy or acetoxy group by ozonolysis or by PCC oxidation, respectively. Following the *N*-alkylation, the vinyloxy group was subjected to ozonolysis to give the corresponding formate, which in the presence of a Lewis acid could be substituted by the oxygen atom of the *p*-methoxybenzyl ether. The *p*-methoxybenzyl group plays the role of protecting group which in the presence of an acid undergoes elimination as a stabilized benzyl cation [134–136].

A comparison of three routes for the formation of 5-oxa-cephams **149** and **150** from glycerols derived from D-glyceraldehyde has been reported (Scheme 40) [105]. The first two routes: the [2 + 2]cycloaddition of CSI to **146** and the condensation of **144** to **147**, are followed by the ring closure step involving *N*-alkylation. The third route consists of *N*-alkylation of **145** by **148** prior to the cyclization step. While the comparison of stereoselectivities of the first and second method reveals unequivocally the advantage of the first one, the third one, which leads to the opposite direction of asymmetric induction, seems to be the most attractive [105].

An excellent stereoselectivity was achieved when a chiral fragment derived from the 1,2-*O*-isopropylidene-α-D-xylofuranose attached to the nitrogen atom of the azetidinone ring **151** was subjected to cyclization (Scheme 41). The tetracyclic cepham **152**, with (*S*) configuration in the azetidinone fragment, was obtained in a good yield and as the sole product [137].

The observed high asymmetric induction is worth emphasizing. In all known cases when both 4-vinyloxy-azetidin-2-one and *p*-methoxybenzyl ether moieties are attached to five or six-membered rings, formation of the

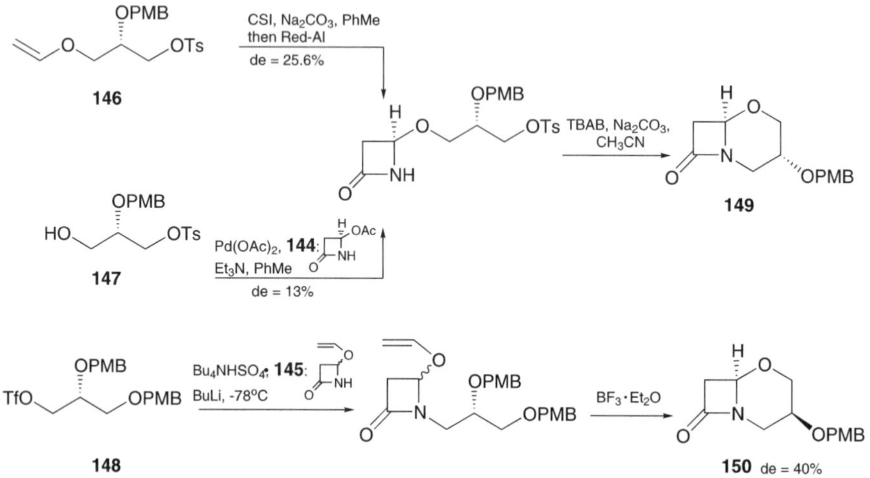

Scheme 40

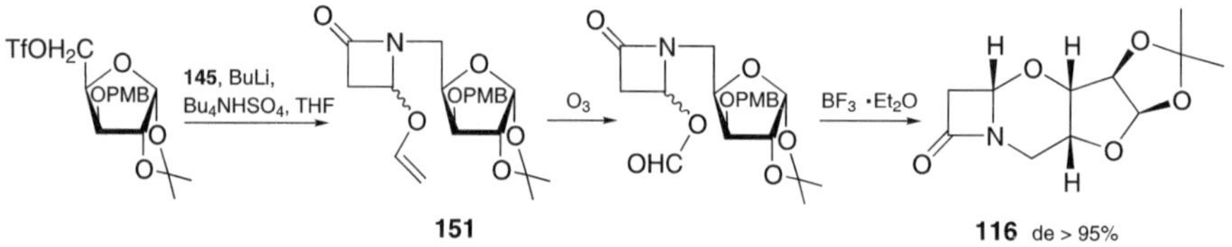

Scheme 41

cepham skeleton proceeds with a very high asymmetric induction [107, 134–138]. Although the synthetic strategy involving **145** has been successfully applied for the 5-oxacepham synthesis, all attempts to synthesize clavams by this method failed. It was found that the 4-vinyloxy group in the presence of a Lewis acid may play the role of leaving group [139]. A combination of the 4-*O*-vinyloxy group and the *p*-methoxybenzyl group, which are both resistant to bases and labile in the presence of acids, simplifies the formation of 5-oxacephams.

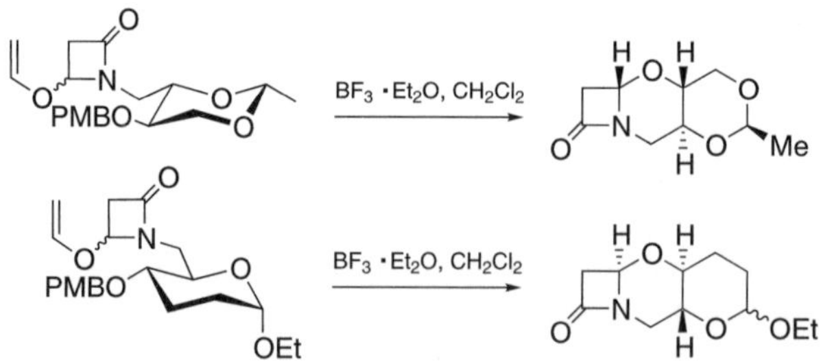

Scheme 42

The effectiveness of the methodology based on azetidinone **145** is well illustrated by the examples shown in Scheme 42 [108, 109]. In every case this methodology provides a diastereomer having the configuration of the bridgehead carbon atom opposite to that obtained by the [2 + 2]cycloaddition of CSI to the corresponding vinyl ether (compare Schemes 32 and 41, as well as Schemes 35 and 42).

7
Solid-Phase Synthesis of β-Lactams

Because of the importance of β-lactam antibiotics, a number of solid-phase approaches to this class of compounds have recently been reported [140]. In particular, [2 + 2]cycloadditions between ketenes and resin-bound imines have attracted the attention of many laboratories [140]. The alternative method, based on [2 + 2]cycloaddition of isocyanates to olefins has been reported by our laboratory only.

Binding of the vinyl ether to the polymer support through a sulfonyl linker seemed to be the most attractive approach, since the intramolecular alkylation of the nitrogen atom in the cycloadduct, would remove the product from the resin by a cyclization/cleavage methodology [141]. The two alternative methods of binding of vinyl ethers to the resin are shown in Schemes 43 and 44 [142]. The first one (Scheme 43) uses the p-oxyphenylsulfonyl linker which can be attached to the Wang resin via the Mitsunobu procedure [142, 144], whereas the second one (Scheme 44) utilizes alkylation of the lithium salt of a mesylate [143] with the terminal of the Merrifield resin [142, 145]. Both procedures do not leave any free sulfonyl acid group on the resin. This was found to be critical, since reactions carried out on sugar vinyl ethers bound to a polymer using chlorosulfonylated resins followed by the cyclization/cleavage methodology do not produce any β-lactam [144].

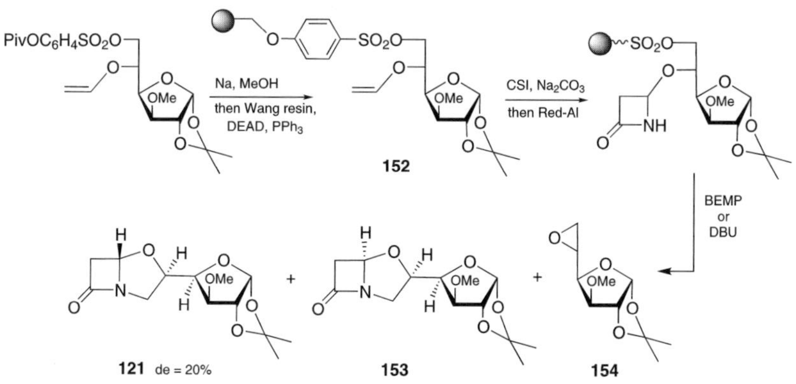

Scheme 43

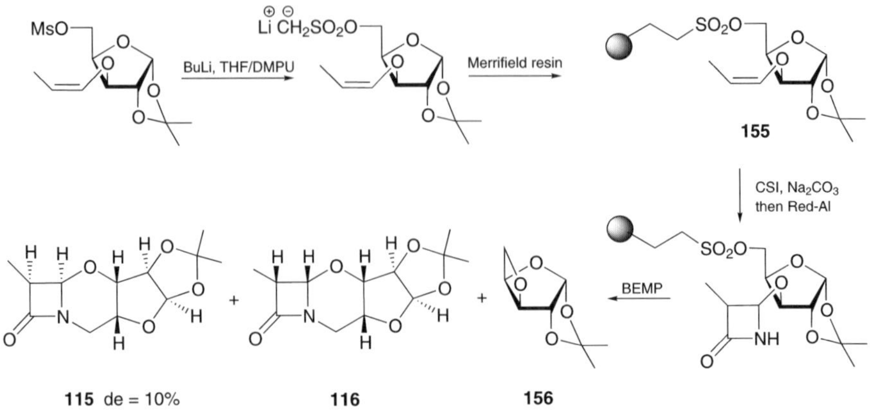

Scheme 44

[2 + 2]Cycloaddition between CSI and polymer bound vinyl ethers, for example **152** and **155**, followed by the intramolecular alkylation provided mixtures of corresponding clavams **121/153** or cephams **115/116** [144]. In all cases, reactions proceeded with a lower yield and stereoselectivity than the corresponding sequences of reactions performed in solution. Moreover, in both cases formation of anhydrosugars **154** and **156** was observed [144, 145]. It should be, however, pointed out that the same reactions performed with soluble polymer proceeded similarly to those carried out in solutions [145].

The strategy based on 4-vinyloxy-azetidinone (**145**) was particularly effective when performed on a solid support. The reaction sequence shown in Scheme 41 was successfully performed on the Wang resin [146, 147]. The terminal of Wang resin acted as the *p*-methoxybenzyl equivalent (Scheme 45). Formation of the cepham skeleton is accompanied by a cleavage of the

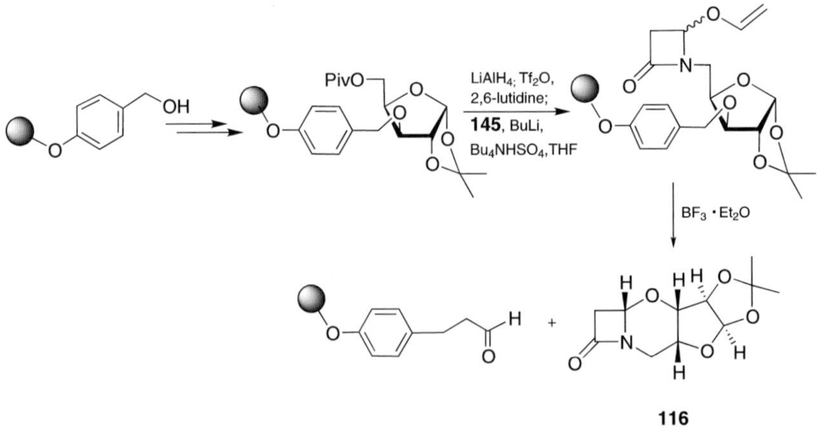

Scheme 45

product from the resin. This is known as a cyclization-cleavage methodology [140]. It should be emphasized that formation of the cepham skeleton in the course of the intramolecular nucleophilic substitution at C-4' of the azetidin-2-one in solution and on the solid phase proceeded with the same direction and magnitude of asymmetric induction [134–136, 145, 146].

Solid-phase and combinatorial synthesis of β-lactams using the Staudinger reaction has widely been studied. Carbohydrates as a chiral pool, however, were not employed except for one report [148]. The polymer supported-imines were employed to prepare several β-lactams by enolate/imine condensation and ketene/imine cycloaddition (Scheme 46) [146]. The reactions carried out on the polymer-bound imines showed a remarkable similarity to those in solution, both in terms of yield and stereoselectivity [54, 55]. β-Lactams were removed from the polymer by CAN oxidation [148].

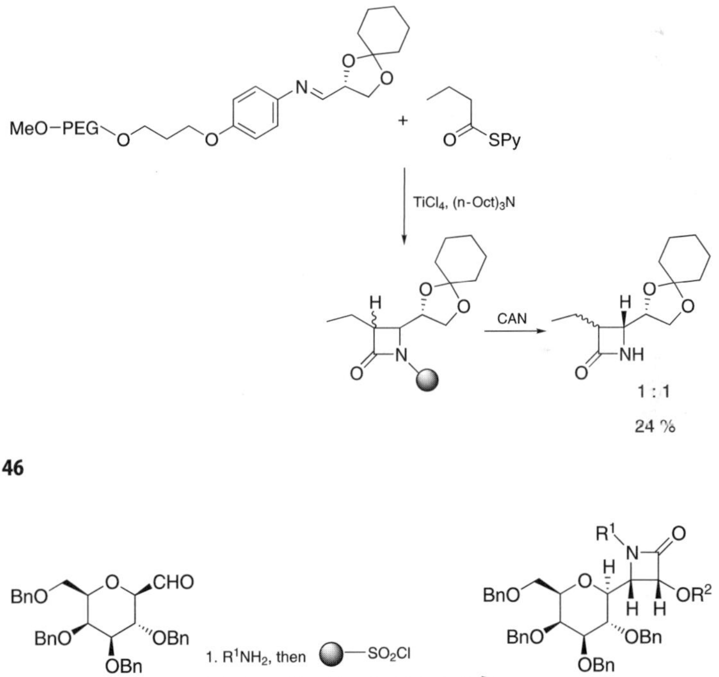

Scheme 46

Scheme 47

An interesting method of purification of Staudinger reaction products was recently reported (Scheme 47) [149]. Sulfonyl chloride and aminomethylated resins were used consecutively to remove excess of components and by-products. Two pure *C*-glycosyl *β*-lactams were transformed into *C*-glycosyl isoserins and a *C*-rybosyl dipeptide via base-promoted heterocycle ring opening by methanol and L-phenylalanine methyl ester, respectively [149].

8
Conclusion

The majority of the approaches presented in this survey show the potential of modern organic synthesis rather than any practical value. Carbohydrates were used either as a chiral pool or chiral auxiliaries. The direct formation of a four-membered *β*-lactam ring via [2 + 2]cycloaddition of ketenes to imines and of isocyanates to olefins dominated over other methods since it offers in most cases an excellent stereoselectivity. *β*-Lactam antibiotics still remain the main tool against bacterial infections but the search for new active compounds has not been the aim of reported investigations. Particularly, in the case of cycloaddition of ketenes or their equivalents to imines, the cycloadducts were used mostly as intermediates for the preparation of other biologically active compounds. In the case of [2 + 2]cycloaddition involving isocyanates, despite low asymmetric induction, cycloadditions to alkoxyallenes seem to be attractive since they provide not only a *β*-lactam ring, but also allow one to introduce a variety of substituents to the carbon atom neighbor to the carbonyl group. Applications of 4-vinyloxy-azetidinone in the synthesis of *β*-lactams looks very promising, particularly because it can be easily alkylated at the nitrogen atom and that the intramolecular nucleophilic displacement at C-4 offers high asymmetric induction. The particular attraction of this new methodology consists of its easy application to solid-phase conditions.

Acknowledgements This work was supported by the Ministry of Science and Higher Education, grant PBZ-KBN-126/T09/08/2004.

References

1. Fraser-Reid B (1975) Acc Chem Res 8:192
2. Hanessian S (1979) Acc Chem Res 12:159
3. Vasella A (1980) In: Scheffold R, Salle A (eds) Modern Synthetic Methods. Otto Salle Verlag, Frankfurt, p 173
4. Hanessian S (1983) Total Synthesis of Natural Product Chiron Approach. Pergamon Press, Montreal
5. Inch TD (1984) Tetrahedron 40:3161

6. Jacobsen EN, Pfaltz A, Yamamoto H (eds) (1999) Comprehensive Asymmetric Catalysis. Springer, Berlin, Heidelberg, New York
7. Katsuki T, Martin VS (1996) Org React 48:1
8. Katsuki T, Sharpless BK (1980) J Am Chem Soc 102:5974
9. Johnson R, Sharpless BK (1993) In: Ojima I (ed) Catalytic Asymmetric Synthesis. VCH, New York, p 103
10. Johnson RA, Sharpless BK (1993) In: Ojima I (ed) Catalytic Asymmetric Synthesis. VCH, New York, p 227
11. Jacobsen EN (1993) In: Ojima I (ed) Catalytic Asymmetric Synthesis. VCH, New York, p 159
12. Evans DA, Chapman KT, Bisaha J (1984) J Am Chem Soc 106:4261
13. Evans DA, Chapman KT, Bisaha J (1988) J Am Chem Soc 110:1238
14. Noyori R (1981) Pure Appl Chem 53:2315
15. Noyori R (1989) Chem Soc Rev 18:187
16. Noyori R, Takaya H (1990) Acc Chem Res 23:345
17. Lichtenthaler FW, Mondel S (1997) Pure Appl Chem 69:1853
18. Lichtenthaler FW (ed) (1991) Carbohydrates as Organic Raw Materials. VCH, Weinheim
19. Kałuża Z, Abramski W, Chmielewski M (1994) Indian J Chem 33B:913
20. Veinberg G, Vorona M, Shestakova I, Kanepe I, Lukevics E (2003) Curr Med Chem 10:1741
21. Hubschwerlen C, Schmid G (1983) Helv Chim Acta 66:2206
22. Hubschwerlen C (1986) Synthesis p 962
23. Wagle DR, Garai C, Chiang J, Montelone MG, Kurys BE, Strohmeyer TW, Hegde VR, Manhas MS, Bose AK (1988) J Org Chem 53:4227
24. Deshmukh ARAS, Bhawal BM, Krishnaswamy D, Govande VV, Sinkre BA, Jayanthi A (2004) Curr Med Chem 11:1889
25. Hassan HH, A. M., Soliman R (2000) Synth Commun 30:2465
26. Doyle MP, Hu W, Phillips IM, Moody CJ, Pepper AG, Slawin AMZ (2001) Adv Synth Catal 343:112
27. Bose AK, Banik BK, Manhas MS (1995) Tetrahedron Lett 36:213
28. Bose AK, Jayaraman M, Okawa A, Bari SS, Robb EW (1996) Tetrahedron Lett 37:6989
29. Bose AK, Banik BK, Mathur C, Wagle DR, Manhas MS (2000) Tetrahedron 56:5603
30. Manhas MS, Banik BK, Mathur A, Vincent JE, Bose AK (2000) Tetrahedron 56:5587
31. Banik BK, Banik I, Samajdar S, Wilson M (2004) Heterocycles 63:283
32. Banfi L, Guanti G, Rasparini M (2003) Eur J Org Chem p 1319
33. Alcaide B, Polanco C, Sierra MA (1998) J Org Chem 63:6786
34. Alcaide B, Almendros P (2004) Curr Med Chem 11:1921
35. Alcaide B, Almendros P, Aragoncillo C (2001) J Org Chem 66:1612
36. Alcaide B, Almendros P, Rodríguez-Vicente A, Ruiz MP (2005) Tetrahedron 61:2767
37. Alcaide B, Sáez E (2005) Eur J Org Chem p 1680
38. Alcaide B, Almendros P, Salgado NR (2000) J Org Chem 65:3310
39. Alcaide B, Almendros P, Alonso JM, Redondo MC (2003) J Org Chem 68:1426
40. Alcaide B, Almendros P, Aragoncillo C (2003) Org Lett 5:3795
41. Alcaide B, Pardo C, Rodríguez-Ranera C, Rodríguez-Vicente A (2001) Org Lett 3:4205
42. Alcaide B, Polanco C, Sierra MA (1998) Synlett, p 416
43. Alcaide B, Almendros P, Redondo MC (2004) Org Lett 6:1765
44. Alcaide B, Almendros P, Aragoncillo C, Rodríguez-Acebes R (2001) J Org Chem 66:5208
45. Alcaide B, Almendros P, Aragoncillo C (2000) Org Lett 2:1411

46. Alcaide B, Almendros P, Rodríguez-Acebes R (2002) J Org Chem 67:1925
47. Alcaide B, Almendros P, Aragoncillo C, Rodríguez-Acebes R (2004) J Org Chem 69:826
48. Alcaide B, Almendros P, Aragoncillo C (2002) Chem Eur J 8:3646
49. Palomo C, Oiarbide M, Esnal A (1997) Eur J Chem Soc Chem Commun p 691
50. Palomo C, Oiarbide M, Esnal A, Landa A, Miranda JI, Linden A (1998) J Org Chem 63:5838
51. Palomo C, Oiarbide M, Landa A (2000) J Org Chem 65:41
52. Palomo C, Oiarbide M, Landa A, Esnal A, Linden A (2001) J Org Chem 66:4180
53. Palomo C, Aizpurua JM, Ganboa I, Oiarbide M (2004) Curr Med Chem 11:1837
54. Annunziata R, Benaglia M, Cinquini M, Cozzi F, Maggioni F, Puglisi A (2003) J Org Chem 68:2952
55. Annunziata R, Benaglia M, Cinquini M, Cozzi F, Puglisi A (2002) Bioorg Med Chem 10:1813
56. Barton DHR, Gateau-Olesker A, Anaya-Mateos J, Cleophax J, Gero SD, Chiaroni A, Riche C (1990) J Chem Soc Perkin Trans 1 p 3211
57. Anaya J, Barton DHR, Gero SD, Grande M, Hernando JIM, Laso NM (1995) Tetrahedron: Asymmetr 6:609
58. Ruano G, Anaya J, Grande M (1999) Synlett, p 1441
59. Ruano G, Grande M, Anaya J (2002) J Org Chem 67:8243
60. Ruano G, Martiáñez J, Grande M, Anaya J (2003) J Org Chem 68:2024
61. Anaya J, Fernández-Mateos A, Grande M, Martiáñez J, Ruano G, Rubio-Gonzalez R (2003) Tetrahedron 59:241
62. Arun M, Joshi SN, Puranik VG, Bhawal BM, Deshmukh ARAS (2003) Tetrahedron 59:2309
63. Deshmukh ARAS, Jayanthi A, Thiagarajan K, Puranik VG, Bhawal BM (2004) Synthesis, p 2965
64. Jayanthi A, Gumaste VK, Deshmukh ARAS (2004) Synlett, p 979
65. Storz T, Bernet B, Vasella A (1999) Helv Chim Acta 82:2380
66. Saul R, Kopf J, Köll P (2000) Tetrahedron: Asymmetry 11:423
67. Shin DG, Heo HJ, Jun J-G (2005) Synth Commun 35:845
68. Shimizu M, Tachi M, Hachiya I (2004) Chem Lett 33:1394
69. Barton DHR, Anaya J, Gateau-Olesker A, Gero SD (1992) Tetrahedron Lett 33:6641
70. García-Martín MG, Paz Báñez MV, Galbis JA (2000) J Carbohydr Chem 19:805
71. Durham TB, Miller MJ (2002) Org Lett 4:135
72. Zegrocka O, Abramski W, Urbańczyk-Lipkowska Z, Chmielewski M (1998) Carbohydr Res 307:33
73. Izquierdo I, Plaza MT, Robles R, Mota AJ (2000) Tetrahedron: Asymmetr 11:4509
74. Valpuesta M, Durante P, López-Herera F-J (1990) Tetrahedron 46:7911
75. Valpuesta M, Durante P, López-Herera F-J (1993) Tetrahedron 49:9547
76. López-Herera F-J, Pino M-S, Sarabia F, Heras A, Ortega J-J, Pedraza M-G (1996) Tetrahedron: Asymmetr 7:2065
77. Banik BK, Manhas MS, Bose AK (1994) J Org Chem 59:4714
78. Banik BK, Manhas MS, Bose AK (1997) Tetrahedron Lett 38:5077
79. Banik BK, Zegrocka O, Bose AK (1997) Heterocycles 46:173
80. Banik BK, Barakat KJ, Wagle DR, Manhas MS, Bose AK (1999) J Org Chem 64:5746
81. Ohba S, Hosomi H, Tanaka K, Miyamoto H, Toda F (2000) Bull Chem Soc Jpn 73:2075
82. Natarajan A, Wang K, Ramamurthy V, Scheffer JR, Patrick B (2002) Org Lett 4:1443
83. Łysek R, Borsuk K, Furman B, Kałuża Z, Kazimierski A, Chmielewski M (2004) Curr Med Chem 11:1813

84. Grodner J, Chmielewski M (1992) J Carbohydr Chem 11:691
85. Grodner J, Chmielewski M (1995) Tetrahedron 51:829
86. Bełżecki C, Urbañski R, Urbañczyk-Lipkowska Z, Chmielewski M (1997) Tetrahedron 53:14153
87. Abramski W, Urbañczyk-Lipkowska Z, Chmielewski M (1997) J Carbohydr Chem 16:63
88. Bełżecki C, Chmielewski M (2000) Polish J Chem 74:1243
89. Kałuża Z, Furman B, Patel M, Chmielewski M (1994) Tetrahedron: Asymmetr 5:2179
90. Furman B, Molotov S, Thürmer R, Kałuża Z, Voelter W, Chmielewski M (1997) Tetrahedron 53:5883
91. Furman B, Kałuża Z, Chmielewski M (1996) Tetrahedron 52:6019
92. Łysek R, Furman B, Kałuża Z, Chmielewski M (2000) Polish J Chem 74:51
93. Effenberger F, Fischer P, Prossel G, Kiefer G (1971) Chem Ber 104:1987
94. Effenberger F, Fischer P, Prossel G (1971) Chem Ber 104:2002
95. Cossio FP, Lecea B, Lopez X, Roa G, Arrieta A, Ugalde JM (1993) J Chem Soc Chem Commun, p 1450
96. Cossio FP, Roa G, Lecea B, Ugalde JM (1995) J Am Chem Soc 117:12306
97. Łysek R, Kałuża Z, Furman B, Chmielewski M (1998) Tetrahedron 54:14065
98. Kałuża Z, Furman B, Chmielewski M (1995) Tetrahedron: Asymmetr 6:1719
99. Brown DB, Evans JR (1979) J Chem Soc Chem Commun, p 282
100. Wanning M, Zahner H, Krone B, Zeeck A (1981) Tetrahedron Lett 22:2539
101. Pruess DJ, Kellett M (1983) J Antibiot 36:208
102. Evans RH Jr, Ax H, Jacoby A, Williams TH, Jenkis E, Scanneli JP (1983) J Antibiot 36:213
103. King HD, Langharig J, Sanglier JJ (1986) J Antibiot 39:510
104. Naegeri HV, Loosli H-R, Nussbaumer A (1986) J Antibiot 39:516
105. Kałuża Z, Furman B, Krajewski P, Chmielewski M (2000) Tetrahedron 56:5553
106. Neuß O, Furman B, Kałuża Z, Chmielewski M (1997) Heterocycles 45:265
107. Furman B, Krajewski P, Kałuża Z, Thürmer R, Voelter W, Kozerski L, Williamson MP, Chmielewski M (1999) J Chem Soc Perkin Trans 2, p 217
108. Borsuk K, Suwiñska K, Chmielewski M (2001) Tetrahedron: Asymmetr 12:979
109. Borsuk K, Kazimierski A, Solecka J, Urbañczyk-Lipkowska Z, Chmielewski M (2002) Carbohydr Res 337:2005
110. Chmielewski M, Kałuża Z, Grodner J, Urbañski R (1992) In: Giuliano RM (ed) Cycloaddition Reaction in Carbohydrate Chemistry. ACS Symposium Series 494:50
111. Chmielewski M, Kałuża Z, Furman B (1996) J Chem Soc Chem Commun, p 2689
112. Furman B, Kałuża Z, Łysek R, Chmielewski M (1999) Polish J Chem 73:43
113. Ruhela D, Chaterjee P, Vishwakarma RA (2005) Org Biomol Chem 3:1043
114. Łysek R, Furman B, Kałuża Z, Frelek J, Suwiñska K, Urbañczyk-Lipkowska Z, Chmielewski M (2000) Tetrahedron: Asymmetr 11:3131
115. Tong TD, Bocian W, Kozerski L, Szczukiewicz P, Frelek J, Chmielewski M (2005) Eur J Org Chem p 429
116. Łysek R, Urbañczyk-Lipkowska Z, Chmielewski M (2001) Tetrahedron 57:1301
117. Buynak JD, Rao MN, Pajouhesh H, Chandrasekaran RY, Finn K, de Meester P, Chu SC (1985) J Org Chem 50:4245
118. Buynak JD, Rao MN (1986) J Org Chem 51:1571
119. Buynak J, Doppalapudi VR, Frotan M, Kumar R, Chambers A (2000) Tetrahedron 56:5709
120. Kawamura Y, Yasuda Y, Mayama M, Tanaka K (1982) J Antibiot 35:10

121. Shoji J, Hinoo K, Sakazaki R, Tsuji N, Nagashima K, Matsumoto K, Takahashi Y, Kozuki S, Hattori T, Kondo E, Tanaka K (1982) J Antibiot 35:15
122. Tsuji N, Nagashima K, Kobayashi M, Shoji J, Kato T, Terui Y, Nakai H, Shiro M (1982) J Antibiot 35:24
123. Murakami K, Doi M, Yoshida T (1982) J Antibiot 35:39
124. Arisawa M, Then RL (1982) J Antibiot 35:1578
125. Brenner DG (1985) J Org Chem 50:18
126. Buynak JD, Wu K, Bachmann B, Khasnis D, Hua L, Nguyen HK, Carver CC (1995) J Med Chem 38:1022
127. Buynak JD, Doppalapudi VR, Frotan M, Kumar R (1999) Tetrahedron Lett 40:1281
128. Buynak J (2004) Curr Med Chem 11:1951
129. Tong TD, Borsuk K, Solecka J, Chmielewski M (2006) Tetrahedron 62:10928
130. Clauss K, Grimm D, Prossel G (1974) Liebigs Ann Chem p 539
131. De Bernardo S, Tengi JP, Sasso GJ, Weigele M (1985) J Org Chem 50:3457
132. Müller JC, Toome V, Pruess DL, Blount JF, Weigele M (1983) J Antibiot 36:217
133. Hoppe D, Hilpert T (1987) Tetrahedron 43:2467
134. Kałuża Z, Park H-S (1996) Synlett, p 895
135. Kałuża Z (1998) Tetrahedron Lett 39:8349
136. Kałuża Z, Kazimierski A, Lewandowski K, Suwińska K, Szczêsna B, Chmielewski M (2003) Tetrahedron 59:5893
137. Kałuża Z, Łysek R (1997) Tetrahedron: Asymmetr 8:2553
138. Borsuk K, Frelek Łysek R, Urbańczyk-Lipkowska Z, Chmielewski M (2001) Chirality 13:532
139. Kałuża Z (1999) Tetrahedron Lett 40:1025
140. Mata IG (1999) Curr Pharm Design 5:955
141. Maarseveen JH (1988) Comb Chem High Throughput Screening 1:185
142. Furman B, Łysek R, Matyjasek Ł, Wojtkielewicz W, Chmielewski M (2001) Synth Commun 31:2795–2802
143. Musicki B, Widlanski TS (1990) J Org Chem 55:4231
144. Furman B, Cierpucha M, Grzeszczyk B, Matyjasek L, Chmielewski M (2002) Eur J Org Chem p 2377
145. Łysek R, Grzeszczyk B, Furman B, Chmielewski M (2004) Eur J Org Chem, p 4177
146. Furman B, Thürmer R, Kałuża Z, Łysek R, Voelter W, Chmielewski M (1999) Angew Chem Int Ed 38:1121
147. Furman B, Thürmer R, Kałuża Z, Voelter W, Chmielewski M (1999) Tetrahedron Lett 40:5909
148. Annunziata R, Benaglia M, Cinquini M, Cozzi F (2000) Chem Eur J 6:133
149. Dondoni A, Massi A, Sabbatini S, Bertolasi V (2004) Adv Synth Catal 346:1355

Top Heterocycl Chem (2007) 7: 133–177
DOI 10.1007/7081_2007_050
© Springer-Verlag Berlin Heidelberg
Published online: 1 June 2007

Azide–Alkyne 1,3-Dipolar Cycloadditions: a Valuable Tool in Carbohydrate Chemistry

Francisco Santoyo-González (✉) · Fernando Hernández-Mateo

Departamento de Química Orgánica, Instituto de Biotecnología.
Facultad de Ciencias, Universidad de Granada, 18071 Granada, Spain
fsantoyo@ugr.es

Abstract Recent applications of the 1,3-dipolar cycloaddition reaction of azides and alkynes in carbohydrate chemistry are summarized in the present review. The efficient catalyzed version of this reaction, also referred to as one of the so-called click chemistry reactions, has joined in a short period of time a select group of the most efficient and useful organic reactions. In particular, its application in the carbohydrate field has shown its value as a synthetic tool allowing the preparation of a wide range of carbohydrate-containing molecular constructs such as glycopeptides, glycoaminoacids, glycoclusters, glycodendrimers, glycopolymers, glycosylated biomolecules, and immobilization of carbohydrates onto solid surfaces (glycoarrays) as the most representative examples. In the majority of cases, the formed 1,2,3-triazole heterocycle plays the role of a linking tether between the coupling partners.

Keywords 1,3-Dipolar cycloadditions · Alkynes · Azides · Click chemistry · Carbohydrates

Abbreviations
Arg   Arginine
BnBr   Benzyl bromide
CPMV   Cowpea mosaic virus

ConA   Concanavaline A
DBU   1,8-Diazabicyclo[5.4.0]undec-7-ene
DIPEA   Diisopropyl ethylamine
DMAP   4-(Dimethylamino)pyridine
DMF   Dimethylformamide
DMSO   Dimethylsulfoxide
DTT   Dithiothreitol
EMC N-(ε-Maleimidocaproyl)
Fuc   Fucose
Gal   Galactose
Glc   Glucose
GlcNAc N-Acetyl glucosamine
Lac   Lactose
LacNAc   Lactosamine
Man   Mannose
MW   Microwave
NDDA N,N'-(Dithiodidecane-10,1-diyl)bispropiolamide
NHS N-Hydroxysuccinimide
PEG   Polyethyleneglycol
PMP p-Methoxyphenyl
QCM   Quartz crystal microbalance
RCA   Ricinus communis agglutinin
rt   Room temperature
SAM   Self-assembled monolayer
SPR   Surface plasmon resonance
TBAI   Tetrabutylammonium iodide
TCEP   Tris(carboxyethyl)phosphine
TsCl   Tosyl chloride
TMS   Trimethylsilyl
Tyr   Tyrosine

1
Introduction

1,3-Dipolar cycloadditions have been recognized as extremely important transformations in organic chemistry, as evidenced by the large number and wide scope of targets than can be prepared by this chemistry. Over the years, this reaction has developed into a generally useful method for the construction of five-membered heterocycles, whose importance is enhanced by the wide selection of 1,3-dipoles and dipolarophiles and the regio- and stereoselectivity during the cycloaddition. The method is also useful for the further transformations of the cycloadducts into a variety of functional molecules. Numerous natural and unnatural products have been prepared by synthetic routes that have a 1,3-dipolar cycloaddition as a crucial step [1].

Synthetic glycochemicals have attracted increasing interest as a consequence of both the functional and the structural diversity of carbohydrates.

Owed to this diversity, these biomolecules are attractive compounds for the development of therapeutic agents. The huge efforts of researchers of different areas in developing carbohydrate mimetics as well as in the preparation and subsequent biological evaluation of carbohydrate conjugates have favored a flourishing of the field of glycobiology and, more recently, the related emerging field of glycomics. In this context, the symbiosis of the 1,3-dipolar cycloaddition reactions and carbohydrates has been long established as a fertile pairing. Some excellent previous reviews [2–4] have covered the successful applications of 1,3-dipolar cycloadditions to carbohydrate derivatives. The majority of the different 1,3-dipoles have been used in different intermolecular and intramolecular chemical strategies but a special preference in the use of nitrones, nitrile oxides, and nitronates is observed.

The discovery at the beginning of this century of the Cu(I) catalytic and regioselective effect in the 1,3-dipolar cycloaddition of azide and alkynes [5, 6] has initiated a golden era for this cycloaddition reaction that has found multiple applications in biomedical science, organic synthesis, and material science [7, 8]. A vast majority of these results involved the use of carbohydrates. The present review is focused on the recent advances concerning the non-catalytic and catalytic azide–alkyne conjugation in the carbohydrate field.

2
The Huisgen 1,3-Dipolar Cycloaddition of Azides and Alkynes

The Huisgen 1,3-dipolar cycloaddition reaction between an azide and an acetylene is the most efficient route for synthesis of 1,2,3-triazole heterocycles. The usefulness of this reaction is due to the relatively easy functionalization of organic molecules with azides and alkynes as well as relative stability of those functions under a variety of conditions. They are essentially inert to most biological and organic conditions, including highly functionalized biological molecules, molecular oxygen, water, and the majority of common reaction conditions in organic synthesis. These two functionalities can be installed when convenient and remain unaffected throughout subsequent transformations. Although azide decomposition is a thermodynamically favorable process, kinetic factors allow aliphatic azides to remain nearly inert until presented with a good dipolarophile. This kinetic stability is directly responsible for the slow rate for the cycloaddition of azides and alkynes, which generally requires elevated temperatures and long reaction times affording, in the majority of the cases, mixtures of the 1,4- and 1,5-regioisomers (Scheme 1). The desirable control of the regioselectivity is only observed in the uncatalyzed Huisgen-type cycloaddition for coupling reactions involving highly electron-deficient terminal alkynes. Additionally, regiocontrol can be achieved in some cases by supramolecular interactions with a macrocycle

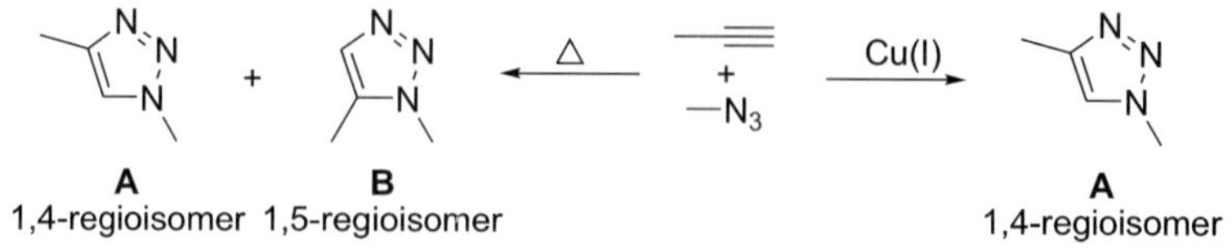

A
1,4-regioisomer
B
1,5-regioisomer
A
1,4-regioisomer

Scheme 1 Thermal and catalyzed Huisgen 1,3-dipolar cycloaddition of azides and alkynes

with guest binding abilities (curcurbituril [9] or resorcinarene [10]) or between the two reactant partners [11]. The elevated temperatures required for the thermal conjugation of azides and alkynes preclude their use in biological systems with a demand of orthogonality.

In spite of the multiple applications that this fusion reaction has found in its thermal version, the benefits of this cycloaddition were only fully expressed after the discovery of the Cu(I) catalytic effect in the reaction of azides and terminal alkynes. Cu(I) increases the reaction rate up to 10^7 times, thus eliminating the need for elevated temperatures. In addition, Cu(I) dramatically improves the regioselectivity affording exclusively the 1,4-regioisomer. This high-yielding reaction tolerates a variety of functional groups and affords the 1,2,3-triazole product with minimal work-up and purification.

The outstanding characteristics mentioned for both the uncatalyzed and the catalyzed reactions have led to the 1,3-dipolar cycloadditions of azide and alkynes being considered the most efficient reaction under the new concept of "click chemistry" [12]. The term coined by Sharpless and coworkers refers to a modular synthetic approach based on a set of the most practical end reliable chemical reactions. These are wide in scope, highly efficient in terms of both conversion and selectivity under very mild and simple reaction conditions, and use simple product isolation procedures. An indicative parameter of the importance of the 1,3-dipolar cycloaddition of azide and alkyne as a prototype of a click chemistry reaction is the exponential number of applications based on this reaction that have appeared in the literature since the description in 2002 of the catalyzed version. Currently, it can be asserted that click chemistry and the Huisgen 1,3-dipolar cycloaddition of azides and alkynes are synonymous.

Although the thermal dipolar cycloaddition of azides and alkynes occurs through a concerted mechanism, a stepwise mechanism has been postulated on the basis of calculations and kinetic studies. These suggest the initial formation of a Cu(I) acetylide but the exact nature of the intermediate species [13, 14] is still not very clear. Since the discovery of the Cu(I) catalytic effect, a variety of conditions and Cu(I) sources have proven to be successful in generating the 1,2,3-triazoles with high yields, in most cases underscoring the robustness and versatility of the reaction. Solution-phase Cu(I)-catalyzed azide–alkyne cycloaddition constitutes, until now, the bulk of the applications of this click chemistry reaction. However, further progress has also allowed the transition to solid-phase chemistry for the azide–alkyne coupling.

Structure 1–3 Heterocyclic chelates as Cu(I) ligands

In traditional solution chemistry, the reactions are performed in water or, more commonly, in an aqueous mixture of an organic co-solvent such as an alcohol or DMSO. The Cu(I)-catalyzed azide–alkyne cycloadditions are normally performed at room temperature but in some cases additional heating has an accelerating effect. Similarly, microwave irradiation and sonication can dramatically reduce the reaction times without affecting the yields or the formation of undesired side reactions. The Cu(I) generation constitutes the factor of major variability and has led to the development of different protocols. In-situ formation of Cu(I) from Cu(II) salts, usually copper sulfate pentahydrate, has been the most widely used protocol and is normally performed by reduction with sodium ascorbate as the favored reducing agent or, in some special applications, by copper-metal comproportionation. Tris(carboxyethyl)phosphine (TCEP) is also another reducing reagent that has found applications, mainly in bioconjugation [15, 16].

The second method of choice is the use of Cu(I) inorganic salts, usually CuI. This option represents a reliable procedure of catalyzing azide–alkyne cycloadditions, particularly in the presence of excess of base (DIPEA or 2,6-lutidine), which probably plays a role in preventing the degradation of Cu(I). A similar action seems to be operative when, by following this protocol, cycloadditions are performed in the presence of certain chelating heterocycles such as compounds 1 [17], 2 [18] and 3 [19]. In this case, the use of other bases is not necessary since the tertiary nitrogen center of these compounds functions both as a donor to Cu(I) and a proton acceptor. This protocol is particularly valuable in bioconjugation where the use of other bases is limited but complicates the purification procedures as it introduces the necessity of removing the ligand. In the case of substrates not soluble in aqueous media, where the use of an organic solvent is a prerequisite, the use of the tandem $CuSO_4$/ascorbate or the exclusive use of CuI proved to be inefficient. To solve this problem, two alternative protocols have been developed. The first made use of organic-soluble catalysts such as the complex $(Ph_3P)_3CuBr$ [20], $(EtO)_3PCuI$ [20], or $Cu(MeCN)_4PF_6$ [17, 18] that solely or in combination with a base (DIPEA or DBU) or in the presence of CuI led to the cycloaddition products in excellent yields. More recently, a CuI polymer-supported catalyst easily prepared from Amberlyst A-21 has been reported as an efficient heterogeneous catalytic system with good activity, stability, and recycling capabilities [21]. Oxidation of copper metal provides the latest alternative protocol in solution chemistry for generating Cu(I) by using copper turnings [5] or preferably Cu(0) nanosize powder in the presence, normally, of an amine hydrochloride [22] or copper nanoclusters [23]. This protocol, however, presents as disadvantages the longer reaction times required when using copper turnings or the acidic environment, and high costs for the Cu(0) nanosize clusters, which limits its applicability. The efficiency and scope of the catalyzed Huisgen 1,3-dipolar cycloaddition of azides and alkynes has been broadened in reported cases by the development of one-pot multistep procedures that avoid the manipulation of azides by their in-situ generation without isolation [24].

Solid-phase Cu(I) catalyzed azide–alkyne cycloadditions have also proven to be a reliable and highly robust procedure [6, 25–27]. Both the azide and the alkyne functions have been incorporated into different resins. These functionalized supports have been applied to the synthesis of triazole- and non-triazole-containing products by using, in the latter case, the triazole functionality as a linker. The cycloadditions proceed easily on the solid phase showing more sensitivity to steric factors than in solution phase and little sensitivity to reaction conditions, resin type, or subsequent transformations. The Cu(II) reduction protocol, as well as the use of Cu(I) salts, work with equal efficiency in various organic solvents (THF, DMF, acetonitrile, DMSO).

The Huisgen cycloaddition has found synthetic applications wherever 1,2,3-triazoles are important. In a broad sense, 1,2,3-triazoles derivates are of interest in industry, agriculture, and in bioscience as bioactive compounds. However, an analysis of the literature since 2002 shows that the special attention given to these systems is based on their capability to act as linking moieties for the connection of different substructures in frameworks of varied complexity, and also on their ability to function as a peptide bond mimic. 1,2,3-Triazoles have an atom emplacement and electronic properties similar to those of an amide function, having the advantage of their resistance to hydrolytic cleavage as well as to oxidation or reduction processes. In addition, the larger dipolar moment that they possess in relation with the amide bond enables both the N(2) and N(3) triazole atoms to act as hydrogen bond donors and acceptors. All these outstanding features have found huge applications in biomedical sciences, organic synthesis, and material science [7, 14].

3
Intermolecular Alkyne–Azide 1,3-Dipolar Cycloadditions

In state-of-the-art carbohydrate chemistry, the introduction of the alkyne or azide group in a sugar is a routine process that allows easy access to mono- as well as polyfunctionalized derivatives. These compounds can be rationally used for the conjugation with other complementary functionalized molecules or scaffolds by means of their intermolecular alkyne–azide 1,3-dipolar cycloaddition reaction. This synthetic strategy enables the rapid preparation of carbohydrates conjugates in which the heterocyclic triazole acts as a tether holding for the carbohydrate decoration. The anomeric position is usually the preferred position for placing the linking azide or alkyne functionality, but the rest of the sugar ring positions have also been used through an adequate protection–deprotection strategy. The versatility of the strategy allows the preparation of a wide variety of glyco-structures that are not limited by the structural requirements of the target structures but mainly by the creativity of the researchers.

3.1
Carbohydrate Conjugates Containing One 1,2,3-Triazole Heterocycle

The reaction of monofunctionalized azide or alkyne carbohydrate derivatives with complementary monofunctionalized counterparts is the simplest approach for the design of carbohydrate conjugates containing triazoles. This strategy was used by Meldal et al. [6] in their pioneering work where CuI was used for the first time as a catalyst in the alkyne–azide 1,3-dipolar cycloaddition reaction. The Cu(I)-catalyzed reaction was applied to the synthesis of peptidotriazole sugar derivatives on solid phase by reacting the 2-azido thioglycoside (4) with two different alkynylated PEGA$_{800}$ resins (5, 6) obtained by standard solid-phase peptide synthesis (Scheme 2). The resin 5 was prepared by acylation of the N-terminus of the linked FGFG tetrapeptide with

Scheme 2 *a* CuI, DIPEA; *b* 0.1 M NaOH; *c* 20% piperidine/DMF; *d* Cu(OAc)$_2$, Na-ascorbate (*F* phenylalanine, *G* glycine) [6, 28]

propargylic acid while the resin **6** was obtained by using propargylglycine as the terminal amino acid. The reactions allowed the preparation of the peptidotriazoles **9** and **10** after basic cleavage of the resins. This approach for the synthesis of triazole-linked amino acids was also validated [28] in solution chemistry by the reaction of propargylglycine (**8**) with the 2-azido functionalized Glc derivative **7** using the $Cu(AcO)_2$/Na-ascorbate system as a source for Cu(I) (Scheme 1). Both reaction conditions proved to be fully compatible with all the amino acids and their protecting groups.

The synthesis of anomeric N-(**14**) as well as C-glycosyl α-amino acids (**17**) has been described by Kuijpers et al. [28], Dondoni et al. [29], and Giguere et al. [30] as an extension of the previous results by reacting glycopyranosyl azides (**12**) or C-glycosyl acetylenes (**15**) with several protected alkyne (**13**) and azide (**16**) functionalized amino acids, respectively, as the partners of the 1,3-dipolar cycloadditions (Scheme 3). The resulting unnatural glycosyl amino acids, featuring a triazole heterocycle moiety between the sugar and the amino acid entities, are valuable as impervious glycoconjugates to chemical and enzymatic degradation. In particular, compound **17** (β-Gal, $n = 1$, Aa = CHCOOMe) has been tested as a new glycomimetic inhibitor of galectis-1 and -3 with modest results by Guigere et al. [30].

R = Ac, Bn; R^1 = OR, (OR)Glc$p\beta$; R^2 = H, Me, Bn; X = OR, NHAc; n = 0,1; m = 1,2,3

Scheme 3 *a* Na-ascorbate [28] or CuI, DIPEA [29, 30]

In an indirect strategy for the synthesis of neoglycopeptides, by means of the alkyne–azide 1,3-dipolar cycloaddition reaction as an assembling tool, MacMillan et al. [31] have used glycopyranosyl azides (**18**) for the synthesis of bromoacetamide sugar derivatives of type **20** by reaction with 2-bromoacetyl propargylamide (**19**). This compound functions as an heterobifunctional adaptor allowing the subsequent coupling with cysteine-containing peptides by nucleophilic displacement (Scheme 4). In particular, it is noteworthy that the strategy is compatible with peptide synthesis on solid phase and in an automated fashion when the bromoacetamides are reacted with the deprotected thiol functions of a linked peptide (**21**). The authors applied this methodology to the synthesis of a peptide similar in sequence to human erythropoietin residues 21–32. This straightforward synthesis of neoglycopeptides **22** has been combined with native chemical ligation by coupling with a peptide thioester corresponding to human erythropoietin residues 1–19.

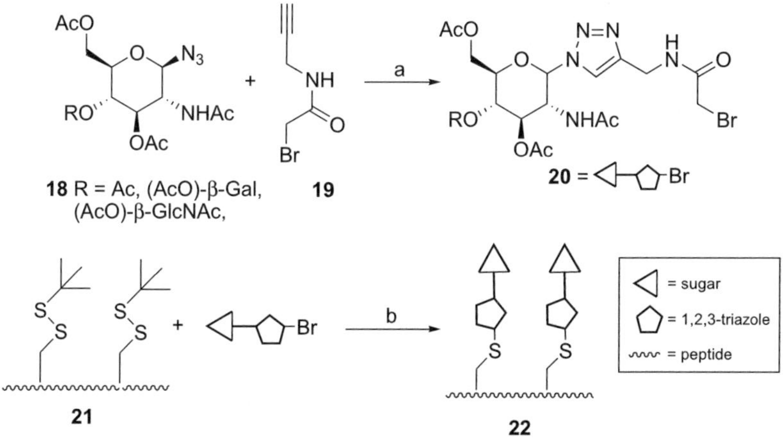

Scheme 4 a CuSO$_4$, Na-ascorbate; b (i) DTT, DIPEA, DMF; (ii) **20** (sugar), Et$_3$N, DMF [31]

The use of various 3-ethynyl-2(1H)-pyrazinones (**24**) as peptidomimetic precursors has been described by Ermolat'ev et al. [32] (Scheme 5). Thus, the Cu(I)-catalyzed ligation of compounds **24** with different β-D-glycopyranosyl azides (**23**) led to the corresponding glycoconjugates **25**. These compounds could allow the access to glycopeptidomimetics after subsequent transform-

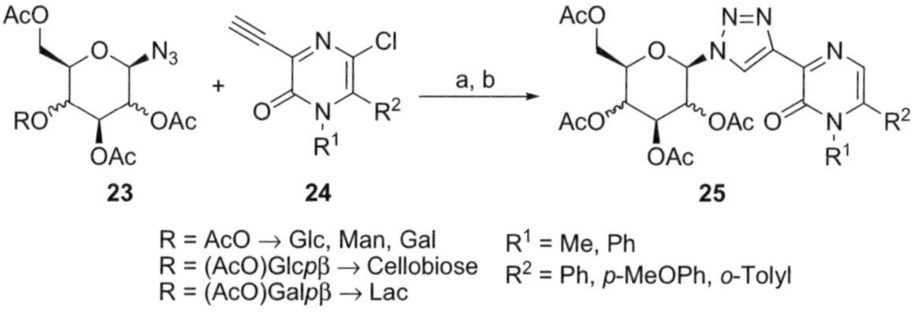

Scheme 5 a Cu (turnings), CuSO$_4$, MW; b H$_2$, Pd/C [32]

ations. This indirect coupling of the 2(1H)-pyrazinone scaffold with the saccharides via a 1,2,3-triazole ring system was performed in a much faster way by the microwave-assisted reaction of the partners. However, in this case the cycloadducts were obtained only in moderate to good yields, probably due to polymerization of the starting acetylenes and/or the thermal instability of the glycoconjugates under the reaction conditions used.

An improvement in the protocol simplicity for the preparation of triazole-linked glycoconjugates (**28**) from azido glycosides was reported by Chittaboina et al. [24] who developed an efficient one-pot glycosylation method starting from both unprotected or peracetylated carbohydrates (**26**). By executing several successive reaction steps (acetylation in the case of unprotected sugars, formation of glycosyl bromides, transformation into the glycosyl azides, and Cu(I)-catalyzed 1,3-dipolar cycloaddition) in one pot this procedure has the advantage of excluding the isolation of all the intermediates, thus reducing significantly the reaction time and improving the overall yield (Scheme 6). In most situations, quantitative transformations were observed, isolating the resulting glycoconjugates (**28**) by a simple filtration operation.

R = R^1 = H → Glc, Gal, Man
R = OAc; R^1 = (RO)Glc$p$α, (RO)Gal$p$β → Cellobiose, Lac
R^2 = Ph, CH$_2$OH, CH$_2$NHCOCH$_2$Br

Scheme 6 *a* Ac$_2$O, iodine (cat), rt; *b* HBr-AcOH; *c* NaN$_3$, Bu$_4$NHSO$_4$, NaHCO$_3$, CuSO$_4$, Na-ascorbate, CHCl$_3$: EtOH : H$_2$O 9 : 1 : 1 (R = H) or CHCl$_3$: EtOH 9 : 1 (R = OAc) [24]

Azidoalkyl **29** as well alkynyl glycosides **32** have been used as easily accessible partners in the preparation of diverse O-glycosyl aminoacids and O-glycopeptidomimetics (**31, 34**) by click ligation with alkyne (**30**) and azide (**33**) amino acids, respectively (Scheme 7). Thus, Hotha et al. [33] reported the preparation of cysteine and phenylalanine derivatives in which the reactive side (alkyne or azide) was separated by a 6-aminocaproic acid linker that in the reaction with 2-azidoethyl and 3-butynyl glucoside and lactoside led to the corresponding amino acid-ligated glycoconjugates. Analogously, in the search for functionalized peptidomimetics with improved affinities for specific receptors and metabolic stability towards endogenous proteases, Arosio et al. [34] performed the conjugation of azabicycloalkane amino acids (**34**), mimicking the homoSer-Pro dipeptide with protected and unprotected propargyl $β$-D-glucopyranoside and other biologically relevant partners.

Scheme 7 *a* CuI, DIPEA [33]; *b* Cu(OAc)$_2$, Na-ascorbate [34]

In the search for new glycomimetic inhibitors of galectins with higher affinities, stabilities, and specificities than the naturally occurring carbohydrate ligands (Lac and LacNAc), Salameh et al. [35] and Guiguère et al. [30] rationally designed a series of 1-thio-β-D-galactosides containing a 1,2,3-triazole moiety at C-3 (**37**, **38**) that were constructed by click chemistry. The choice of the position C-3 was based on the crystal structure of a LacNAc-galectin-3 complex that showed the HO-3 of the Gal moiety as the most suitable position for chemical modifications (Scheme 8). Thioglycosides (**35**) were chosen for the reason that they are more hydrolytically stable than the corresponding *O*-glycoside analogs, which possess low physiological stability due to their acid-sensitive glycosidic bonds. For construction of the triazole ring, the 3-azido function, as well as the 3-*O*-propargyl group, were placed at the C-3 position of Gal (compounds **35** and **39**, respectively) following standard protocols in carbohydrate chemistry, preparing the 3-*O*-propargyl derivative **39** of higher synthetic economy than that of the corresponding 3-azido derivative **35**.

Scheme 8 *a* CuI, DIPEA, toluene, 40 °C or toluene, 80 °C (for R = H); *b* R^1 = Ac → R^1 = H, NaOMe/MeOH; *c* RNH$_2$, MeOH; *d* CuI, DIPEA, THF; *e* CuI, DIPEA, acetonitrile [30, 35, 36]

This compound was reacted with methyl propiolate and a series of aliphatic- and aromatic-substituted acetylenes **36** whereas the alkynyl intermediate **39** was conjugated with the azides **40**. In the case of the 1*H*-[1,2,3]-triazol-1-yl derivatives, the collection was extended by reaction with different amines leading to a series of the corresponding 4-carbamoyltriazoles **38** in good yields. In evaluating the inhibitory properties of these carbohydrate-based 1,2,3-triazoles it was found that some of the prepared compounds exhibited high inhibitory potencies against Gal-3 suggesting the 4-substituted triazole moiety as a promising structural motif with an affinity-enhancing effect. In a complementary study, Tejler et al. [36] also prepared some lactose derivatives with triazole-containing aglycon moieties **44** by reaction of 2-azidoethyl β-lactoside (**42**) with acetylene derivatives of phenylalanine and phenyl ethylamine (**43**). Compounds **44** were evaluated in a competitive fluorescence polarization assay against different galectins. The assays gave some insights regarding the influence of the aglycon structure in their blocking capability of these monovalent lactose derivatives. In particular, it was found that galectin-1 showed a preference for ligands with a carbamate linker.

In a more elaborate and elegant strategy, Griffith et al. [37] used in vitro chemoenzymatic glycorandomization in tandem with the azide–alkyne 1,3-dipolar Huisgen cycloaddition to diversify the glycosylation pattern of the natural glycopeptide vancomycin (Scheme 9). In the first stage, chemoenzymatic glycorandomization was applied to nucleotide diphosphosugar donor libraries, obtained by incubating the corresponding sugar 1-phosphate **46** with nucleotidyl transferasa (Ep), as substrates for the inherent promiscuous glycosyltransferases (GtfE), thus transferring the sugar moiety to the complex vancomycin aglycon. The extensive synthetic manipulations typically required for chemical glycosylation strategies are overcome with these multienzyme single-vessel reactions. The combination of this procedure with the inclusion of a clickable azido group in the carbohydrate sugar allowed the downstream diversification, via chemoselective ligation in an aqueous environment, of a series of selected terminal alkynes. The obtained vancomycin analogs were tested for their antibacterial activity, revealing that significantly diverse substitutions are tolerated and that two of the 39 members of the library rival the natural vancomycin.

R¹ | N₃ | O | OPO₃⁻² | a, b, c | R¹ | R | N | O | OAg
R | HO | OH | OH | | R | N=N | HO | OH | OH
45 | **46** | | | | | **47**

Ag = Vancomycin Aglycon

Scheme 9 *a* Ep/dTTP; *b* GtfE vancomycin aglycon; *c* CuI [37]

In addition to monofunctionalized (azido and alkyne) amino acid derivatives as the substrate of choice in the majority of the syntheses of carbohydrate conjugates bearing one 1,4-disubstituted 1,2,3-triazole tethers, the self-coupling of carbohydrate molecules by the click reaction of their monoazido and monoalkyne derivatives has also been exploited for the synthesis of divalent neoglycoconjugates (Table 1). In this case, the versatility of the coupling strategy allows the reliable preparation of homo- (entries 1, 3, 6, 9) as well as heteroneoglyconjugates (entries 2, 4, 5, 7, 8, 10) bear-

Table 1 Divalent neoglycoconjugates

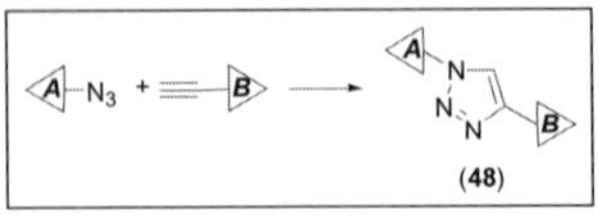

$$\boxed{\;\text{A}-N_3 \; + \; \equiv\!\!-\text{B} \;\longrightarrow\; \text{triazole (A, B)}\;} \quad (48)$$

Entry	 A—	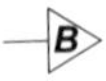 —B	Conditions	Refs.
1	(AcO)Man$p\beta$OCH$_2$CH$_2$ –	(AcO)Man$p\beta$XCH$_2$ – (X = O, S, NAc)	A (X=O) B (X=S) C (X=NAc)	[20]
2	(AcO)LacβOCH$_2$CH$_2$ –	(AcO)Glc$p\beta$OCH$_2$CH$_2$ –	D	[33]
3	(AcO)LacβOCH$_2$CH$_2$ –	(AcO)LacβOCH$_2$CH$_2$ –	D	[33]
4	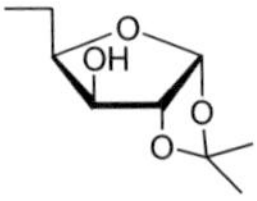 *(sugar structure)*	(AcO)Glc$p\beta$OCH$_2$	D	[33]
5	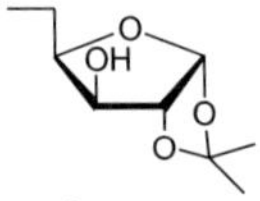 *(sugar structure)*	(AcO)LacβOCH$_2$	D	[33]
6	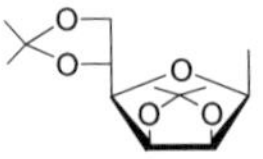 *(sugar structure)*	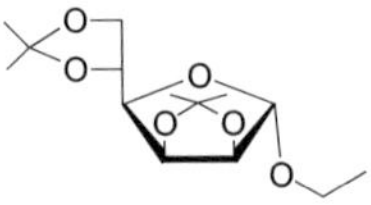 *(sugar structure)*	E	[38]
7	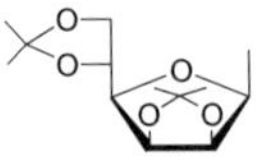 *(sugar structure)*	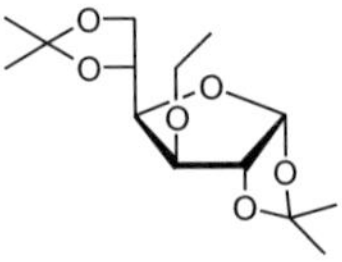 *(sugar structure)*	E	[38]
8	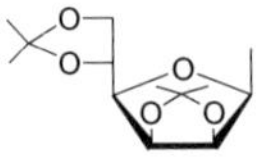 *(sugar structure)*	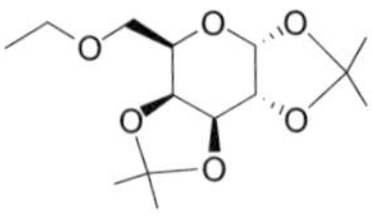 *(sugar structure)*	E	[38]
9	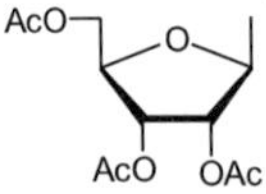 *(sugar structure)*	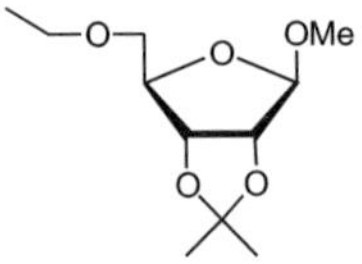 *(sugar structure, OMe)*	F	[24]
10	(AcO)Glc$p\beta$ –	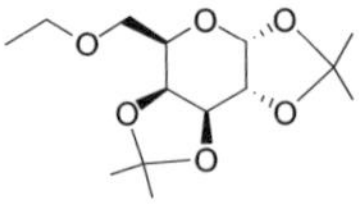 *(sugar structure)*	F	[24]
11	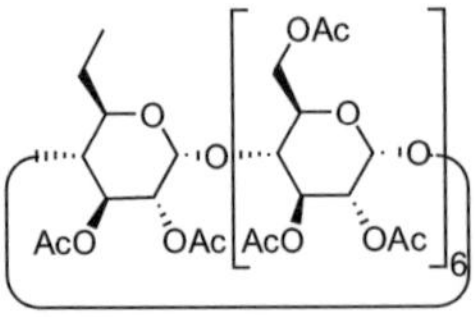 *(cyclodextrin structure)*	X-Sugar(OAc), X = O,S, Sugar = (AcO)Man, (AcO)Gal, (AcO)Lac	B	[20] [39] [a]

A: (Ph$_3$P)$_3$CuBr, DIPEA, MW; B: (EtO)$_3$P · CuI, DIPEA, MW; C: (Ph$_3$P)$_3$CuBr, DBU, MW;
D: CuI, DIPEA, acetonitrile; E: (Ph$_3$P)$_3$CuBr, DBU; F: CuSO$_4$, Na-ascorbate
[a] Ortega-Muñoz et al. submitted

ing two structurally equal or different carbohydrate moieties, respectively. Thus, not only were monosaccharides derived from furanosyl (entries 4–9) or pyranosyl substrates assembled between them but also oligosaccharides (entries 2, 3, and 5) were covalently linked to monosaccharides or other oligosaccharides. In addition, the strategy has been also applied to more complex platforms such as the 6-monoazide derivative of β-cyclodextrin and propargyl *O*- and *S*-glycosides (entry 11). In the starting building blocks, the clicking functionalities were placed in the anomeric positions but also in other carbohydrate ring positions. In the majority of the collected cases in Table 1, the reactions were performed in a proof-of-concept of the utility of the click chemistry coupling for the synthesis of multivalent carbohydrate structures or in the search for new reaction conditions. This is the case for the work reported by Pérez-Balderas et al. [20] (entries 1 and 11) who first introduced the use of copper complex as a useful organic-soluble catalyst in refluxing solvents or with the assistance of microwave irradiation. Also, Chitaboina et al. [24] (Scheme 6) developed the one-pot protocol for the construction of well-defined chiral structures. However, although there has not been any report up to the present concerning the evaluation of their biological activity, these carbohydrate homodimers and heterodimers have a potential as glycomimetics, representing appealing tools for evaluating distances between carbohydrate binding sites in polyvalent molecular recognition, to act as potent reversible cross-linking reagents, or to be used as transporters in oriented drug delivery (entry 11) [20].

Other non-biologically active molecules have also been conjugated with monofunctionalized carbohydrates. Thus, the click ligation strategy has been used in the authors' laboratory by Casas-Solvas et al. [40] for the preparation of ferrocene–carbohydrate conjugates in studies directed towards the development of carbohydrate-based biosensors (Scheme 10). The 1-azidomethyl-(49) and 1-ethynyl-ferrocene (50) are easily accessible and their regiospecific coupling with the propargyl and 2-azidoethyl glycosides of Glc and Man led, after *O*-deprotection, to the corresponding water-soluble metallocene derivatives (51, 52) bearing binding sites with potential in molecular recognition process. The electrochemical behavior of the synthesized glycoconjugates was

Scheme 10 *a* (EtO)$_3$P · CuI, toluene, reflux; *b* NaOMe, MeOH [40]

investigated and showed that these compounds exhibit reversible oxidation and reduction of the Fe^{2+} center.

On the other hand, an easy access to *S*-neoglycoconjugates containing a triazole heterocyclic ring in the anomeric position has been described by Zhu et al. [41] as an illustrative application for the development of a general methodology for accessing acid-resistant glycoconjugates (Scheme 11). The authors easily prepared glycosylthiomethyl chlorides as flexible building blocks for the construction of *S*-linked glycoconjugates by reaction of *O*-acyl-protected thiosugars with dichloromethane. Nucleophilic substitution of the chlorine of those compounds by azide gave the corresponding azido-containing carbohydrates (**53**). Subsequent thermal cycloaddition reaction of **53** with diverse symmetric and non-symmetric acetylene derivatives (**54**) led to 1-*N*-glycosylthiomethyl-1,2,3-triazoles (**56**), which are ring-substituted. The authors also explored the reaction with norbornadiene (**55**) as dipolarophile. Under thermal conditions, norbornadiene works as an acetylene equivalent by a tandem 1,3-dipolar cycloaddition/retro–Diels–Alder reaction providing a general and convenient means for accessing the parent 1-*N*-glycosylthiometyl-1,2,3-triazoles (**56**). These compounds are troublesome to obtain by direct triazolylmethylation of glycosyl thiols. In this respect, it is also noticeable the use made by Hager et al. [42] of 1,2-disubstituted phenylsulfonyl-ethenes (**57**) as dienophiles in the synthesis of fluoroalkyl-substituted 1,2,3-triazoles linked to the C-6 atom of monosaccharides (**59**), which are conceived as reverse nucleosides (Scheme 11). In this cycloaddi-

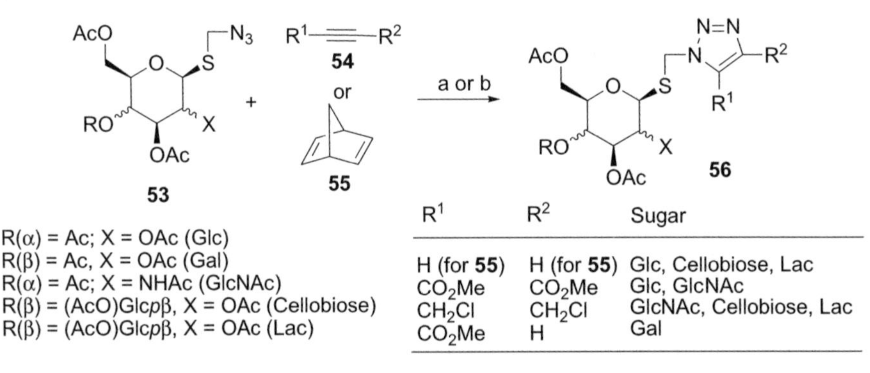

R^1	R^2	Sugar
H (for **55**)	H (for **55**)	Glc, Cellobiose, Lac
CO_2Me	CO_2Me	Glc, GlcNAc
CH_2Cl	CH_2Cl	GlcNAc, Cellobiose, Lac
CO_2Me	H	Gal

R(α) = Ac; X = OAc (Glc)
R(β) = Ac, X = OAc (Gal)
R(α) = Ac; X = NHAc (GlcNAc)
R(β) = (AcO)Glc*p*β, X = OAc (Cellobiose)
R(β) = (AcO)Glc*p*β, X = OAc (Lac)

R = C_4F_9, C_6F_{13}

Scheme 11 *a* Dioxane, 100 °C; *b* toluene, reflux [41, 42]

tion reaction, the vinyl sulfones (**57**) (easily accessible by ultrasound-assisted Wittig–Horner olefination) act likewise as terminal acetylenic equivalents in the reaction with 6-azido-6-deoxy-galactopyranosides (**58**). This is due to the spontaneous aromatization that follows the cyclization step as a consequence of the sulfonyl elimination. The reported reactions showed completely regio-selectivity, only the 1,4-disubsituted regioisomers being isolated.

In an effort to expand the click chemistry concept, Ijsselstijn et al. [43] have reported the use of two N-protected ynamides (**60**) as dienophiles in the Cu(I)-catalyzed cycloaddition reaction with a variety of azides, including glycosyl azides (Scheme 12). The ynamides were successfully clicked under the standard conditions for these reactions [Cu(OAc)$_2$/Na-ascorbate] giving rise to 1-substituted 4-amino 1,2,3-triazole-linked sugars (**61**). These results are noteworthy as they allow direct access to amino triazole rings, which are found in some bioactive molecules. However, additional studies are required in order to evaluate the scope and applicability to more complex ynamides.

Scheme 12 *a* Cu(OAc)$_2$, Na-ascorbate [43]

Some recent improvements in the reaction conditions and protocol of the click ligation reported by different research teams are noteworthy especially because they have been developed or validated in the preparation of carbohy-drate conjugates containing 1,2,3-triazoles. First, Lee et al. [44] introduced the use of CH$_2$Cl$_2$ as a co-solvent with H$_2$O in the Cu(I)-catalyzed 1,3-dipolar cy-cloaddition of organic azides and alkynes as an alternative when other more popular co-solvents (*t*-BuOH, EtOH, CH$_3$CN, DMSO, THF) failed or gave only poor yields after long reaction times. These reaction conditions eliminate the need for ligands such as **1** and simplifies the reaction protocol, having been applied to diverse azido-alkyl *O*-glycosides. Secondly, Guezguez et al. [45] performed the synthesis of new 2′-deoxynucleosides featuring a function-alized 1,2,3-triazoyl ring as aglycone moiety by using microwave activation coupled with Cu(I) catalysis under solvent-free conditions by using silica gel as a solid in open vessels. This expanded the previous use of microwaves in solution chemistry. These conditions not only reduce dramatically the reac-tion times required for the coupling of azido-2′-deoxyribose and terminal alkynes but also afford the corresponding 4-substituted 1,2,3-triazoyl nu-cleosides in almost quantitatively yields. Finally, Deng et al. [46] applied sonication as a means of facilitating organic reactions. Sonication has been

$R^2 = C_6H_{13},\ C_8H_{18},\ CH_2NHBoc;\ R^3 = alkyl, sugar$

Scheme 13 *a* Cu(OAc)$_2$, Na-ascorbate, MeOH – THF, H$_2$O, sonication [46]

demonstrated to be a powerful way of enhancing the efficiency of the most important representative reactions in carbohydrate synthesis, including the 1,3-dipolar cycloaddition of a series of alkynes and glycosyl azides (**62, 64**) which can, in turn, be prepared by sonochemistry (Scheme 13). The corresponding triazole derivatives (**63, 65**) were thus obtained in high yields.

3.2
Carbohydrate Conjugates Containing Multiple 1,2,3-Triazole Heterocycles

The efficient construction of molecular systems bearing multiple carbohydrate appendages has become a necessity in the fields of glycobiology and glycomics. The main reason for this requirement is the prevalence of the multivalent principle in the ligand–receptor interactions in which carbohydrates are involved. This effort to emulate Nature has led to the development of multivalent glycomimics with different architectures (glycoclusters [47], glycodendrimers [47–51], glycopolymers [52], liposomes [53]) and carbohydrate densities to endow them with binding power greater than that of their monovalent analogs. These glycomimics have attracted attention because of their great potential not only to gain insights into molecular recognition events in which carbohydrates participate but also for their capabilities in biotechnology, pharmaceutical, and medical applications. Another reason is the demand for reliable techniques for the immobilization of carbohydrates in the preparation of carbohydrate-based surfaces. This driving force has led to the development of a variety of glyco-functionalized solid supports (microarrays, microbeads, biosensor chips, and self-assembled monolayers) as valuable tools with important applications in the "omic" sciences.

Among the different reported methodologies for the covalent assembling of the constitutive modules of these polyvalent carbohydrate systems, the 1,3-dipolar cycloaddition of azides and alkynes have reached a relevant place in recent times because of the reliability, efficiency, and robustness of the copper-catalyzed version. In common with other ligating methodologies, monofunctionalized sugars are normally used to be grafted to polyfunctionalized scaffolds or solid supports giving rise to molecular glyco-systems containing multiple 1,2,3-triazole rings with a tether function.

In addition, bifunctionalized carbohydrate derivatives containing simultaneously one azido and alkyne groups have been intermoleculary reacted, affording cyclic carbohydrate-containing molecule of interest that also possess more that one triazole ring in their structures.

3.2.1
Glycoclusters

For the synthesis of multivalent glycomimetics with cluster architectures by means of click chemistry reactions, diverse polyalkyne- as well as polyazide-functionalized scaffolds have been used with complementary monoazide or monoalkyne sugar derivatives. The number of functions expressed in the scaffolds is the variable that allows access to glycoclusters with different sugar densities and topologies. The scaffold structure and the nature of the connecting spacer between the core and the clicking functions are normally the factors that provide the variability for reaching glycoclusters with diverse cluster presentation and lengths between the sugar moiety and the core. These variables are essential for gaining fundamental insights into the spatial factors regulating the binding of carbohydrates.

Tables 2, 3, and 4 summarize some of the more representative contributions showing that dimeric (**66, 67**), trimeric (**68, 69**), and tetra-, penta- and heptaantenary clusters (**70, 71**) have been the preferred structures. In the majority of cases, the reported glycoclusters are aromatic-centered compounds. In other cases, simple multivalent aliphatic compounds (Table 2, entry 1, and Table 4, entries 3 and 4) or carbohydrates (Table 4 entries 2, 5, 6, and 7) have been used as the core. Carbohydrates have been selected to play this structural role because of their easy availability, biocompatibility, low toxicity, polyfunctionality, and intrinsic chirality. The anomeric position has been almost exclusively the position of choice to functionalize the carbohydrates with azido and alkyne clicking groups but the methodology works equally well when other carbon positions are selected for the ligation (Table 2, entry 5). In spite of the more structural complexity of glycoclusters, the cycloaddition reactions work as well as those with the simpler carbohydrate conjugates containing one 1,2,3-triazole heterocycle, leading to the desired multivalent neoglycoconjugates with equal efficiency and yields. A range of different reaction conditions have been used but the use of copper organic-soluble complexes appears to be the best option when protected carbohydrate sugars are used.

In some cases, the reactions were performed in a proof-of-concept of the utility of click chemistry in the construction of multivalent neoglyconjugates and constitute a logical extension of already commented on works for the

Table 2 Divalent neoglycoconjugates

Entry	◁—	Esp	Conditions	Refs.
1	$(AcO)Manp\beta OCH_2 -$	CH_2CH_2	A	[20]
2	$(AcO)Manp\beta XCH_2 -$ (X=O,S)	(phenyl)	A, B	[20]
3	$(AcO)Manp\alpha OCH_2 -$ $(AcO)Glcp\beta OCH_2 -$	(ferrocene)	C	[40]
4	(sugar with OAc, OC_2H_5)	(para-xylylene)	D	[55]
5	$(HO)Lacp\beta O(CH_2)_2 -$	MeO_2C...CO_2Me (bis-acetamido)	C	[36]
6	$(HO)Lacp\beta O(CH_2)_2 -$	MeO_2C...CO_2Me (bis-ethylcarbamate)	C	[36]

A: $(Ph_3P)_3CuBr$, DIPEA, MW; B: $(EtO)_3P \cdot CuI$, DIPEA, MW; C: $(Ph_3P)_3CuBr$, DBU, MW; D: CuI, DIPEA, acetonitrile

synthesis of carbohydrate conjugates containing one 1,2,3-triazole ring. However, in other cases oriented-design synthesis was the guideline for the preparation of the multivalent neoglyconjugates. Thus, bis(glycosyl) ferrocene derivatives (Table 2, entry 3) were designed by Casas-Solvas et al. [40] in the search for carbohydrate-based sensors. Dimeric deoxystreptamine derivatives were sought by Thomas et al. [54] to obtain a library of these compounds, with which they searched for size-specific ligands for RNA hairpin loops. The compounds showed in Table 2, entry 4, exhibited the higher affinity. Tejler et al. [36] prepared multivalent lactose derivatives in the search for selective galectin inhibitors and demonstrated that a significant cluster effect relative to corresponding β-lactoside monomers can be observed in the mul-

Table 3 Trivalent neoglycoconjugates

Entry ▷—	(N$_3$)$_3$-core or (HC≡C)$_3$-core		Conditions	Refs.
	▷—≡ + (N$_3$)$_3$–Core ⟶ Core–(triazole–▷)$_3$ (68)			
1	(AcO)ManpβXCH$_2$ – (X=O,S)	triazine core, R = O(CH$_2$)$_2$N$_3$	A (X=O) / A (X=S)	[20] / [39]
2	(AcO)ManpβSCH$_2$ –	methyl/Me benzene core, R = CH$_2$N$_3$	A	[39]
	▷—N$_3$ + Core–(≡)$_3$ ⟶ Core–(triazole–▷)$_3$ (69)			
3	(AcO)Glcpβ	R = OCH$_2$≡CH	B	[24]
4	(AcO)Lacβ(CH$_2$)$_2$	R = C(O)NH–CH$_2$–C≡CH	C	[30]
5	(AcO)Lacβ	R = NH–CH(CO$_2$Me)–C(O)–C≡CH	C	[36]

A: (Ph$_3$P)$_3$CuBr, DIPEA, MW; B: CuSO$_4$, Na-ascorbate; C: CuI, DIPEA

tivalent ligands indicated in Table 2, entries 5 and 6, and Table 3, entry 5, for galectin-1 and to a lesser extent for galectin-4N. Similar results were found by Guiguère et al. [30] with the C$_3$-symmetrical lactoside shown in Table 3, entry 4. Finally, the pentavalent Glc-centered (Table 4, entry 6) and the heptavalent β-cyclodextrin-centered clusters bearing Man, Gal, and Lac appendages (Table 4, entries 2 and 7) were designed in the authors' laboratory by Ortega-Muñoz et al. in a study (submitted for publication) that has showed the capability of some of these compounds to promote the cell adhesion and stimulation of monocytes acting as lipopolysacharide mimics. These glycocyclodextrin derivatives also have the potential to act as transporters in oriented drug delivery.

Table 4 Multivalent neoglycoconjugates

Entry	$\triangleleft$—	$(N_3)_n$-core or $(HC\equiv C)_n$-core	Conditions	Refs.
	$\triangleleft\!\!=\!\!=\ +\ (N_3)_n\!\!-Core\ \longrightarrow\ Core\!\!\left(\!\!N\genfrac{}{}{0pt}{}{}{}N\!\!=\!\!N\!\!-\!\!\triangleright\right)_n$ **(70)**	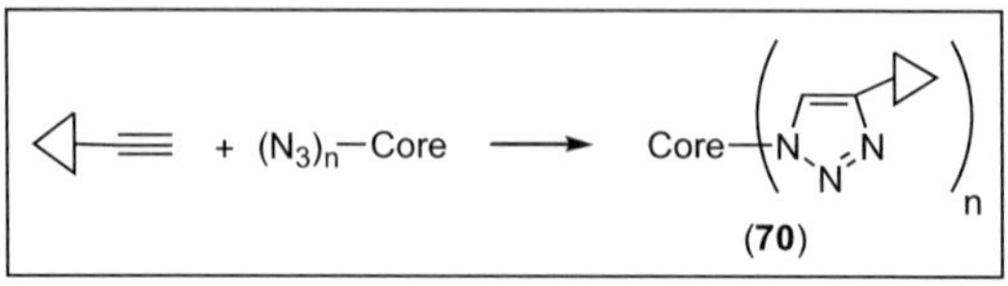		
1	$(AcO)Manp\beta XCH_2 - (X=O,S)$	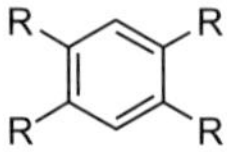 $R = CH_2N_3$ (benzene core, R at 1,2,4,5)	A (X=O) B (X= S)	[20] [39]
2	$(AcO)Manp\beta OCH_2 -$ $(AcO)Galp\beta OCH_2 -$ $(AcO)Lacp\beta OCH_2 -$ $(AcO)Galp\beta XCH_2 -$	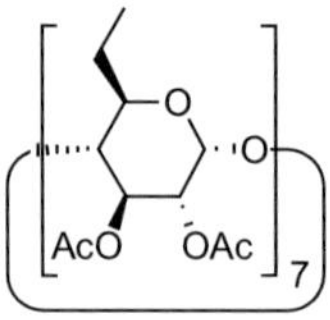 (cyclodextrin core, AcO, OAc)$_7$	C	a [39]
	$\triangleleft\!\!-N_3\ +\ Core\!\!=\!\!\!=\!\!)_n\ \longrightarrow\ Core\!\!\left(\!\!\genfrac{}{}{0pt}{}{}{}N\!\!=\!\!N\!\!-\!\!\triangleright\right)_n$ **(70)**	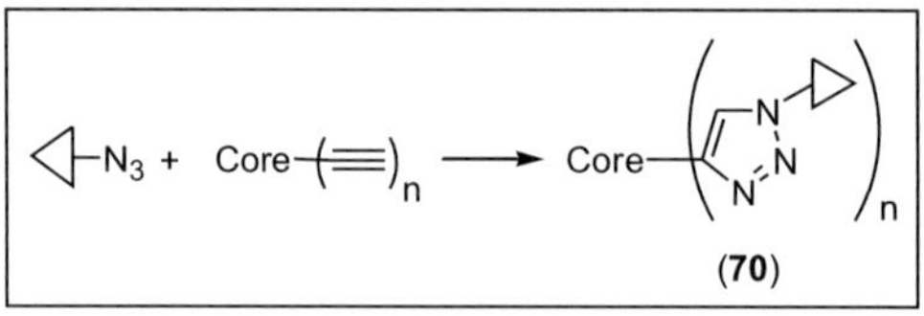		
3	$(AcO)Glcp\beta$	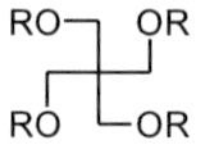 $R = CH_2\equiv CH$ (pentaerythrityl core, RO, OR)	B	[24]
4	$(AcO)Glcp\beta$	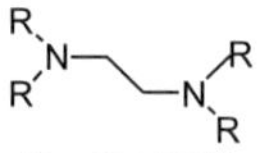 $R = CH_2\equiv CH$ (ethylenediamine core)	D	[24]
5	$(HO)Lac\beta$	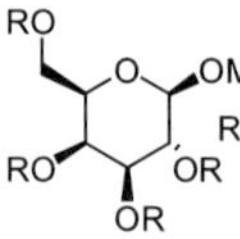 $R = CH_2\equiv CH$ (methyl glycoside core, RO, OR, OMe)	D	[24]
6	$(AcO)Galp\beta OCH_2CH_2 -$ $(AcO)Lac\beta OCH_2CH_2 -$	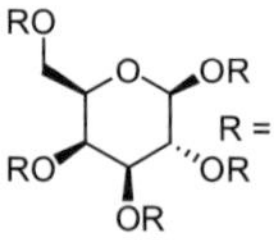 $R = CH_2\equiv CH$ (glycoside core, RO, OR)	C	a
7	$(HO)Galp\beta OCH_2CH_2 -$ $(HO)Lac\beta CH_2CH_2 -$	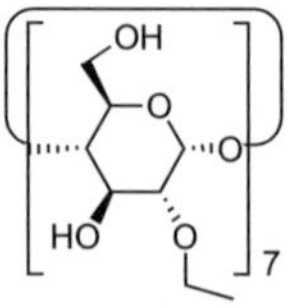 (cyclodextrin core, OH, HO)$_7$	C	a

A: $(Ph_3P)_3CuBr$, DIPEA, MW; B: $(EtO)_3P \cdot CuI$, DIPEA; C: $(EtO)_3P \cdot CuI$, MW; D: $CuSO_4$, Na-ascorbate; [a] Ortega-Munoz et al. submitted

3.2.2
Glycodendrimers

The high fidelity and efficiency of click chemistry has been exploited success-fully in the synthesis of carbohydrate end functionalized dendritic macro-molecules. Glycodendrimers are especially attractive glycoconjugates due to their high functionality, monodisperse nature, and the possibility of controlling their size and the number of carbohydrate motifs at the periphery. These characteristic are particularly appealing in the case of monodendritic architectures versus globular-like dendrimeric structures as they can mimic the complex multiantennary glycans found at the tips of natural glycoproteins. Thus, glycodendrimers are object of studies to evaluate their binding interaction with biological receptors and cell surfaces mainly in the construction of targeted drug delivery systems and in the preparation of therapeutic agents and inhibitors of adhesion.

Both azide- and acetylene-terminated dendritic macromolecules (72, 73) have been prepared for functionalization by click chemistry as an initial step in dendrimer design (Scheme 14). Convergent as well as divergent strategies have been used for the synthesis of clickable dendrons with high efficiency. Thus, dendritic acetylenes of type 73 have been prepared by Malkoch et al. [56], who reported a convenient synthesis of alkyne-functionalized poly(benzyl ether) in a convergent approach, and the divergent growth of PAMAM/DAB, bis-MPa, and polyester-based dendrimers. In addition, Joosten et al. [57] have also prepared different dendritic aromatic-centered acetylenes of type 73. Alternatively, Fernández-Megía et al. [58] prepared several generations of azide-terminated dendrimers of type 72 in a divergent approach from a core comprising gallic acid and triethylene glycol den-

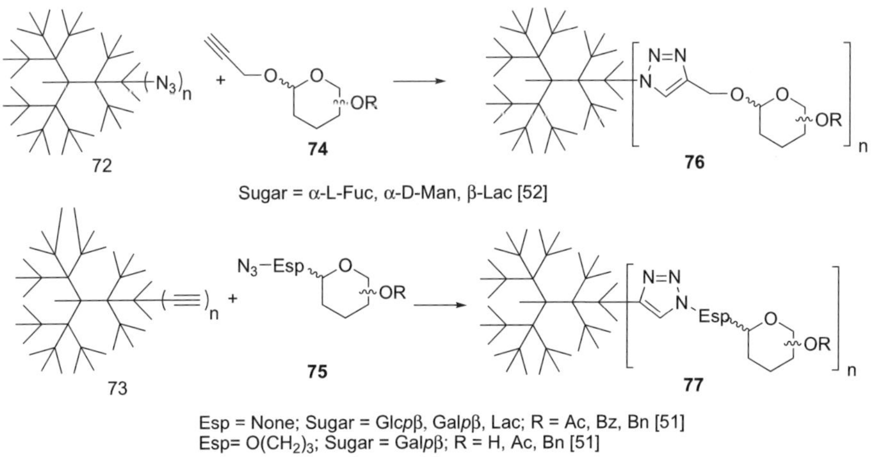

Scheme 14 Synthesis of glycodendrimers [56–58]

drimers, Different anomeric (**75**) as well as non-anomeric azide and alkyne carbohydrate derivatives (**74**) have been used to perform the sugar decoration of the synthesized clickable dendrimers 72 and 73. The reactions conditions were optimized in each of the reported cases, exploiting the flexibility of the 1,3-dipolar cycloadditions of azides and alkynes. Microwave irradiation notably facilitated the coupling reactions of these multifunctional systems giving the glycodendrimers (**76, 77**) in short reaction times and in high yields. Additionally, the incompatibility observed in some cases with the standard aqueous conditions ($CuSO_4$/sodium ascorbate in aqueous solutions), mainly due to insolubility of the reacting compounds, could be overcome by employing organic solvents and organic solvent-soluble catalysts (($PPh_3)_3 \cdot CuBr$ or $(EtO)_3P \cdot CuI$).

One of the advantages of click chemistry for the glyco-functionalization of dendrimers is that it allows the quick, efficient, and reliable multivalent conjugation of unprotected azide- and alkyne-derived carbohydrates as opposed to the lengthier process employing protected glycosides. The high efficiency is especially manifested not only for the excellent yields that are normally obtained in these reactions but also for the atom economy that they exhibit as they allow the use of stoichiometric amounts of the reagents or fair excess of the clickable carbohydrates. In addition, the very mild conditions of the cycloaddition permits a wide functional group tolerance during the derivatization of the dendrimeric scaffolds and only minimal purification steps are normally required, making the preparation of glycodendrimers a routine process in comparison with previous traditional approaches.

The power of the Cu(I)-catalyzed azide–alkyne cycloaddition as an assembling tool in the preparation of glycodendrimers and the potential of these compounds has been nicely illustrated by Wu et al. [59]. These authors reported the synthesis of non-symmetrical dendrimers (**80**) containing both Man-binding units and coumarin fluorescent units as agents with a dual function: recognition and detection. The compounds were prepared using a modular synthetic strategy that allowed the efficient introduction of functional groups in the dendritic structures at different stages of the process in a stepwise manner. Thus, from dendrons with alkyne groups at the focal point (**78**), the coumarin azide derivative was grafted by their click reaction followed by sequential differentiation to the dendritic chain end groups that led to the corresponding alkyne-terminated dendrimer (**79**). This compound was then coupled with unprotected 2-azidoethyl mannopyranoside by means of their cycloaddition reaction, yielding the mannosylated triazole glycodendrimer (**80**) whose inhibitory properties were evaluated by standard hemagglutination assays.

In a different context, Ortega-Muñoz et al. [60] prepared tailored glycosilicas containing Man (**84**) from functionalized silica gel (**81, 82**) by click immobilization of azido and alkyne mannosides and Man-terminated dendrons (**83**), previously obtained also by click cycloadditions. This demon-

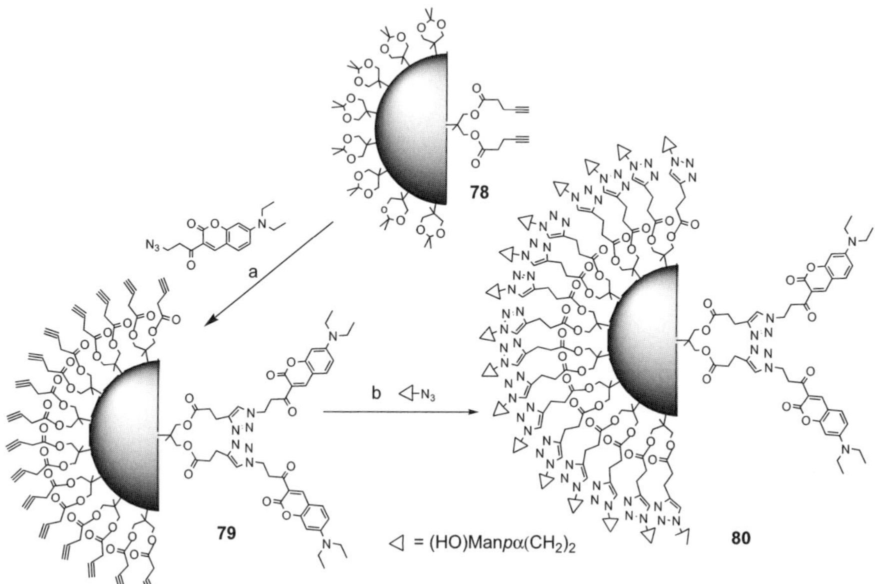

Scheme 15 *a* (i) DOWEX, MeOH, (ii) azido cumarine derivative, $CuSO_4$, Na-ascorbate, THF/H_2O; (iii) $(HC\equiv CCH_2CH_2CO)_2O$, DMAP/pyridine; *b* azido sugar, $CuSO_4$, Na-ascorbate, THF/H_2O [59]

Scheme 16 *a* $(EtO)_3Si(CH_2)_3N_3$, toluene, reflux; *b* $(EtO)_3Si(CH_2)_3NHCOC\equiv CH$, toluene, reflux; *c* $(EtO)_3P\cdot CuI$, DIPEA, MW [60]

strates that the Cu(I)-catalyzed 1,3-dipolar cycloaddition of alkyne and azides works efficiently as a general, versatile, and efficient synthetic methodology for the immobilization of carbohydrates in a solid support. These new materials have shown to be valuable bioselective affinity supports in which the unspecific interactions that make silica unsuitable for affinity chromatography are passivated, turning raw silica into a support with a clear potential in glycomic research. The characterization of such materials, performed by evaluating their binding capacity to a suitable lectin (ConA), demonstrated that the structure of the glyco appendages plays a role in the chromatographic properties of those glyco-silicas that show a better performance in a dendritic architecture than in a monovalent structure. The utility of the prepared glyco-silicas is manifested not only for the purification of a protein mixture but also for the possibility of the fluorescent labeling of the attached ConA in a single step that allows the preservation of the carbohydrate recognition domains of such lectins. The chemical stability of the glyco-silicas and their reproducibility and recycling characteristics as affinity supports are also remarkable characteristics, indicating the viability of silica as a universal and inexpensive scaffold in the immobilization of biomolecules.

3.2.3
Glycopolymers

The copper(I)-catalyzed Huisgen 1,3-dipolar cycloaddition of alkynes and azides is also a particularly attractive route for the preparation of new synthetic glycopolymers featuring well-defined macromolecular architectures and has expanded the synthetic repertory of techniques for accessing these compounds. Glycopolymers have emerged, as have other glycoforms, in the glycobiology field with the main aim of creating synthetic tools for investigating glycopolymer–protein interactions. They have also attracted the attention of other related areas wanting to develop carbohydrate-containing structures aimed at biomedical, pharmaceutical, and medical applications. The chemical strategies for accessing these neoglyconjugates involve the chemical modification of a polymer backbone, the polymerization of sugar-containing monomers, and chemo-enzymatic procedures. Up to now, click chemistry has found applications as a linking tool in the first mentioned strategy but also has the potential to be used in the latter approach as a polymerization reaction, as illustrated below in the synthesis of trehalose click polymers.

For the construction of glycopolymers through click coupling, alkyne polymers have been used more extensively than azide polymers, normally from a safety perspective as this reduces the number of azide groups in the same molecule. In spite of this precaution, the reported synthesis of polysaccharide-based functional materials by Hasegawa et al. [61] and Ortega-Muñoz et al. (submitted for publication) are representative examples of the utility that azido-appended polymers also have (Scheme 17). The first research team described [61] the synthesis of (1,3)-β-D-glucan having Lac functional appendages (**86**) as a potential gene carrier. This

▷ = Glc(OH), Gal(OH), Lac(OH); Fph = Fluorophore

Scheme 17 *a* CuBr$_2$, ascorbic acid, propylamine, rt, Me$_2$SO (EtO)$_3$Si(CH$_2$)$_3$N$_3$, toluene, reflux [61]; *b* (EtO)$_3$P·CuI, DMF, MW (Ortega-Munoz et al. submitted)

compound has been obtained in almost quantitative yield by reaction of 6-azido-6-deoxy-curdlan (**85**) with propagyl lactoside. With the efficient introduction of these bulky appendages, the authors argued the advantage of this chemical approach over chemical/enzymatic strategies previously used. The reported methodology is thus a suitable approach for obtaining polysaccharide derivatives having functionalities at desired positions, avoiding tedious synthetic routes. On the other hand, Ortega-Muñoz et al. performed the synthesis of polystyrene-based glycopolymers (**88**) by reaction of poly[*p*-(azidomethyl)styrene] (**87**) and propagyl *O*-glycosides in a study (submitted for publication) that demonstrated the capability of synthetic multivalent neoglycoconjugates to promote the cell adhesion and stimulation of monocytes. By exploiting the modularity of the click grafting strategy, the fluorescent counterparts of these glycopolymers (**89**) were also prepared by the sequential reaction of the clickable polymer with a stoichiometric defect of an alkynyl dansyl derivative and subsequent reaction with monoalkyne sugar derivatives.

In relation to the use of alkyne polymers, a novel series of heteroglycopolymers (**91**) for lectin binding studies have been prepared by Ladmiral et al. [62] from appropriate sugar azides and poly(methacrylates) bearing terminal alkyne functionalities (**90**). These compounds were obtained by living radical polymerization of (trimethylsilyl)propargyl methacrylate with excellent control over the polymer properties (Scheme 18). After removal of the TMS protecting groups, the click grafting of protected and unpro-

90

91

◁ = (OH)Man$p$αO(CH$_2$)$_3$ ⬭ = (OH)Gal$p$βO(CH$_2$)$_3$

92

93

Scheme 18   *a* (PPh$_3$)$_3$CuBr, Et$_3$N, DMSO [61]; *b* (PPh$_3$)$_3$CuBr, DIPEA [63]

tected carbohydrates via either a C-6 or an anomeric azide onto the polymer backbone was successfully performed, giving rise to a number of Man- and Gal-containing multidentate ligands (**91**) by simultaneously reacting different sugar azides onto the polyalkyne backbone. This co-clicking approach successfully couples the advantages of controlled radical copolymerization with a highly efficient post-functionalization process, leading to a library of multivalent displays that only differ in their epitope density. In addition, fluorescent ligands could be also introduced into the polymeric structure by simply adding a visibly fluorescent coumarin azide tag to the reaction mixture. The binding behavior of the glycopolymers was tested with the lectins ConA and RCA (Ricinus communis agglutinin), showing in the first case that both the clustering rate and the stoichiometry of the polymer–protein conjugates depend on the epitope density of the displays employed. The synthetic strategy proposed is quite general, opening the way for the synthesis of a wide range of precision materials. It appears to be an extremely powerful tool for the synthesis of libraries of materials that differ only in the nature of the sugar moiety presented on a well-defined polymer scaffold.

In a further and elegant step ahead, Malkoch et al. [63] developed an orthogonal approach for the simultaneous and cascade functionalization of polymers by combining copper-catalyzed cycloaddition chemistry of azides and alkynes with other synthetic transformations. These single-step strategies exploit the diverse array of functional groups present in many polymeric systems. In the case of the simultaneous approach, the preparation of the glycopolymer **93** is illustrative of the efficiency of the concept (Scheme 18). The water-soluble (N,N-dimethyl)acrylamide-based terpolymer (**92**) was the selected starting material as it bears multiple copies of different functional groups along the backbone. This characteristic qualifies this polymer to undergo distinct and independent reactions when reacted with different functional reagents. The wide compatibility of the click azide–alkyne coupled with a broad range of functional groups and reaction conditions allows the highly efficient and modular construction of diverse decorated polymers by this strategy, representing a significant advance when compared with traditional multistep approaches.

The versatility of the click conjugation of azides and alkynes has also allowed its application as a polymerization reaction by reacting diazide and dialkyne monomers. Srinivasachari et al. [64] have reported the synthesis of a new family of glycopolymers in which 6,6′-diazido-trehalose (**95**) has been used as a co-monomer in the reaction with a series of dialkyne-oligoamine monomers (**94**) (Scheme 19). The resulting tailored structures (**96**) were designed to contain: (1) a trehalose moiety to promote biocompatibility, water solubility, and stability against aggregation; (2) amide-triazole groups to enhance DNA binding affinity; and (3) an oligoamine unit to facilitate DNA encapsulation, phosphate neutralization, and interactions with cell surfaces. The trehalose click polymers proved to be very effective nucleic acid carriers in both the absence and the presence of serum, particulary in the case of the vector **96** (x = 3) that revealed exceptionally stable polyplex formation, high cellular delivery, and relatively low toxicity. Finally, Marmusse et al. [65] reported the efficient preparation of synthetic well-defined fragments of amylopectin (**98**) incorporating α-(1 → 6)-branch points as potential tools for physicochemical and biochemical studies (Scheme 19). These compounds were obtained by using 6,6′-dipropargylated p-methoxyphenyl maltoside (**97**) as templates for the assembly of linear carbohydrate chains containing an azido group in the reducing terminal anomeric position.

Scheme 19 *a* (i) CuSO$_4$, Na-ascorbate, 1 : 1 *t*-BuOH : H$_2$O; (ii) NaOMe/MeOH [64]; *b* (PPh$_3$)$_3$CuBr, DIPEA, toluene, rt [65]

3.2.4
Cyclodextrin Analogous and Glycosylated Cyclic Oligopeptides

The catalyzed 1,3-dipolar cycloaddition of azides and alkynes has also found applications in developing efficient synthetic strategies for the preparation of carbohydrate-containing macrocycles. Macrocycles are important building blocks in supramolecular chemistry with diverse applications as molecular pores, artificial receptors, and other components in complex supramolecular architectures. Among the extensive array of macrocyclic structures reported, cyclodextrin analogous, macrocyclic carbohydrate/amino acid hybrids, and glycosylated cyclic oligopeptites have encountered in click chemistry a valuable tool for the intermolecular assembly of differently functionalized modules through the formation of multiple heterocyclic triazoles as connecting tethers.

The synthesis of cyclodextrin-like triazoles and macrocyclic carbohydrate/amino acid hybrids has been reported by Bodine et al. [66, 67] and Billing et al. [68], respectively, through a common strategy based in the convergent cyclization of bifunctional azide/alkyne carbohydrate derivatives. Thus, Bodine et al. [66, 67] prepared 2,3,6-tri-*O*-benzyl propargyl-α-D-mannopyranosyl azide and protected di- and trisachharide mannose derivatives consisting of 1,4-linked Man units with an anomeric azide and a 4-propargyl ether at the opposing termini following conventional carbohydrate procedures. The click chemistry reaction conditions employed by the authors allowed the cyclodimerization or cyclotrimerization of these components over unproductive linear oligomerization reactions, giving rise to the corresponding cyclic carbohybrids (structures **99–101**). The exclusive high-yielding formation of the cyclodimer (**100**) is observed in the case of the disaccharide substrate although in the case of the monosaccharide and trisaccharide substrates the corresponding cyclotrimer (**99**) and cyclodimer (**101**) coexist with the formation of other minor cyclic byproducts. The synthetic protocol offers a convenient, highly convergent methodology for the preparation of cyclooligosaccharides. This methodology also has the potential of allowing the modular preparation of functionally diverse cyclodextrin analogs, enlarging and improving other available synthetic methodologies for these compounds. Stepwise oligosaccharide synthesis followed by cycloglycosylation or cyclooligomerization have been previously used with major drawbacks such us long synthetic sequences, low cyclization yields, anomeric mixtures, and, for cyclooligomerization, mixtures of macrocyles of different sizes. These disadvantages are partially diminished by using click chemistry as the ligation reaction. Analogously, Billing et al. [68] performed the preparation of macrocyclic carbohydrate/amino acid

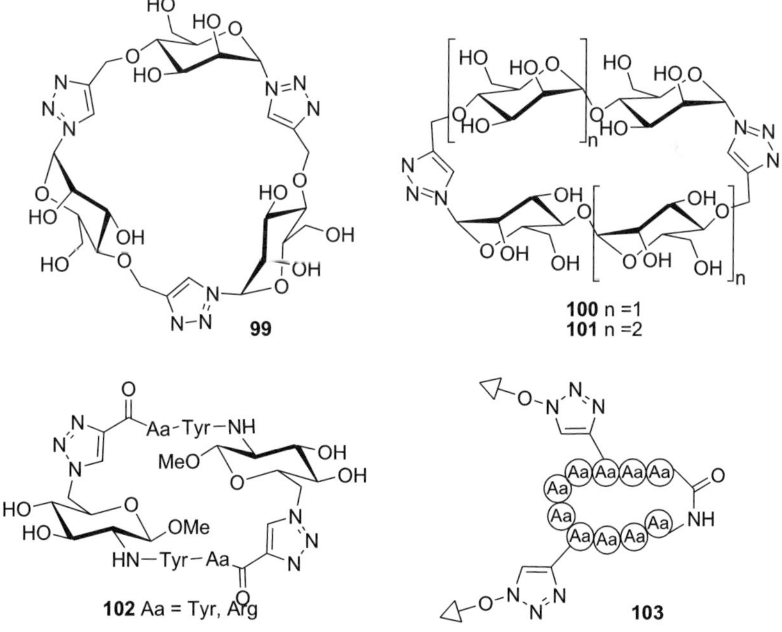

Structure 4 Cyclodextrin analogous and glycosylated cyclic oligopeptides

hybrids (**102**) by the click macrocyclization of bifuntional 2-amino-6-azido-3,4-di-*O*-benzoyl-2,6-dideoxy-β-D-glucopyranoside coupled with *N*-propiolyl dipeptides (propiolyl-Tyr-Tyr-OH and propiolyl- Arg(Mtr)-Tyr-OH) at the amino group of the sugar. The click cyclodimerization of these compounds, using CuI and *N*,*N*-diisopropylethylamine in CH_3CN as the most optimal condition, led to C_2-symmetric rigid macrocycles, which are promising candidates for artificial receptors in water.

In this context, the contribution of Lin et al. [69] should be also mentioned. They described a chemoenzymatic approach to make carbohydrate-modified cyclic peptide antibiotics (**103**). Their chemical strategy was planned through a two-step sequence that began in the preparation of alkyne-containing cyclic peptides by enzymatic macrocyclization and was followed by the click conjugation of these compounds to azido sugars. Using a thioesterase domain from the decapeptide tyrocidine synthetase, 13 head-to-tail cyclized tyrocidine derivatives were obtained with one to three propargylglycines incorporated at positions 3–8. These cyclic peptides were then conjugated to 21 azido sugars via Cu(I)-catalyzed cycloaddition, giving rise to a library of 247 glycopeptides. Antibacterial and hemolytic assays allowed the identification of two components of this library having a sixfold better therapeutic index than that of the natural tyrocidine.

3.2.5
Glycosylated Biomolecules

The outstanding reaction profile of the catalyzed click ligation of azides and alkynes has been rapidly adopted by bioconjugation chemistry, whose aim is the ligation of two or more biomolecules by means of chemoselective processes that are orthogonal to biological species (bioorthogonality). With this purpose, click chemistry has been applied to most bioplatforms with different purposes and with high success.

Concerning the chemoselective ligation of carbohydrates to other bioplatforms, one of the best examples in the click bioconjugation of proteins has been reported by Gupta et al. [70, 71] who employed cowpea mosaic virus (CPMV) as a prototypical protein polyvalent scaffold (Scheme 20). The viral capsids were labeled with azides at surface-exposed lysine residues (**104**) using standard NHS ester chemistry. Simultaneously, azide–alkyne cycloaddition with a chromophoric dialkyne derived from fluorescein (**105**) served to label the azide-containing carbohydrate partner (**106**) with a single dye molecule, allowing convenient monitoring of further manipulations. Either an azide sialyl Lewis X derivative or a Glu-containing glycopolymer, previously obtained by an end-functionalization strategy using an azide-containing initiator in a living polymerization process, were chosen for the preparation of the dye–alkyne derivatives (**106**). Bioconjugation of the click partners was efficiently performed, yielding the glycosylated virus (**107**) only

Sugar = Sialily Lewix X

Sugar = Glycopolymer $R, \triangleright = Glc(OH)$, n = 125± 12

Scheme 20 *a* $CuSO_4$, Na-ascorbate; *b* CuOTf or $[Cu(MeCN)_4](OTf)$, ligand **1** [70, 71]

when the water-soluble sulfonated bathophenantroline ligand **3** was used instead of the most widely utilized tris(triazolyl)amine ligand **1**. The ligand **3** accelerates the reaction and maintains the Cu(I) oxidation state although the reaction has to be performed in an inert atmosphere owing to its oxygen sensitivity. The methodology outperforms other described bioconjugation procedures for polymer attachment to proteins, allowing in the present case the production of molecular structures with a dendrimer-like display with polyvalent lectin-binding properties.

Bouillon et al. [72] have demonstrated that the strategy also works well in the case of oligonucleotides (Scheme 21). These authors describe the mul-

$\triangleleft = \beta\text{-D-Gal-}(OCH_2CH_2)_3$

$\triangleleft = \alpha\text{-D-Man(OH)}$

Scheme 21 *a* (i) $CuSO_4$, Na-ascorbate, MW; (ii) NH_4OH; *b* $CuSO_4$, Na-ascorbate, ligand **3** [72, 73]

tiple click labeling of solid-supported oligonucleotides with anomeric azido galactosyl derivatives. The cyanoethyl-protected dodecathymidine (T_{12}) was prepared on solid support using phosporamidite chemistry and transformed into the alkyne-bearing oligonucleotide (**108**) following well-established chemistry. The cycloaddition of this trivalent alkyne oligonucleotide with the monovalent azide worked efficiently, especially when the reaction was assisted by microwave irradiation, affording the cluster **109**. The reactions could also be performed in solution under similar conditions but the solid-support protocol offers outstanding advantages. The methodology thus appears as a simple and robust strategy for anchoring multiple carbohydrate derivatives to an oligonucleotide. On the other hand, Hassane et al. [73] reported the conjugation of an unprotected α-D-mannosyl derivative (**110**), carrying a spacer arm functionalized with an azide group, to the surface of preformed liposomes (**111**), presenting a synthetic lipid carrying a terminal alkyne function (Scheme 21). This function was selected as anchor because of its chemical "inertness" in vivo and its absence of reactivity with the (bio)molecules that could be entrapped within the liposomes. The relatively long PEG spacer arm of the carbohydrate derivative provides an optimal accessibility of the liposomal Man ligands (**112**) to their receptors. The use of $CuSO_4$/ascorbate as the catalyst system failed in this system; a limitation that was circumventing by the simultaneous use of the sulfonated bathophenantroline ligand **3**. The conjugation conditions were demonstrated to be harmless for the assayed liposomes although the authors pointed out the necessity of alternative copper-free ligation reactions in the light of reported oxidation of vesicles made of unsaturated phospholipids.

Finally, the contribution of Ballell et al. [74] should also be highlighted. This group reported the rational design of a new chemical probe suitable for studying the carbohydrate-binding protein interactions of galectin-1 (Scheme 22). The synthetic strategy was based on a sophisticated and efficient

Scheme 22 *a* hν; *b* $CuSO_4$, Cu wire, tris(triazolyl)amine ligand **3**. *P* photoaffinity label [74]

combined application of photoaffinity labeling, chemoselective ligation, and affinity-tag visualization. The probe was constructed using Lac as the sugar of choice to be recognized by the galectin. The placement of a photoaffinity label, which proved to be a crucial factor, was performed at the 3-OH position of the Gal unit (**113**) based on the crystal structures of the Lac and LacNAc complexes of galectin-1 and galectin-3. The ethylene glycol linker was incorporated to confer water-solubilizing properties to the molecule and to introduce an azide function in its terminus. The purpose of this was to take advantage of the chemoselective click ligation with alkynes, thus allowing indirect attachment of a rhodamine dye (**114**) to the targeted protein. An advantage of this methodology is its versatility, as the approach is amenable to other proteins given that only a photoreactive version of a ligand or inhibitor is needed. Another advantage is its simplicity, as the use of large molecular constructs is minimized as a result of the use of click chemoselective ligation to incorporate the reporter group.

3.2.6
Glycoarrays

The array technology, which pursues displaying biomolecules on a surface with the aim of mapping their biological interactions, has also been a rich beneficiary of click chemistry thanks to its ability to make molecular assemblies with high efficiency and its capacity to preserve the activity of the biomolecules. For these reasons, click chemistry is emerging as a robust immobilization technique. In particular, glycoarrays, in which carbohydrate probes are covalently or non-covalently attached in a spatially addressable manner to a solid substrate, represent valuable devices with wide applications in the emerging field of glycomics, especially for the large-scale high-throughput analysis of sugar–receptor interactions (microarrays) [75–78]. Covalent linkages are normally preferred over non-covalent assemblies as they produce a defined and uniform orientation of the carbohydrates probes that enhances their accessibility towards binding. Click chemistry has been exploited in both approaches, expressing its full potential as a chemoselective and bioorthogonal reaction.

The first contribution in the development of sugar arrays based on click chemistry was reported by Fazio et al. [79] who described the use of Cu(I)-catalyzed triazole formation for the synthesis and in-situ non-covalent attachment of saccharides to the surface of microtiter plates through lipid tethers (Scheme 23). The alkyne clickable counterpart was prepared from a C_{14} aliphatic chain that was conjugated with a series of 2-azidoethyl β-D-glycosides on micromolar scale by using an experimental protocol that avoided purification procedures. In an effort to extend this approach to the covalent attachment of saccharides to the microtiter plate surface, Bryan et al. [80] reported a strategy based on bifunctional linkers (**115 and 116**).

$\triangleleft$ = (OH)Sugar-O(CH$_2$)$_2^-$

115

116

Scheme 23 *a* C$_{14}$H$_{29}$HNC(O)C ≡ CH; *b* CuI, DIPEA; *c* **115**, DIEA; *d* **116**, DIEA; *e* DTT [79–81]

These compounds allowed the previous orthogonal attachment to a functionalized microtiter plate surface while leaving a terminal alkyne group for conjugation to the saccharide counterpart (Scheme 23). For covalent attachment of the linkers, both NHS- and amine-coated surfaces were used to be coupled, respectively, with the complementary amine (**115**) and isothiocyanate linkers (**116**). Remarkably, these linkers were designed to contain a cleavage site by the introduction of a disulfide bond, which allows characterization and quantitative analysis of the array. The click conjugation of the resulting alkyne-functionalized plates with azide-containing sugars allows the efficient preparation of the glycoarrays. Additionally, the methodology can be exported to other surfaces. In both of these reports, binding studies have proven that these microarrays are functional in biological screening and, therefore, applicable in ELISA-type formats. The non-covalent approach has been also exploited by Huang et al. [81] who demonstrate the validity of quantum dots attached to ConA as fluorescent label for the detection of carbohydrate–protein interactions.

Clicked carbohydrate self-assembled monolayers (SAMs) on gold, obtained by an indirect coupling on a preformed template, has been reported by Zhang et al. [82] (Scheme 24). Carbohydrate SAMs are of interest as they offer extensive control over the presenting pattern, density, and orientation of the ligands. They also allow the use of surface-based real-time and label-free analysis methods such as surface plasmon resonance (SPR) and quartz crystal microbalance (QCM). The fabrication was based on preformed SAM templates incorporated with alkyne terminal groups, which could further anchor the azido sugars to form well-packed, stable and rigid sugar SAMs. The rationale behind the structural motifs selected for the synthetic design was based on the incorporation of a long chain alkyl (C_{10}) disulfide linker NDDA with a terminal alkyne group in the preformed SAMs. An amide linkage was selected to connect the alkyl chain and the alkyne group as it provides activated alkyne functions to be readily coupled to the azido sugars. In addition, a tris(ethylene oxide) glycol was incorporated into the azido sugars to eliminate non-specific adsorptions. The click coupling of these partners occurs with high efficiency and the resulting carbohydrate SAM sensors kept the specificity for the corresponding lectins. This indirect coupling strategy offers advantages over the direct coupling through a thiol anchor as it significantly reduces the synthetic labor needed for preparation of carbohydrate thiolates and promises rapid and flexible construction of arrays of carbohydrate SAMs. Moreover, the efficient combination of these arrays with QCM or SPR detection techniques established the basis for the development of a potent platform for high-throughput characterization of carbohydrate–protein interactions.

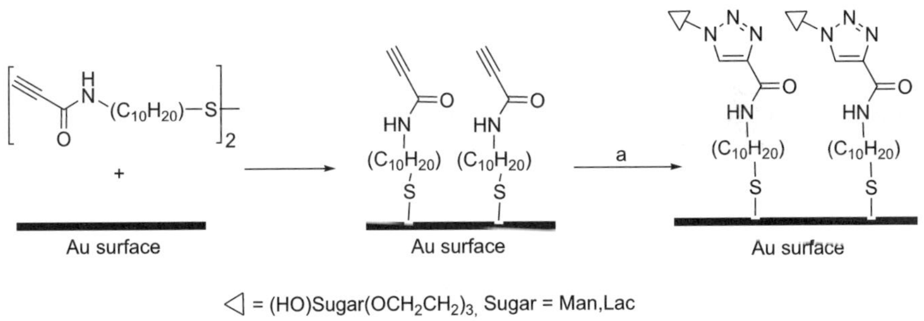

Scheme 24 *a* Azide–sugar, $CuSO_4$, Na-ascorbate [82]

Glycoarrays constructed on glass slides have been reported by Sun et al. [83], who designed an appropriate bifunctional PEG linker and used sequential Diels–Alder and azide–alkyne cycloaddition reactions for the immobilization of carbohydrates and proteins, following a strategy with resemblances to that developed by Bryan et al. [80] (Scheme 25). An α,ω-poly(ethylene glycol) (PEG) linker carrying alkyne and cyclodiene terminal groups was synthesized to facilitate surface bioconjugation. The immobiliza-

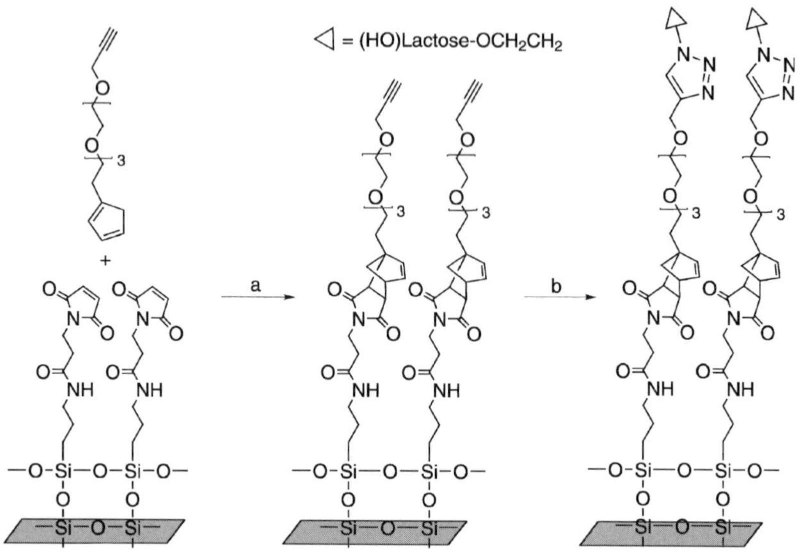

◁ = (HO)Lactose-OCH₂CH₂

Scheme 25 *a* H₂O, *tert*-BuOH, rt; *b* Azidoethyl lactoside, CuSO₄ · 5H₂O, TECP, tris(triazol)amine, PBS-*tert*-BuOH, 4 °C [83]

tion onto an *N*-(ε-maleimidocaproyl) (EMC)-functionalized glass slide via an aqueous Diels–Alder reaction provided an alkyne-terminated PEGylated surface for the ulterior conjugation of azide-containing biomolecules, including carbohydrates. The click coupling proceeded efficiently at low temperature and in aqueous solvent. The developed methodology shows that azide, alkyne, cyclopentadiene, and EMC groups can coexist under bioconjugation conditions. This fact will probably inspire new design schemes directed toward the fabrication of microarrays. These results and the previous advances described herein show the great potential of click chemistry in material chemistry.

4
Intramolecular Alkyne–Azide 1,3-Dipolar Cycloadditions

The simultaneous presence of an azido and an alkyne group on a scaffold is a prerequisite for performing the 1,3-dipolar cycloaddition of such functions in its intramolecular version. In the case of carbohydrates, the sequential introduction of such functions is always feasible but lengthy and laborious synthetic strategies, normally based on protection and deprotection of the hydroxyl sugar groups, are required. For this reason, the intramolecular alkyne–azide of 1,3-dipolar cycloadditions have found more limited application than the intermolecular reaction between these functions. However, this approach is irreplaceable for the oriented preparation of molecular architectures in which the 1,2,3-triazole ring is fused or integrated with other structural components. This comprehensive section summarizes the scarce contributions made in the period that it covers. For a more extensive view of the fields in which these intramolecular cycloadditions have found applications, previous reviews should be consulted [4].

Intramolecular uncatalyzed and catalyzed 1,3-dipolar cycloaddition of azide and alkynes has been envisaged as a valuable strategy for accessing azoles fused to furanoses and pyranoses compounds. These sugar derivatives are of interest as some of them have been identified as good, selective, and potent glycosidase and glycosyl transferase inhibitors. Although the synthesis of these azasugars can also be performed using alternative strategies, the preparative methods based on cycloaddition reactions normally compete in efficiency. Dolhem et al. [84] reported an expedient synthesis of 1,2,3-triazoles-fused bicyclic compounds (**121**) from naturally occurring aldohexoses (Glc, Man, Gal, and 2-deoxy-Glc **118**) and aldopentoses (**117**) via glyco-ynitols (**119**) (Scheme 26). The alkyne function was first introduced starting from tritylated aldoses by exploiting the reactivity of the anomeric position through a one-carbon chain-elongation reaction without creation of a new stereogenic center. The subsequent activation and azido substitution of the primary hydroxyl group allowed direct access to the correspond-

Scheme 26 *a* Ph₃PCHBr₂, Zn, 1,4-dioxane, reflux; *b* nBuLi, THF, −70 °C; *c* (i) BnBr, TBAI, NaH, DMF; (ii) HCl conc., CHCl₃/MeOH; *d* TsCl, Py; *e* NaN₃, DMF, 80 °C; *f* toluene, reflux; *g* NaN₃, DMF, 120 °C; *h* (i) methyl-(triphenylphosphoranylidene)-acetate, toluene, 80 °C; (ii) TsCl or MsCl, DMAP, pyridine; *i* DBU, 80 °C; *j* air oxidation. *P* protective group [84, 85]

ing bycyclic triazoles (**121**). This was possible because the introduction of the azido group and the thermal cyclization reaction could be performed in a one-pot procedure without isolation of the intermediate **120**. The chiral compounds **121** possessing a controlled and predefined stereochemistry were obtained in satisfactory overall yields by this versatile strategy. On the other hand, Flessner et al. [85] reported a related strategy in which adequately protected D-Man and L-Fuc (**118**) were elongated by two carbons at the anomeric position through a Horner–Wittig–Emmons reaction that led to α,β-unsaturated esters (**122**) (Scheme 26). The manipulation of the terminal hydroxyl group allowed its conversion to a tosylate or mesylate leaving group that in a three-step one-pot reaction (nucleophilic substitution, thermal cycloaddition, aromatization by air oxidation) afforded the corresponding 1,2,3-triazole-containing compounds (**124**). Unfortunately, in this case the reported yields were poor.

Hotha et al. [86] and Ray et al. [87] have reported a strategy for the synthesis of 1,2,3-triazole-fused tetracylic compounds (Scheme 27). The methodology was developed using the well-known 1,2-isopropylidene furanose skeleton (**125**) as scaffold for the placement of an azido function at the primary position of the sugar and an alkyne at the C-3 position. While Hotha et al. [86] used the *O*-propargyl derivative, Ray et al. [87] placed at this position more elaborated appendages containing ester, peptides, or aromatic-containing tethers between the terminal alkyne and the sugar ring. In the first case, the 1,3-dipolar cycloaddition were carried out under reagent-free thermal conditions leading to the 1,2,3-triazoles fused to furanosyl-derived seven-membered rings (**127**, R = i) in a excellent yield. These satisfactory results allowed the authors to extend the methodology to other azido–alkyne substrates comprising glucose-, allo-, xylo-, ribo-, and arabino-derivatives. Conversely, the results reported by Ray et al. [87] showed that although the strategy allowed access to 12- to 17-membered monomeric triazolophanes (**127**, R = ii, iii) under Cu(I)-catalyzed cycloadditions these compounds are only obtained in low yields, in contrast with the high yields normally obtained in click chemistry. In both cases the monocyclization was the only

Scheme 27 *a* Toluene, 100 °C for R = i [85]; *b* CuSO$_4$, Na-ascorbate or CuI, DIPEA for R = ii and iii [87]

detected cycloaddition reaction. Cyclodimerization of the substrates was not observed in either of the two cases.

In this context, it is worth noting the contribution of Marco-Contelles et al. [88] who described the synthesis of fused-azole piperidinoses, annulated onto furanose templates (Scheme 28). The basis of this strategy is the introduction of an N-triazole at C-3 (**129**) in a hexofuranose starting azido material **128** by intramolecular thermal cycloaddition with mono- and disubstituted acetylenes. Subsequent halogenation of the primary hydroxyl group gave radical precursors that were submitted to a free-radical cyclization process by reaction with tributyltin hydride or tris(trimethylsilyl)silane. The 5-*exo-trig* or the 6-*exo-trig* cyclization of a carbon-centered (at C-5 or C-6) radical species onto the corresponding azaannulated sugars (**130**) occurs in yields that are dependent on the strain associated with the resulting ring systems and on the presence of activating groups in the heterocyclic ring.

R = H, COOR^1; R^1 = Me, Et

Scheme 28 *a* (i) RC≡COOR^1, toluene, 110 °C; (ii) AcOH, H$_2$O; (iii) TsCl, pyridine; (iv) Nal, DMF; *b* (TMS)$_3$SiH, Bu$_3$SnCl, AIBN or (TMS)$_3$SiH, AIBN; *P* protective group [88]

Structure 5 Cyclodextrin analogous

Finally, the synthesis of cyclodextrin analogs (**131**) has been reported by Hoffman et al. [89] through the intramolecular thermal cyclization reaction of the azidoalkyne derivative of a protected maltohexaoside. In particular, the azidoetyl and the propargyl group were placed, respectively, at the anomeric position and at the C-4 position of the non-reducing unit of the oligosaccharide. This strategy elegantly complements the click methodology developed by Bodine et al. [66, 67] for the preparation of cyclodextrin-like triazoles by the cyclodimerization or cyclotrimerization of the simplest azide–alkyne mono-, di-, and trisaccharide derivatives (structures **99–101**). However, in this case, the thermal conditions used did not allow control over the regioselectivity and a mixture of the two possible regioisomers was obtained.

5
Concluding Remarks

The well-established and recognized synthetic utility of 1,3-dipolar cycloadditions for the coupling of azides with alkynes has experienced a flourishing interest in the present century owing to the discovery of the catalytic and regiocontrol effect of Cu(I) in this reaction. The outstanding characteristics of this catalyzed cycloaddition, which allows its inclusion in the click chemistry concept, means that it has been rapidly embraced by a diverse spectrum of fields. The results described here show that, for the particular case of carbohydrate chemistry, this reaction has gained in popularity because of its impressive scope, reliability, and robustness. These qualities enable the use of a wide variety of mild reaction conditions, which is highly appreciated especially in glycobiology and bioconjugation. The intrinsic chemoselectivity and bioorthogonality of the Huisgen 1,3-dipolar cycloaddition of azides and alkynes, together with the favorable physicochemical properties of the 1,2,3-triazole heterocycle, have made this transformation an invaluable linking reaction for the construction of natural and synthetic complex architectures in which carbohydrates are present. The reaction has shown its full capabilities in the preparation of sophisticated structures and devices (such as glyco- and microarrays) that require the simultaneous grafting of carbohydrate motifs to a scaffold in an effort to mimic Nature, which on numerous occasions uses multivalent carbohydrate displays.

The relatively easy preparation of tailored azide- and alkyne-functionalized carbohydrate derivatives, as well as the efficiency of the click reaction with complementary counterparts, determine that this reaction could be considered at the present as a valuable synthetic tool. In spite of the fact that the reported examples summarized here reveal the potential of this cycloaddition reaction, their applications in the carbohydrate field are just beginning to be explored. A fruitful future can be anticipated, limited only by the level of creativity of the research community.

References

1. Padwa A, Pearson WH (2003) Synthetic applications of 1,3-dipolar cycloaddition chemistry toward heterocycles and natural products. Wiley, Chichester, UK
2. Gallos JK, Koumbis AE (2003) Curr Org Chem 7:397
3. Koumbis AE, Gallos JK (2003) Curr Org Chem 7:585
4. Koumbis AE, Gallos JK (2003) Curr Org Chem 7:771
5. Rostovtsev VV, Green LG, Fokin VV, Sharpless KB (2002) Angew Chem Int Ed Engl 41:2596
6. Tornoe CW, Christensen C, Meldal M (2002) J Org Chem 67:3057
7. Kolb HC, Sharpless KB (2003) Drug Discov Today 8:1128
8. Wang Q, Chittaboina S, Barnhill HN (2005) Lett Org Chem 2:293
9. Mock WL, Irra TA, Wepsiec JP, Adhya M (1989) J Org Chem 54:5302
10. Chen J, Rebek J (2002) Org Lett 4:327
11. Howell SJ, Spencer N, Philp D (2001) Tetrahedron 57:4945
12. Kolb HC, Finn MG, Sharpless KB (2001) Angew Chem Int Ed 40:2005
13. Himo F, Lovell T, Hilgraf R, Rostovtsev VV, Noodleman L, Sharpless KB, Fokin VV (2005) J Am Chem Soc 127:210
14. Bock VD, Hiemstra H, van Maarseveen JH (2006) Eur J Org Chem 51
15. Wang Q, Chan TR, Hilgraf R, Fokin VV, Sharpless KB, Finn MG (2003) J Am Chem Soc 125:3192
16. Speers AE, Adam GC, Cravatt BF (2003) J Am Chem Soc 125:4686
17. Chan TR, Hilgraf R, Sharpless KB, Fokin VV (2004) Org Lett 6:2853
18. Meng JC, Fokin VV, Finn MG (2005) Tetrahedron Lett 46:4543
19. Lewis WG, Magallon FG, Fokin VV, Finn MG (2004) J Am Chem Soc 126:9152
20. Perez-Balderas F, Ortega-Muñoz M, Morales-Sanfrutos J, Hernandez-Mateo F, Calvo-Flores FG, Calvo-Asin JA, Isac-Garcia J, Santoyo-Gonzalez F (2003) Org Lett 5:1951
21. Girard C, Onen E, Aufort M, Beauviere S, Samson E, Herscovici J (2006) Org Lett 8:1689
22. Orgueira HA, Fokas D, Isome Y, Chan PCM, Baldino CM (2005) Tetrahedron Lett 46:2911
23. Pachon LD, van Maarseveen JH, Rothenberg G (2005) Adv Syn Cat 347:811
24. Chittaboina S, Xie F, Wang Q (2005) Tetrahedron Lett 46:2331
25. Lober S, Rodriguez-Loaiza P, Gmeiner P (2003) Org Lett 5:1753
26. Lober S, Gmeiner P (2004) Tetrahedron 60:8699
27. Collman JP, Devaraj NK, Chidsey CED (2004) Langmuir 20:1051
28. Kuijpers BHM, Groothuys S, Keereweer AR, Quaedflieg PJLM, Blaauw RH, van-Delft FL, Rutjes FPJT (2004) Org Lett 6:3123
29. Dondoni A, Giovannini PP, Massi A (2004) Org Lett 6:2929
30. Giguere D, Patnam R, Bellefleur MA, St-Pierre C, Sato S, Roy R (2006) Chem Commun 2379
31. Macmillan D, Blanc J (2006) Org Biomol Chem 4:2847
32. Ermolat'ev D, Dehaen W, Van der Eycken E (2004) QSAR Comb Sci 23:915
33. Hotha S, Kashyap S (2006) J Org Chem 71:364
34. Arosio D, Bertoli M, Manzoni L, Scolastico C (2006) Tetrahedron Lett 47:3697
35. Salameh BA, Leffler H, Nilsson UJ (2005) Bioorg Med Chem Lett 15:3344
36. Tejler J, Tullberg E, Frejd T, Leffler H, Nilsson UJ (2006) Carbohydr Res 341:1353
37. Griffith BR, Langenhan JM, Thorson JS (2005) Curr Opin Biotech 16:622
38. Miner PL, Wagner TR, Norris P (2005) Heterocycles 65:1035
39. Perez-Balderas F (2005) PhD Thesis. University of Granada

40. Casas-Solvas JM, Vargas-Berenguel A, Capitan-Vallvey LF, Santoyo-Gonzalez F (2004) Org Lett 21:3687
41. Zhu XM, Schmidt RR (2004) J Org Chem 69:1081
42. Hager C, Miethchen R, Reinke H (2000) J Fluor Chem 104:135
43. IJsselstijn M, Cintrat JC (2006) Tetrahedron 62:3837
44. Lee BY, Park SR, Jeon HB, Kim KS (2006) Tetrahedron Lett 47:5105
45. Guezguez R, Bougrin K, El Akri K, Benhida R (2006) Tetrahedron Lett 47:4807
46. Deng S, Gangadharmath U, Chang CWT (2006) J Org Chem 71:5179
47. Roy R (2003) Trends Glycosci Glyc 15:291
48. Jayaraman N, Nepogodiev SA, Stoddart JF (1997) Chem Eur J 3:1193
49. Bezouska K (2002) Rev Mol Biotechnol 90:269
50. Turnbull WB, Stoddart JF (2002) Rev Mol Biotechnol 90:231
51. Roy R, Baek M-G (2002) Rev Mol Biotech 90:291
52. Ladmiral V, Melia E, Haddleton DM (2004) Eur Polym J 40:431
53. Nishikawa M, Kawakami S, Yamashita F, Hashida M (2003) Methods Enzy 373:384
54. Thomas JR, Liu X, Hergenrother PJ (2005) J Am Chem Soc 127:12434
55. De Oliveira RN, Sinou D, Srivastava RM (2006) Synthesis 467
56. Malkoch M, Schleicher K, Drockenmuller E, Hawker CJ, Russell TP, Wu P, Fokin VV (2005) Macromolecules 38:3663
57. Joosten JAF, Tholen NTH, El Maate FA, Brouwer AJ, van Esse GW, Rijkers DTS, Liskamp RMJ, Pieters RJ (2005) Eur J Org Chem 3182
58. Fernandez-Megia E, Correa J, Rodriguez-Meizoso I, Riguera R (2006) Macromolecules 39:2113
59. Wu P, Malkoch M, Hunt JN, Vestberg R, Kaltgrad E, Finn MG, Fokin VV, Sharpless KB, Hawker CJ (2005) Chem Commun 5775
60. Ortega-Munoz M, Lopez-Jaramillo J, Hernandez-Mateo F, Santoyo-Gonzalez F (2006) Adv Synth Cat 348:2410
61. Hasegawa T, Umeda M, Numata M, Li C, Bae AH, Fujisawa T, Haraguchi S, Sakurai K, Shinkai S (2005) Carbohydr Res 341:35
62. Ladmiral V, Mantovani G, Clarkson GJ, Cauet S, Irwin JL, Haddleton DM (2006) J Am Chem Soc 128:4823
63. Malkoch M, Thibault RJ, Drockenmuller E, Messerschmidt M, Voit B, Russell TP, Hawker CJ (2005) J Am Chem Soc 127:14942
64. Srinivasachari S, Liu Y, Zhang G, Prevette L, Reineke TM (2006) J Am Chem Soc 128:8176
65. Marmuse L, Nepogodiev SA, Field RA (2005) Org Biomol Chem 3:2225
66. Bodine KD, Gin DY, Gin MS (2004) J Am Chem Soc 126:1638
67. Bodine KD, Gin DY, Gin MS (2005) Org Lett 7:4479
68. Billing JF, Nilsson UJ (2005) J Org Chem 70:4847
69. Lin H, Walsh CT (2004) J Am Chem Soc 126:13998
70. Gupta SS, Kuzelka J, Singh P, Lewis WG, Manchester M, Finn MG (2005) Bioconjugate Chem 16:1572
71. Gupta SS, Raja KS, Kaltgrad E, Strable E, Finn MG (2005) Chem Commun 4315
72. Bouillon C, Meyer A, Vidal S, Jochum A, Chevolot Y, Cloarec JP, Praly JP, Vasseur JJ, Morvan F (2006) J Org Chem 71:4700
73. Hassane FS, Frisch B, Schuber F (2006) Bioconjugate Chem 17:849
74. Ballell L, Alink KJ, Slijper M, Versluis C, Liskamp RMJ, Pieters RJ (2005) ChemBioChem 6:291
75. Wang D (2003) Proteomics 3:2167
76. Ortiz MC, Garcia Fernandez JM (2002) ChemBioChem 3:819

77. Love KR, Seeberger PH (2002) Angew Chem Int Ed 41:3583
78. Feizi T, Fazio F, Chai W, Wong CH (2003) Curr Opin Struct Biol 13:637
79. Fazio F, Bryan MC, Blixt O, Paulson JC, Wong CH (2002) J Am Chem Soc 124:14397
80. Bryan MC, Fazio F, Lee HK, Huang CY, Chang A, Best MD, Calarese DA, Blixt O, Paulson JC, Burton D, Wilson IA, Wong CH (2004) J Am Chem Soc 126:8640
81. Huang GL, Liu TC, Liu MX, Mei XY (2005) Anal Biochem 340:52
82. Zhang Y, Luo S, Tang Y, Yu L, Hou KY, Cheng JP, Zeng X, Wang PG (2006) Anal Chem 78:2001
83. Sun XL, Stabler CL, Cazalis CS, Chaikof EL (2006) Bioconjugate Chem 17:52
84. Dolhem F, Al Tahli F, Lievre C, Demailly G (2005) Eur J Org Chem 5019
85. Flessner T, Wong CH (2000) Tetrahedron Lett 41:7805
86. Hotha S, Anegundi RI, Natu AA (2005) Tetrahedron Lett 46:4585
87. Ray A, Manoj K, Bhadbhade MM, Mukhopadhyay R, Bhattacharjya A (2006) Tetrahedron Lett 47:2775
88. Marco-Contelles J, Rodriguez-Fernadez M (2001) J Org Chem 66:3717
89. Hoffmann B, Bernet B, Vasella A (2002) Helv Chim Acta 85:265

Top Heterocycl Chem (2007) 7: 179–248
DOI 10.1007/7081_2007_056
© Springer-Verlag Berlin Heidelberg
Published online: 19 May 2007

Synthesis of Glycoporphyrins

José A. S. Cavaleiro (✉) · João P. C. Tomé · Maria A. F. Faustino

Department of Chemistry, University of Aveiro, 3810-193 Aveiro, Portugal
jcavaleiro@dq.ua.pt

Abstract Porphyrins and carbohydrates are two groups of natural compounds that are vital to our life. However, only a few examples of glycoporphyrins have been isolated from a natural source. The attachment of saccharide units to porphyrin macrocycles gives rise to derivatives which might be of great significance for certain medicinal and other applications. Synthetic methodologies published during the last 20 years leading to such glycoporphyrin derivatives have been reviewed. Glycoporphyrin derivatives can be obtained by chemical transformation of natural porphyrins or by synthesis from adequate precursors. Such precursors can be porphyrins adequately substituted for the glycoderivatization to take place or can be pyrrolic and glycoaldehyde species, which give rise to target glycoporphyrins by acid catalyzed condensations. Glycochlorins, another group of derivatives of great biological significance, have been synthesized in recent years. The recently reported 1,3-dipolar cycloaddition reactions of porphyrins with glyconitrones and glyco-azomethine ylides have provided a new avenue for the synthesis of such compounds.

Keywords Carbohydrates · Glycochlorins · Glycoporphyrins · PDT · Porphyrins

Abbreviations

BOP	benzotriazol-1-yloxy*tris*(dimethylamino)phosphonium hexafluorophosphate
BuOH	butanol
DMAP	4-*N*,*N*-dimethylaminopyridine
EDC	1-ethyl-3-(3-*N*,*N*-dimethylaminopropyl)carbodiimide
NMP	*N*-methylpyrrolidinone
NMR	nuclear magnetic resonance
OEP	octaethylporphyrin
Pbf	2,2,4,6,7-pentamethyldihydrobenzofuran-5-sulfonyl
PDT	photodynamic therapy
PPIX	protoporphyrin-IX
PS	photosensitizer
RGD	tripeptide arginine-glycine-aspartate
SOD	superoxide dismutase
TBDMSOH	*tert*-butyldimethylsilanol
TPP	*meso*-tetraphenylporphyrin
TPPS$_4$	*meso*-tetraphenylporphyrin tetrasulfonic acid

1
Introduction

Porphyrin and sugar derivatives constitute two groups of natural compounds which play key roles in many vital functions of life. Over the last few centuries Man has begun to investigate the natural occurrence of such compounds. Walter Raleigh, in his "History of the World: Preface (1614)", stated "Man can not give a true reason for the grass under his feet why it should be green rather than red or any other color". Almost two centuries later Verdeil (1844) was able to convert chlorophyll into a red pigment and this prompted him to suggest a structural relationship between chlorophyll and heme, the blood red pigment. This was confirmed later (1879) by Hoppe-Seyler. Intensive studies were being carried out by several groups at the time. Willstäter was awarded the Nobel prize for his chlorophyll work but it was Hans Fischer's fantastic work who definitively set up the tetrapyrrolic macrocyclic feature of such compounds with the synthesis, in 1929, of hemin, a Fe(III) complex of protoporphyrin-IX, the porphyrin present in mammals in the form of Fe(II) and Fe(III) complexes [1]. This was a landmark which contributed to the Nobel prize award to Fischer in 1930. During the following decades several research groups devoted all their efforts to the establishment of new, efficient synthetic methodologies and to the total structural identification and mode of action of such vital compounds. It is known that certain vital functions (e.g., respiration, xenobiotics detoxification and enzymatic actions) are played by porphyrin iron complexes [2–4]. Also the chlorophylls, mainly the *a* and *b*, are the "catalytical forces" in the photosynthetic process, forming a cycle linking animals and plants, in which carbohydrates also play a key role [5]. Among those who dedicated their lives to the study of carbo-hydrates, Emil Fischer deserves special mention. It was Fischer who in 1888 undertook to determine the total structure of glucose and so the configuration of each center. This was a true masterpiece of organic chemistry and Fischer was awarded with the Nobel prize in 1902.

It would be difficult to overestimate the importance of carbohydrates to human beings; they can be oxidized to yield energy and so they can act as energy-storage molecules, and polymeric derivatives are found in cell walls and protective coatings of many organisms. Carbohydrates attached to cell membranes are involved in cell recognition and in cell-to-cell communication. Sugar derivatives are present in a number of biological entities and also they incorporate the structures of several drugs. The search for other applications in the medical field is an exciting area of research for quite of number of research groups [6].

Aiming at medicinal applications, porphyrins have been intensively studied since the mid-1970s as photosensitizers (PS) in photodynamic therapy (PDT) of oncological, cardiovascular, dermatological, ophthalmic and infectious diseases [7–11], and in clinical trials as photomarkers in cancer diagnosis by fluorescence [12–14] and NMR imaging applications [15–18].

New and highly efficient synthetic routes have generated many porphyrin-like compounds with unique characteristics for their uses in several other applications like oxidative catalysis [19, 20] and as biomimetic model systems of the primary processes of photosynthesis [21, 22]. Presently, the interest includes also the supramolecular units, including molecular recognition in chemical receptors and sensors [23–25], use as light-harvesting devices [26–29], and as materials for advanced technologies, mainly in nanosciences [30, 31].

PDT is certainly the most promising application. Its concept is based on the PS concentration in target cells and, upon subsequent irradiation with visible light in the presence of oxygen, with specific destruction of the target cells/tissues. There is a general agreement that the main mechanism by which the cell destruction occurs involves mainly the disruption of the cellular, mitochondrial, or nuclear membranes by cytotoxic agents such as singlet oxygen generated by the PS action [7, 32–35].

Photofrin®, the first PDT formulation, has been approved in several countries (USA, Canada, Netherlands, France, Germany, Japan and others) for treatment of bladder, esophageal, gastric, cervical and lung cancers and several other diseases [35, 36]. Although Photofrin® has been effectively used against a number of malignancies, it is not an ideal photosensitizer; it is not a well-defined single agent, it has a weak absorbance in the red region of the visible spectrum and it induces long-lasting skin photosensitivity. Mainly because of this, several other compounds were synthesized and are under clinical trials [7, 36].

PDT treatments of various skin disorders (psoriatic plaques, impetigo, atopic dermatitis), localized microbial infections (periodontal diseases, acne vulgaris) and viruses are also taking place [37–39]. Also the Visudyne® formulation being used against age-related macular degeneration gives another significant application for humans [40–42].

Despite significant efforts to find an ideal photosensitizer for a specific PDT purpose, there are few PDT agents, thus there remains a need to further develop the chemistry and the understanding of the structure-activity relationships of this class of compounds.

The porphyrin skeleton is essentially hydrophobic; this feature may be an important factor affecting the preferential accumulation in cellular hydrophobic loci since such molecules must be able to get into cells by crossing lipid membranes. That brings insolubility in physiological fluids; to overcome this fact adequate formulations have to be used, such as incorporation into liposomes, biopolymers and cyclodextrins [43–45].

However, the design and synthesis of water-soluble molecules, avoiding the use of a delivery system in biological fluids, are present targets for many pharmaceutical chemists. The structural modification of the porphyrinic core can provide the required peripheral substituents (like the carbohydrate residues) used to control water solubility biodistribution, pharmacokinetics and affinity/selectivity for cancer cells [10, 46–49].

The conjugation of carbohydrates to porphyrin derivatives will bring increased water solubility and ideal hydrophilic/hydrophobic ratios for such glycoporphyrins. Porphyrins with sugar moieties have not only better solubility in aqueous solutions, which is important in the biodistribution, but also possible specific membrane interactions that can affect the plasmatic life time of the drug [50].

Porphyrins appended with a variety of carbohydrate moieties have been recently synthesized and their biological evaluation revealed that they are efficient photosensitizer agents in PDT [46, 48–51]. Glycoporphyrins efficacy as antibiotic and antiviral agents is another area under investigation [52–54].

Other promising applications for porphyrin glycoconjugates are centered on the catalysis and molecular recognition areas.

Over the last three decades it has become known that certain synthetic metalloporphyrins can also act as oxidative catalysts thus mimicking the action of cytochrome P_{450}. Transformations concerning substrates epoxidation and hydroxylation have been widely studied [19, 55, 56].

Glyco-metalloporphyrin conjugates, being chiral species with significant solubility in several solvents, might improve the development of oxidative processes, particularly in asymmetric transformations, especially relevant when the desired compound has pharmaceutical or agrochemical interest. Glyco-metalloporphyrins 1–5 substituted in one *ortho* or *meta* position of each phenyl group by mono- or disaccharide moieties have been used to catalyze asymmetric oxidation reactions (Fig. 1) [57–61]. The introduction

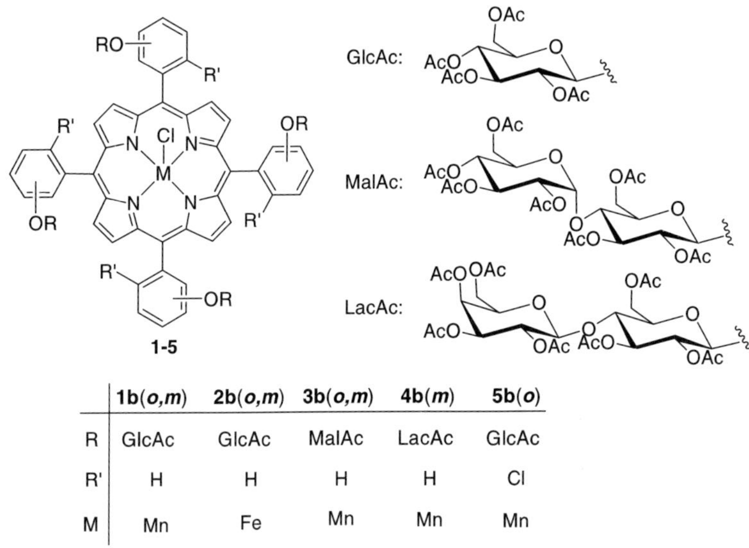

	1b(o,m)	2b(o,m)	3b(o,m)	4b(m)	5b(o)
R	GlcAc	GlcAc	MalAc	LacAc	GlcAc
R'	H	H	H	H	Cl
M	Mn	Fe	Mn	Mn	Mn

Fig. 1 Glycosylated metalloporphyrins which have been used in catalytic systems

of the chlorine atom, an electron-withdrawing substituent, in the *ortho*-aryl position of the glycosylated metalloporphyrin **5** protects the catalyst against oxidative degradation [62].

Studies already carried out have revealed that the catalytic turnover numbers of acetylglycosylated metalloporphyrins were doubled when compared with those of the same metalloporphyrins without sugar groups. Certainly the improvement of such catalytic methods requires the development of new synthetic methodologies.

Molecular recognition is one of the most interesting topics in modern chemistry and several host systems have been described as receptors for the recognition of biologically important substrates. Numerous systems are based on the combination of different covalent and noncovalent binding modes [63–65]. Several porphyrins have already been tested. If the compound tends to aggregate in aqueous media a linkage of saccharide-boronic acid complexes will bring increased hydrophilic properties. Shinkai et al. used the boronic-acid group appended to protoporphyrin-IX derivative **6** to detect sugar molecules in aqueous systems. In this case, the porphyrin acts as an aggregative chromophore and the two boronic acid residues as the sugar interfaces (Fig. 2) [66].

The same research group has also exploited the sugar-boronic acid interactions as shown with **7**, to create oriented supramolecular assemblies [67–75].

This methodology was extended to other energy-transfer systems giving results pointing out the importance of the sugar unit connected to the electron-donor and electron-acceptor to control the electron-transfer efficiency [76].

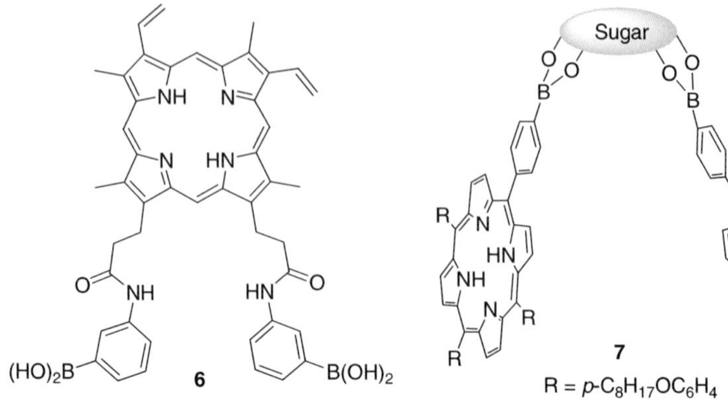

Fig. 2 Boronic acid porphyrin conjugates

Glycosylated porphyrins have also been tested as drug sensors. As examples, porphyrins **1b (o)** and **8b** (Fig. 3) have been used for the amperometric responses to metronidazole (MTZ) and dimetridazole (DNZ), two clinically useful drugs with selective toxicity to anaerobic bacteria and protozoarium [77, 78].

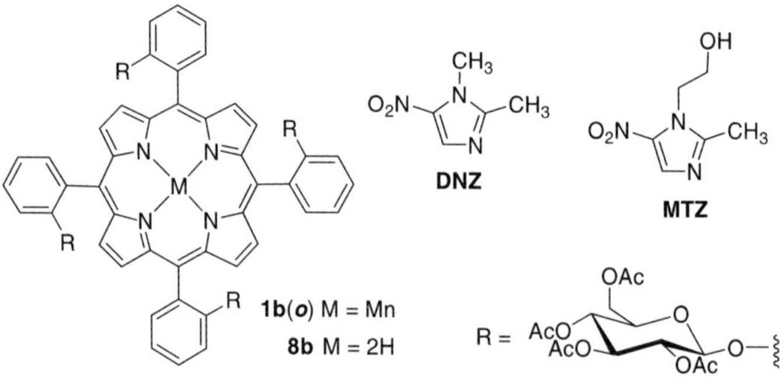

Fig. 3 Glycosylated porphyrins tested as drug sensors and the structures of metronidazole (MTZ) and dimetridazole (DNZ)

Recently, studies on β-lactosylated 5,15-diphenylporphyrinatoiron(III) chloride derivatives have shown unique specific colorimetric responses to calcium ion [79]; no other cations induced such colorimetric behavior. This response is dependent on pH values and appeared only under basic conditions (pH > 8.5). The μ-oxo-bis[porphyrinatoiron(III)] **9** (Fig. 4), is the species responsible for this calcium ion colorimetric behavior [79].

More recently Kawakami et al. reported that manganese porphyrin-carbohydrate conjugates **10** and **11** (Fig. 5) can be considered as new superoxide dismutase (SOD) mimics with cellular recognition [80, 81]. The role of the

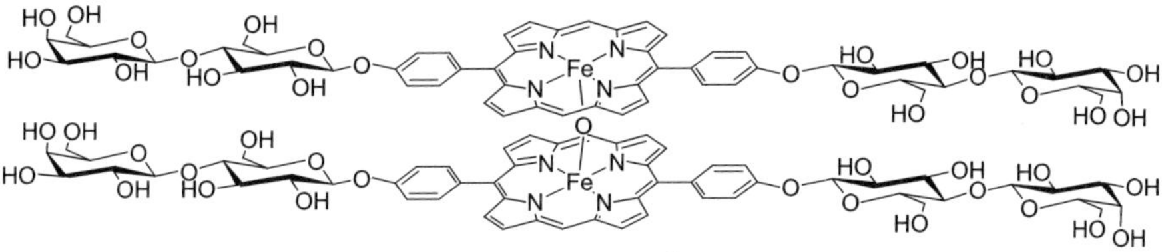

Fig. 4 Structure of μ-oxo-bis[5,15-bis(β-lactosylphenyl)porphyrinatoiron(III)] **9**

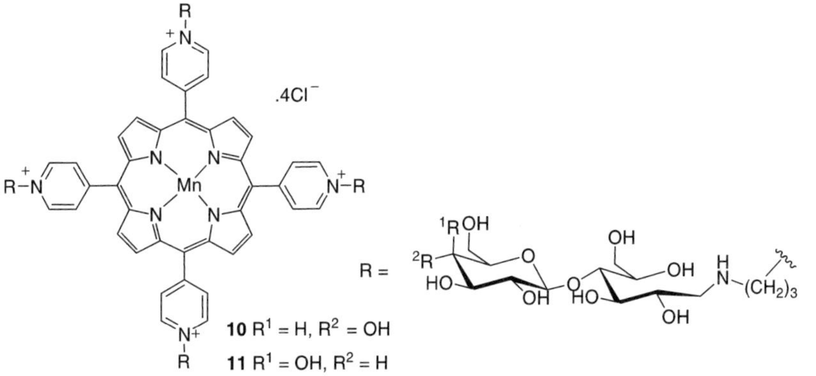

Fig. 5 Manganese(III) glycoporphyrin conjugates for new superoxide dismutase mimics with cellular recognition

glycosylated moiety is to facilitate the receptor binding with active targeting. In fact, the manganese porphyrin-lactose **11** showed significant SOD activity, cellular recognition, and negligible cytotoxicity [81].

2
Natural Glycoporphyrins: the Tolyporphins

Carbohydrate and porphyrin derivatives are two groups of naturally occurring compounds of great significance to all existing organisms. However, only a few examples of natural glycoporphyrins are known. These have been mainly isolated from the blue-green alga *Tolypothrix nodosa* [82, 83]. In 1992 Moore et al. reported that a lipophilic extract component of this microalga can become an anticancer agent [82, 84]. In the first publication, they accounted that most of this activity is related to the presence of unusual porphyrin-like compounds which were designated as tolyporphins [82]. However, such compounds are glycobacteriochlorins; the main component of that mixture, isolated from the extract in 0.1% yield, was called tolyporphin A (**13a**), an unsymmetrical dioxobacteriochlorin core with quaternary centers containing β-linked C-glycosides [82]. The structure was, however, later revised to structure **12a** (Fig. 6) [85] after the total synthesis meanwhile carried out by Kishi et al. for isomers **12b** and **13b** [86, 87]. This synthetic work did establish unambiguously the configuration of the natural product **12a**.

Prinsep et al. isolated and characterized ten additional tolyporphins in the same extract. Six of them contain one or two carbohydrate moieties—

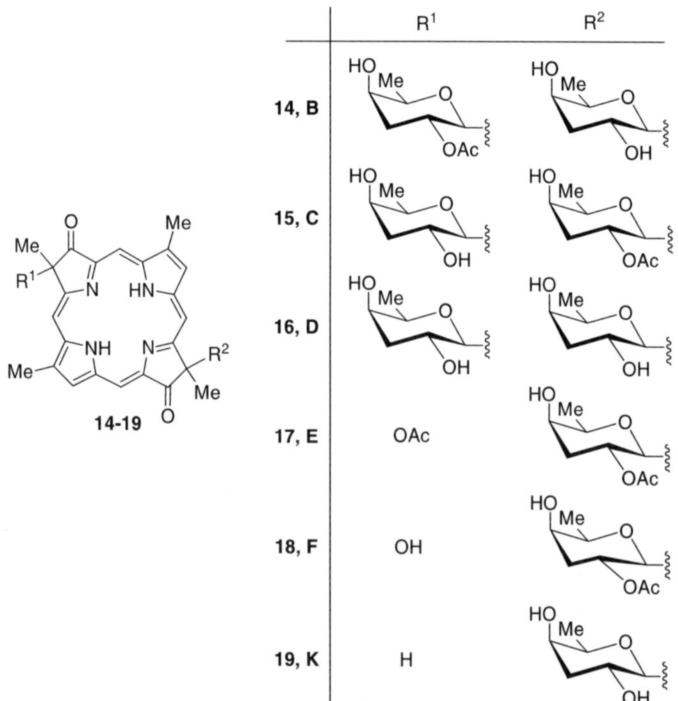

Fig. 6 Tolyporphin A (**12a**) and the two O,O-diacetylated isomers **12b** and **13b**

Fig. 7 Tolyporphins B-F, K found in lipophilic extracts of blue-green alga *Tolypothrix nodosa*

tolyporphins B–F, K (**14–19**), (Fig. 7), which appear in much lower amounts than the tolyporphin A [83, 88]. Interesting biological activity results were found with some of these new tolyporphins [83]. The natural tolyporphin extract was also found to be a very potent photosensitizer against EMT-6 tumor cell growth, both in vitro and in vivo [84].

Another natural glycoporphyrin-like compound is the chlorophyll c_2-monogalactosyldiacylglyceride **20** (Fig. 8), a derivative isolated from *Emiliania huxleyi* (prymnesiophyceae) [89]. Chlorophylls *c* are a group of natural pigments found in the photosynthetic algae, where they act as light-harvesting pigments in the aquatic environment. Garrido et al. have described [89] the chromatographic and spectral properties of four different chlorophylls *c*: chlorophyll c_2, chlorophyll c_3, monovinyl chlorophyll c_3 and a nonpolar chlorophyll *c*. In 2000, they characterized the nonpolar chlorophyll *c*, by mass spectrometry studies, as a chlorophyll c_2 moiety esterified to a monogalatosyldiacylglyceride, a lipid typical of thylakoid membranes.

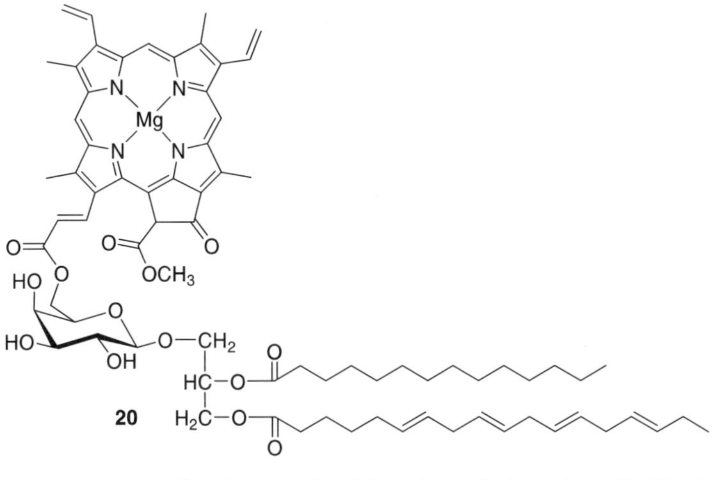

Fig. 8 Structure proposed for the nonpolar chlorophyll *c* isolated from *Emiliania huxleyi*

3
Synthesis of Glycoporphyrins

The biological significance of glycoporphyrins, their scarce natural occurrence and their potential applications have made the availability feature of such compounds a scientific target for several research groups.

Adequate synthetic methodologies leading to differently substituted porphyrins have been established mainly during the first half of the last century [1, 90]. Intensive studies have also been carried out concerning the chemistry of saccharide derivatives [5]. The synthesis of a glycoporphyrin derivative should thus take into consideration the established knowledge about synthesis or chemical modification of a porphyrin and also about the linkage of a sugar species to another compound.

Synthetic approaches to glycoporphyrins can consider the direct glycosylation of adequate functionalized porphyrins obtained from natural sources (Sect. 3.1) or from synthesis (Sect. 3.2) or the chemical synthesis from adequate reagents already fulfilling the glyconjugated requirement (Sect. 3.3).

Many glycoporphyrin derivatives were then prepared with amide or ester, ether or thioether and amine linkages between the porphyrin and the sugar moieties [47]. In all cases sugar hydroxyl functions are protected during the synthesis by suitable groups. These groups allow an easier purification of the products and can be cleaved at a later stage affording, in many cases, water soluble glycoporphyrins.

3.1
Glycoporphyrin Derivatives from Protoporphyrin-IX

Considering that amphiphilic porphyrins may either form micellar fibers or vesicles or they may be integrated into corresponding host systems, Fuhrhop et al. [91] prepared several *N*-glycosylamide protoporphyrin-IX derivatives (Fig. 9).

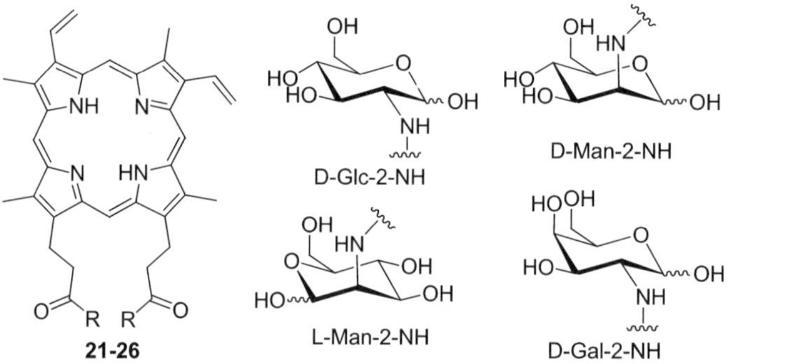

	21	22	23	24	25(D)	25(L)	26
R	OH	OCH$_3$	OC(O)OEt	D-Glc-2-NH	D-Man-2-NH	L-Man-2-NH	Gal-2-NH

Fig. 9 *N*-glycosylamide protoporphyrin-IX derivatives

The protoporphyrin-IX glycosamides were prepared by slow addition of 2-deoxy-2-aminopyranosides, dissolved in ethanol/Et$_3$N/water, to anhydride derivative **23**. This derivative **23** was prepared by treatment of a suspension of protoporphyrin-IX **21** in THF/Et$_3$N at 0 °C with ethyl chloroformate which was added dropwise under stirring. The glycosamide protoporphyrin-IX derivatives **24–26** were obtained in 70–90% yields. To facilitate the products purification and their full characterization the corresponding bis(tetra-acetate) derivatives were prepared.

Other derivatives containing glycosidic and thioglycosidic moieties have also been prepared from protoporphyrin-IX [51]. Esterification of protoporphyrin-IX **21** was performed with methanol containing 5% sulfuric acid [90]. The obtained dimethyl ester **22** was directly converted into isohemato-porphyrin dimethyl ester **27B** (Scheme 1); treatment of protoporphyrin-IX dimethyl ester **22** with thallium(III) nitrate in methanol gave the bis(2,2-dimethoxyethyl)deuteroporphyrin which was then reacted with formic acid.

27A R = H

27B R = CH$_3$

28b R^1 = H, R^2 = OAc

29b R^1 = OAc, R^2 = H

30

28a R^1 = H, R^2 = OH

29a R^1 = OH, R^2 = H

Scheme 1 i) Ag$_2$CO$_3$, Na$_2$SO$_4$, CH$_2$Cl$_2$, 20 °C, 1–7 d, Argon; ii) KOH, MeOH, reflux, 24 h

The obtained product was allowed to react with sodium borohydride in methanol/dichloromethane giving **27B** [51]; this procedure avoids ester cleavage and the need for reesterification.

Considering the PDT applications of hematoporphyrin **30** and its derivatives, Franck et al. [92] synthesized the diglucosides **28** and the digalactosides **29** from isohematoporphyrin dimethyl ester **27B** (Scheme 1). The great advantage of using compounds **27A,B** relatively to the use of hematoporphyrin **30** or its derivatives is due to the fact that **27A,B** are chemically and stereoisomerically pure owing to their primary hydroxyethyl groups. Diastereoselective *O*-glycosylation was accomplished in nonpolar solvents with a solid, surface-active silver catalyst affording stable β-diglycosides. The glycosyla-

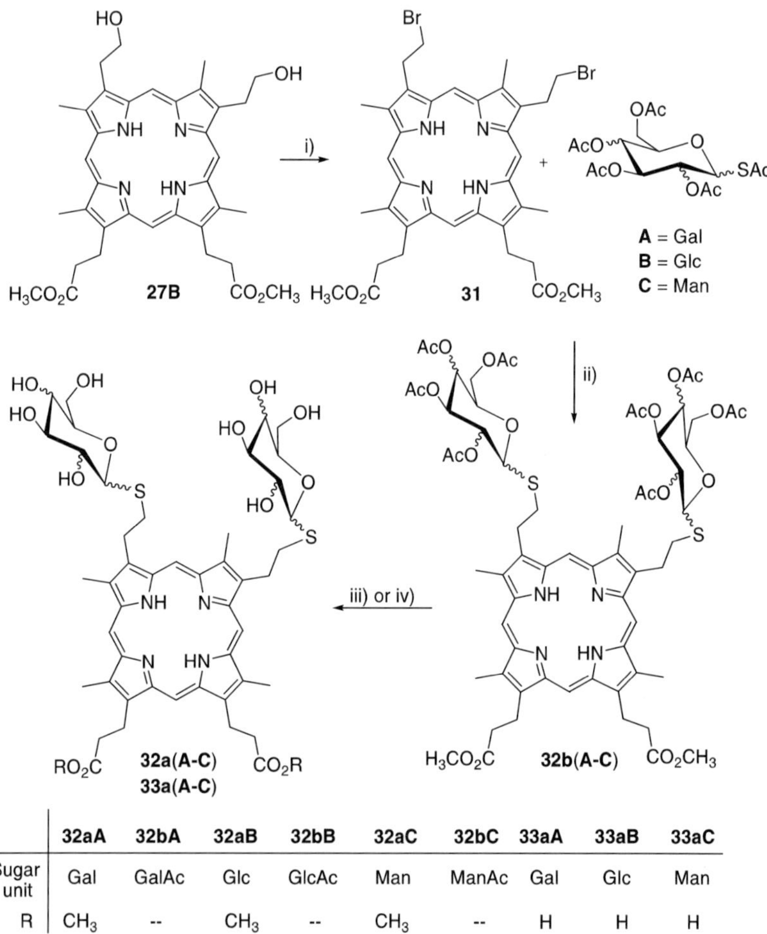

	32aA	32bA	32aB	32bB	32aC	32bC	33aA	33aB	33aC
Sugar unit	Gal	GalAc	Glc	GlcAc	Man	ManAc	Gal	Glc	Man
R	CH₃	--	CH₃	--	CH₃	--	H	H	H

Scheme 2 i) SOBr₂, K₂CO₃, rt, 5 h; ii) DMF, Et₂NH, N₂, rt; iii) NaOMe, MeOH; iv) KOH, CH₂Cl₂/MeOH, reflux

tions of isohematoporphyrins **27** with the 2,3,4,6-tetra-*O*-acetyl-1-bromo-α-D-glucopyranose or 2,3,4,6-tetra-*O*-acetyl-1-bromo-α-D-galactopyranose were carried out at room temperature and in good yields. The corresponding deprotected diglycosides **28a** and **29a** were obtained by alkaline treatment of the acetylated derivatives **28b** and **29b**.

The thioglycosidic analogues **32–33(A–C)** (Scheme 2) have also been obtained from isohematoporphyrin dimethyl ester **27B**. Bromination of **27B** was carried out at room temperature by reaction with thionyl bromide in the presence of potassium carbonate. The dibromoderivative **31** was obtained in 75% yield after purification and demetallation with HCl. Each one of the thioglycosides 2,3,4,6-tetra-*O*-acetyl-1-thioacetyl-β-D-galactopyranose **A**, 2,3,4,6-tetra-*O*-acetyl-1-thioacetyl-β-D-glucopyranose **B** and 2,3,4,6-tetra-*O*-acetyl-1-thioacetyl-β-D-mannopyranose **C**, was added to 2,4-bis(2-bromoethyl)deuteroporphyrin dimethyl ester **31**, in a mixture of DMF and diethylamine, giving rise to compounds **32b(A–C)** in good yields (Scheme 2). The acetyl protecting groups in **32b** can be cleaved by treatment with sodium methoxide in methanol, giving rise to compounds **32a(A–C)**; the acetyl groups and the methyl ester ones can also be cleaved with potassium hydroxide/methanol, giving rise to the water soluble derivatives **33a(A–C)**. The new derivatives **32a(A–C)** and **33a(A–C)** were evaluated for their photocytotoxic activities in comparison with Photofrin®, the commercial anti-cancer hematoporphyrin derivative formulation. Compounds **32a(A–C)** have been revealed to be more PDT efficient than Photofrin®, but surprisingly compounds **33a(A–C)** have been shown to be inactive.

Recently, Hayashi et al. [93] described a very interesting approach to preparing a prosthetic group having carbohydrate moieties. The galactohemin-type **34**, (Fig. 10), contains four galactose moieties appended by a flex-

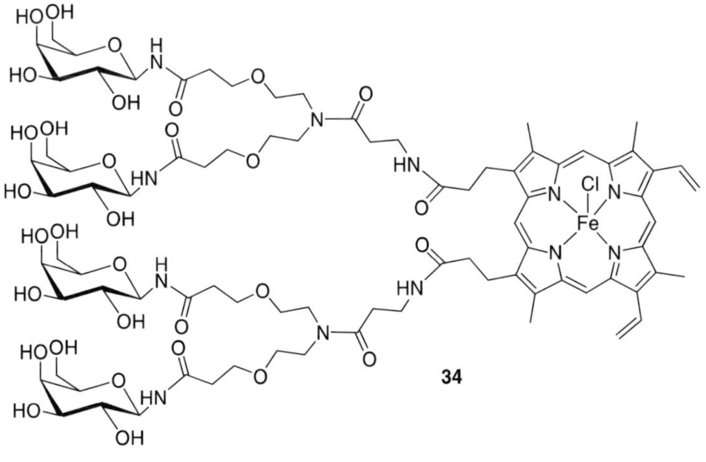

Fig. 10 Structure of galactohemin

ible and branched linker through *N*-glycosyl bonds. Preliminary binding studies performed with lectins (sugar binding proteins), indicated that the galactose units in that reconstituted "myoglobin analogue" might work as the interface for forming the **34**-lectin complex [93].

The synthetic pathway (Scheme 3) for the branched linker galactohemin **34** starts from *N*-benzyl-2,2′-aminodiethanol **35** which, after reaction with *tert*-butyl acrylate followed by hydrogenation/hydrogenolysis processes has afforded the secondary amine **36** in 71% yield. This amine derivative **36**, in the presence of *N,N′*-dicyclohexylcarbodiimide (DCC), was condensed with *N*-benzyloxycarbonyl-*β*-alanine and, after removal of the benzyloxycarbonyl protective group, the linker **37** was obtained in 87% yield. The reaction between **37** and the propionate groups of protoporphyrin-IX **21**, in the presence of 1-ethyl-3-(3-dimethylaminopropyl)carbodiimide (EDC·HCl) gave the derivative **38** (92%). Removal of the *tert*-butyl groups gave the tetracarboxylic acid **39**, which was then coupled with 2,3,4,6-tetra-acetyl-1-amino-*β*-D-galactopyranose using benzotriazol-1-yloxy*tris*(dimethylamino)phosphonium hexafluorophosphate (BOP), affording the bis(amide) derivative **40** in 42% yield. Complexation of **40** with Fe(III), followed by cleavage of the

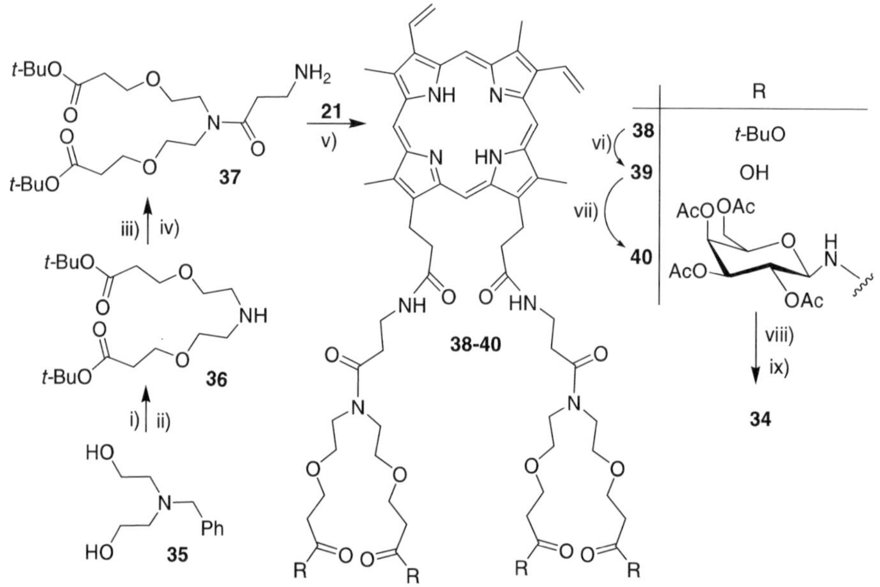

Scheme 3 i) *tert*-butyl acrylate, NaOH, dioxane, rt, 24 h; ii) HCO₂NH₄, 10% Pd/C, MeOH, 50 °C, 7 h; iii) *N*-benzyloxycarbonyl-*β*-alanine, DCC, CH₂Cl₂, 3 h, 0 °C and 16 h rt; iv) HCO₂NH₄, 10% Pd/C, MeOH, 50 °C, 5 h; v) PPIX (**21**), Et₃N, BuOH·H₂O, EDC·HCl, DMF, 2.5 h, 0 °C, and 22 h, rt; vi) CF₃CO₂H, HCO₂H, rt, 22 h; vii) 2,3,4,6-tetra-acetyl-1-amino-*β*-D-galactopyranose, BOP, Et₃N, DMF, 3 h 0 °C, and 40 h rt; viii) FeCl₂, K₂CO₃, CHCl₃/THF, 50 °C, 24 h; ix) NaOH, MeOH, rt, 4 h

acetyl groups on the galactosyl units by sodium methoxide/methanol and treatment with diluted aqueous HCl gave rise to galactohemin **34**.

3.2
Direct Glycosylation of Preliminary Functionalized Porphyrins

As already mentioned, the direct glycosylation of adequately substituted *meso*-arylporphyrins is another synthetic strategy. Several *O*- and *S*-glycoporphyrins have been prepared in such a way.

3.2.1
O-Glycoconjugate Porphyrins

Derivatization of substituted *meso*-tetraphenylporphyrins involving the attachment of acetylated carbohydrate residues at the phenyls *ortho* [57, 94], *meta* [60] and *para* [54, 95] positions has been reported.

Using such a strategy the glucoporphyrins **42** have been synthesized as is depicted in Scheme 4 [54]. The 5-*p*-hydroxyphenyl-10,15,20-triphenylporphyrin **41** was prepared from crossed-Rothemund reaction using the appropriate benzaldehydes and pyrrole in refluxing acetic acid and nitrobenzene. The reaction of porphyrin **41** with 2,3,4,6-tetra-*O*-acetyl-1-bromo-α-D-glucopyranose afforded porphyrin glucoside **42b**. The reaction was carried out at room temperature in the presence of freshly prepared silver carbonate, affording a mixture of the desired porphyrin **42b** with the corresponding silver complex **42c**. Treatment of this mixture with trifluoroacetic acid (TFA) gave, after purification, the pure porphyrin **42b** in 66% yield. The carbohy-

	M	R
42a	2H	H
42b	2H	Ac
42c	Ag	Ac

Scheme 4 i) Ag$_2$CO$_3$, dry CH$_2$Cl$_2$, 24 h, rt; ii) TFA, 5 min; iii) NaOMe, MeOH/CH$_2$Cl$_2$, N$_2$, 6 h

drate protection groups were removed by treating **42b** with sodium methoxide in dry methanol/dichloromethane and the desired glucoporphyrin **42a** was obtained in 80% yield.

Instead of using the mono-hydroxy-TPP **41**, Maillard at al. decided to use 5,10,15-*tris*(*p*-hydroxyphenyl)-20-phenylporphyrin **43** to build and to study the biological and photobiological properties of a novel glycoporphyrin series, in which a diethylene glycol linker makes the bridge between the porphyrin and the sugar moieties. The desired *O*-glycosylated diethylene glycol porphyrin derivatives **46** and **47** (Scheme 5) were obtained by the reaction of **43** with the corresponding bromo-substituted glycosides **44(α,β)** and **45α** (Scheme 6). The *O*-deprotection of the intermediates **46b** and **47b** was performed with sodium methoxide in methanol [49].

The carbohydrates **44** and **45** were prepared, respectively by the reaction of *per*-acetyl galactose and mannose with 2-(2-bromoethoxy)ethanol in the presence of BF$_3$-etherate (Scheme 6) [49].

Scheme 5 i) K$_2$CO$_3$, dry DMF, 60 °C, 15 h; ii) NaOMe, MeOH

Scheme 6 i) 2-(2-bromoethoxy)ethanol, BF$_3 \cdot$OEt$_2$, CH$_2$Cl$_2$, 0 °C to rt

The same principle of nucleophilic substitution has been applied to prepare other derivatives. With porphyrins like *meso*-tetrakis(pentafluorophenyl)porphyrin **48**, the reactivity of the *para*-fluoro atoms towards nucleophilic substitutions has been well established. Thus the galactoconjugate porphyrin derivative **49b** has been obtained from the reaction of porphyrin **48** with 1,2:3,4-di-*O*-isopropylidene-α-D-galactopyranose (Scheme 7). Acidic

Scheme 7 i) NaH, dry toluene, reflux, 24 h, N$_2$; ii) TFA/H$_2$O (9:1), rt, 30 min cleavage of the isopropylidene protective groups gave rise to compound **49a** [54].

Other galactoconjugate porphyrins **55–59** have also been prepared from porphyrinic carboxylic esters [95]. For example, the esters **50–54** achieved transesterification processes when treated with methyl lithium and 1,2:3,4-di-*O*-isopropylidene-α-D-galactopyranose (Scheme 8). The deprotected compounds **55a–59a** have been obtained, as mixtures of α and β anomers, in excellent yields (92–98%).

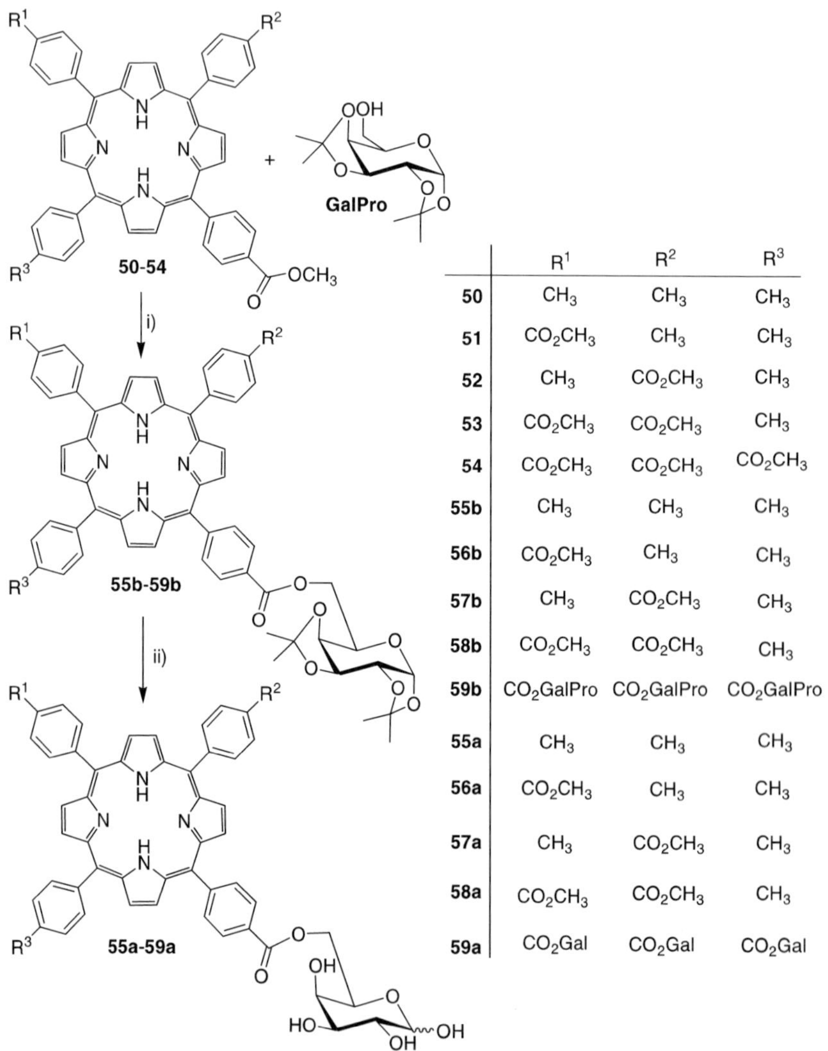

	R^1	R^2	R^3
50	CH_3	CH_3	CH_3
51	CO_2CH_3	CH_3	CH_3
52	CH_3	CO_2CH_3	CH_3
53	CO_2CH_3	CO_2CH_3	CH_3
54	CO_2CH_3	CO_2CH_3	CO_2CH_3
55b	CH_3	CH_3	CH_3
56b	CO_2CH_3	CH_3	CH_3
57b	CH_3	CO_2CH_3	CH_3
58b	CO_2CH_3	CO_2CH_3	CH_3
59b	$CO_2GalPro$	$CO_2GalPro$	$CO_2GalPro$
55a	CH_3	CH_3	CH_3
56a	CO_2CH_3	CH_3	CH_3
57a	CH_3	CO_2CH_3	CH_3
58a	CO_2CH_3	CO_2CH_3	CH_3
59a	CO_2Gal	CO_2Gal	CO_2Gal

Scheme 8 i) (a) MeLi, THF, 0 °C to rt, N$_2$, (b) reflux, 8 h; ii) TFA/H$_2$O (9 : 1), 20 min

Another strategy to obtain porphyrin glyco-esters implied the esterification of a porphyrin carboxylic acid with a halosugar derivative [54].

The attempted esterification of the porphyrin carboxylic acid **60** with galactopyranose, in the presence of DCC and 4-pyrrolidinopyridine, did not lead to the expected compound **61b** but to the corresponding *N*-acylurea derivative [96]. However, glycoporphyrin **61b** was obtained in good yield from the reaction of **60** with 6-iodo-1,2:3,4-di-*O*-isopropylidene-α-D-galactopyranose in the presence of potassium carbonate (Scheme 9). Derivative **61a** was obtained in high yield after cleavage of the sugar protecting groups in **61b**.

Scheme 9 i) K$_2$CO$_3$, dry DMF, 100 °C, 30 h, N$_2$; ii) TFA/H$_2$O (9 : 1), 30 min

Shinkai et al. have used a different methodology to prepare a series of *meso*-tetra-glycoarylporphyrin derivatives **64–67** [97–99]. The sugar-appended porphyrin compounds, **64a–67a** were synthesized in accordance with Scheme 10. The target compounds were obtained after deprotection of the hydroxyl groups of porphyrin precursors **64b–67b**. These were prepared from the condensation of aminophenyl-2,3,4,6-tetra-*O*-acetyl-monosaccharide **63(o,p)** with the acid-chloride derivative **62** [99].

Aiming to prepare *O*-glycosylated porphyrins having the sugar moiety not directly linked to a *meso*-phenyl group, the *ortho* and *para*-hydroxyphenyl derivatives **68(o,p)** were converted into **73b(o,p)** and **73a(o,p)** (Scheme 11). This was done by treating **68(o,p)** with an excess of the spacer reagent **69**

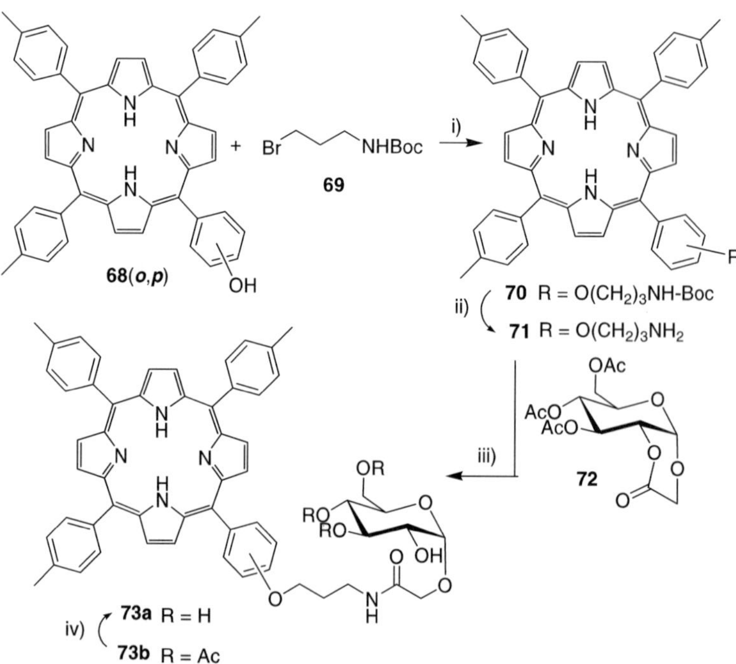

Scheme 10 i) Et$_3$N, THF; ii) NaOMe, MeOH, THF

Scheme 11 i) K$_2$CO$_3$, DMF, rt, 18 h; ii) TFA/CH$_2$Cl$_2$, rt, 3 h; iii) DMAP, CH$_2$Cl$_2$, rt, 16 h; iv) NaOMe, MeOH/CH$_2$Cl$_2$

to generate **70(o,p)**, which by treatment with trifluoroacetic acid gave rise to the aminopropyloxyphenylporphyrins **71(o,p)** in 95% and 82% yield, respectively. Then by reaction of porphyrins **71(o,p)** with the lactone **72**, the porphyrin products **73b(o,p)** were obtained in 99% and 95% yield, respectively; these glycoderivatives, by cleavage of the hydroxyl protecting groups, gave the desired final porphyrins **73a(o,p)** [100].

The derivatization of adequately substituted porphyrins has been applied to the synthesis of polymeric glycoconjugates [101, 102]. Daub et al. have reported the synthesis of a porphyrin functionalized cellulose [101]. The idea was to prepare multichromophore assemblies, organized along regular chains, for studying their optical, electronic, and photophysical properties [101, 103]. Also spectroscopic and electrochemical studies were performed in order to obtain information concerning porphyrin–porphyrin $\pi-\pi$ stacking interactions.

The syntheses of the glucopolymer-porphyrin conjugates **76** and the corresponding zinc(II) complexes **77** were carried out according to the methodology depicted in Scheme 12. A mixture of 2,3-di-*O*-methylcellulose **75** and porphyrin acids **60** or **74**, in a cocktail of EDC, DMAP and pentafluorophenol in dry THF/chloroform, turned out to give the best results. The reaction mixtures were stirred for several days, and after such periods the porphyrin-cellulose conjugates **76** and **77** were obtained by precipitation with methanol [101].

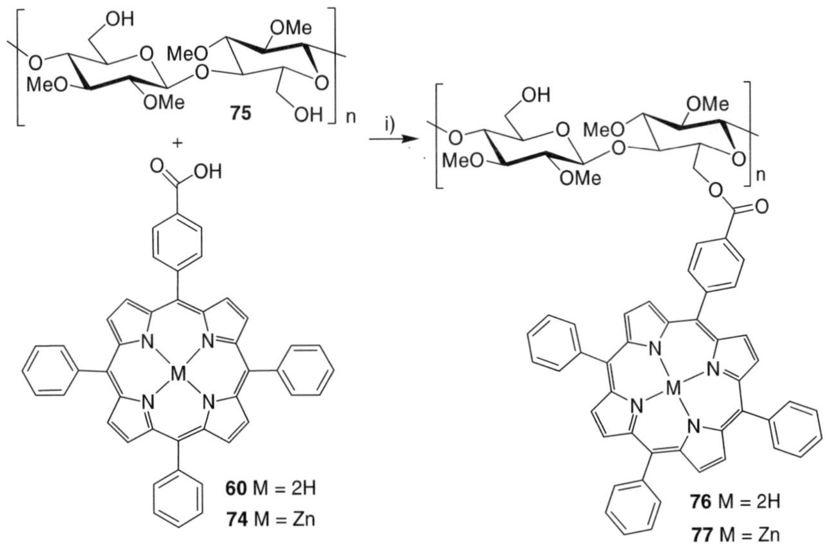

Scheme 12 i) EDC, DMAP, THF/CHCl₃, pentafluorophenol, **60** or **74**

Very recently, Hasegawa et al. prepared interesting polysaccharide-based materials [102]. Using a Cu(I)-catalyzed chemoselective coupling reaction between organic azides and terminal alkynes [104], several functional moieties, such as porphyrins, have been appended to $(1 \rightarrow 3)$-β-D-glucans such as curdlan, **78** (Scheme 13). Curdlan **78** was transformed into curdlan derivative **80** by bromination followed by azidation. The [3 + 2] cycloaddition reactions between **80** and **81** were carried out in NMP, containing copper(II) bromide, ascorbic acid and propylamine. The polymer conjugates **82** were purified by dialysis (against water) followed by lyophilization and solvents washing.

Scheme 13 i) PPh$_3$, DMF, LiCl, rt, 3 h and then CBr$_4$, 60 °C, 24 h; ii) sodium azide, DMSO, 80 °C, 36 h; iii) CuBr$_2$, ascorbic acid, propylamine, NMP, rt, 12 h

3.2.2
S-Glycoconjugate Porphyrins

Porphyrin *S*-glycosides can be considered to be good mimics of the *O*-glycoside derivatives with enhanced stability toward enzymatic hydrolysis. The tetrakis(thioglucosyl) **83b** and tetrakis(thiogalactosyl) **84** derivatives of 5,10,15,20-tetrakis(pentafluorophenyl)porphyrin **48** have been synthesized in high yields [10]. The synthesis of both families can be accomplished using either the protected sugar (ii) followed by deprotection (iv) or directly by using the unprotected sugar (iii) (Scheme 14). The overall yield for compound **83a** after the three steps is 88% and the yield for **84** is 92% [10].

Because of their great triplet quantum yields, it was postulated that these fluorinated porphyrin derivatives can act as potent photosensitizers in their potential medicinal applications [10].

Scheme 14 i) AcSK, dry acetone, rt, overnight; ii) Et$_2$NH, DMF, rt; iii) DMF, rt, overnight; iv) NaOMe, CH$_2$Cl$_2$/MeOH

Other thioglycosylated *meso*-tetra-arylporphyrins **86(o,p)** (**A–C**) (Scheme 15) were synthesized and evaluated for their photocytotoxicities against K562 chronic leukaemia cell lines [51].

The thioglycosylated sugar derivatives 2,3,4,6-tetra-O-acetyl-1-thioacetyl-β-D-galactopyranose **A**, 2,3,4,6-tetra-O-acetyl-1-thioacetyl-β-D-glucopyranose **B** and 2,3,4,6-tetra-O-acetyl-1-thioacetyl-α-D-mannopyranose **C** were condensed with the monobrominated porphyrins **85** in dry DMF, giving rise to the expected compounds **86b(o,p)** in good yields and from these, by cleavage of the acetyl groups, the desired compounds **86a(o,p)** were obtained [105]. The hydroxyporphyrin derivatives **68(o,p)** were converted into **85(o,p)**, in dry DMF, using a large excess of 1,3-dibromopropane and potassium carbonate.

These thioglycosylated derivatives **86a(o,p)** have been shown to resist the enzymatic hydrolysis of the β-thioglucosydic bonds by β-glucosidase. Also all the *ortho*-isomers **86a(o)** (**A–C**) and the *para*-isomer **86aB** showed significant photocytotoxic activity. This phototoxic effect was lower than the one observed with the commercial formulation Photofrin®, when the dead cells were counted immediately after irradiation.

Maillard et al. [49] described a synthetic methodology to prepare the thiosugar-diethylene glycol porphyrin derivatives **88** and **89** (Scheme 16). Condensation of 2,3,4,6-tetra-O-acetyl-1-thio-β-D-galactopyranose or 2,3,4,6-tetra-O-acetyl-1-thio-α-D-mannopyranose with the 5,10,15-*tris*(p-bromo-ethoxyethoxyphenyl)-20-phenylporphyrin **87**, previously prepared from **43**

	86aA	86bA	86aB	86bB	86aC	86bC
(o,p)	Gal	GalAc	Glc	GlcAc	Man	ManAc

Scheme 15　i) $Br(CH_2)_3Br$, K_2CO_3, DMF; ii) Et_2NH, dry DMF, N_2, rt; iii) NaOMe, $CH_2Cl_2/MeOH$

and 1-bromo-2-(2-bromoethoxy)ethane, gave rise to derivatives **88b** and **89b**. These by deacetylation gave compounds **88a** and **89a** quantitatively [49].

Scheme 16 i) K_2CO_3, dry acetone, 20 °C, 15 h; ii) NaOMe, MeOH, rt

3.3
Synthesis of *meso*-Glycoaryl Substituted Porphyrins from Pyrrolic Precursors

Synthetic methodologies leading to *meso*-aryl substituted porphyrins which do not have substituents at the β-peripheral positions are well known in the literature [59, 94, 106–109]. The main ones consider the acid-catalyzed condensation of pyrrole with an aldehyde or a mixture of aldehydes or even with dipyrrylmethane-aldehyde mixtures. In this way a wide range of *meso*-(mono-, di-, tri-, tetra-aryl) substituted porphyrins, can be obtained. For the synthesis of *meso*-glycosylated porphyrins those methodologies can be followed. In this way two main approaches for obtaining *meso*-aryl substituted glycoporphyrins have been described: one implies the condensation of glycosylated aldehydes with pyrrole or dipyrrylmethanes and the other one is based on the direct glycosylation of conveniently functionalized porphyrins.

Good and well-established protecting groups for the carbohydrate moieties during these synthetic procedures are the acetyl groups, which can later be removed by sodium methoxide in dry methanol, and the isopropylidene groups, which can also be removed by TFA/water treatment. Also, according to the type of bond between the aldehyde/porphyrin and the carbohydrate moieties, O, S and C-glycosides can be considered.

3.3.1
Synthesis of *O*-Glycosylarylporphyrins
from Condensation of Pyrrole or Dipyrromethanes with Aldehydes

The acid-catalyzed condensation of pyrrole with benzaldehyde substituted by a carbohydrate residue gave rise to symmetric *meso*-tetrakis(glycosylaryl)porphyrins.

Glycosylbenzaldehydes, such as those containing tetra-acetyl-glucosyl, -galactosyl, -glucosamino, -maltosyl and -lactosyl moieties, (**90–94**) shown in Fig. 11, were prepared according to the method of Halazy et al. (Scheme 17) [110].

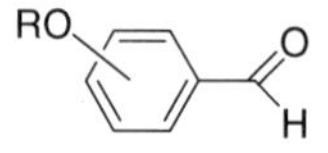

(*o,m,p*)	90a	90b	91a	91b	92a	92b	93a	93b	94a	94b
R	Glc	GlcAc	Gal	GalAc	Glc-2-NH	GlcAc-2-NH	Mal	MalAc	Lac	LacAc

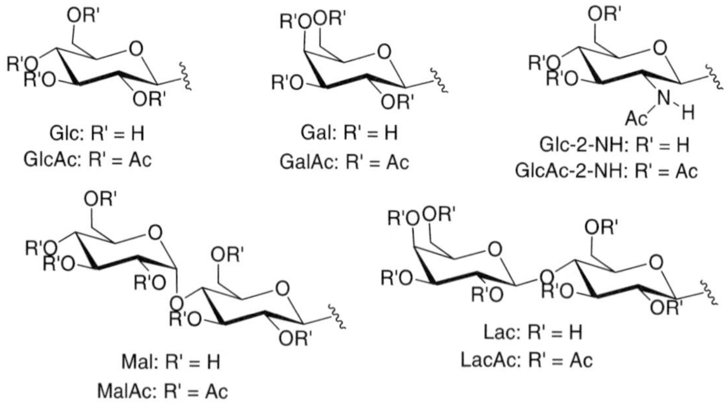

Glc: R' = H
GlcAc: R' = Ac

Gal: R' = H
GalAc: R' = Ac

Glc-2-NH: R' = H
GlcAc-2-NH: R' = Ac

Mal: R' = H
MalAc: R' = Ac

Lac: R' = H
LacAc: R' = Ac

Fig. 11 Glycosylbenzaldehydes prepared according to the method of Halazy

For example, the 2-(2,3,6,2′,3′,4′,6′-hepta-acetyl-β-D-maltosyloxy)benzaldehyde **93b(o)** can be prepared in 36% yield by coupling 2-hydroxybenzaldehyde with α-bromoacetylmaltose **97** (Scheme 17) in a heterogeneous phase (aqueous sodium hydroxide, dichloromethane and tetrabutylam-

Scheme 17 i) Ac_2O, py, 0 °C, 3 d; ii) HBr, AcOH, rt, 3 h; iii) NaOH (aq), Bu_4NBr, CH_2Cl_2, rt, 3 d

monium bromide). This acetyl derivative **97** can be obtained by treating maltose **95** with a mixture of acetic anhydride and pyridine at 0 °C for 3 days, giving the product **96** which is then transformed into α-bromoacetylmaltose **97** by reaction with HBr/acetic acid [111].

Condensation of glycosylated benzaldehydes with pyrrole can take place in acetic acid/pyridine [112] or under Lindsey's conditions [113] to afford glycosylated porphyrins. The Lindsey method considers a high-dilution condensation of aldehydes and pyrrole, in the presence of a Lewis acid catalyst ($BF_3 \cdot OEt_2$), in dichloromethane at room temperature, to afford the corresponding cyclic tetrapyrrolic porphyrinogens, which are oxidized in situ by treatment of the reaction mixture with 2,3-dichloro-5,6-dicyano-1,4-benzoquinone (DDQ), or tetra-chloro-1,4-benzoquinone (p-chloranil) [114]. The reaction between $ortho$-tetra-acetyl-β-D-glucosyloxybenzaldehyde **90b(o)** or -hepta-acetyl-β-D-maltosylbenzaldehyde **93b(o)** with pyrrole gave 5,10,15, 20-tetrakis(2-glycosylphenyl)porphyrins **8b** and **98b** (Scheme 18) [106, 111, 115].

Substituents, such as carbohydrate moieties, in $ortho$ positions, induce strong steric hindrance on both faces of the porphyrin macrocycle, and give rise to the formation of three atropisomers, $\alpha\beta\alpha\beta$ **8bA**, $\alpha\alpha\beta\beta$ **8bB** and $\alpha\alpha\alpha\beta$ **8bC**, isolated in relative ratios 1 : 2 : 4. These derivatives, after treatment with sodium methoxide in dry methanol, are converted into the free glycosylated porphyrins **8a(A–C)**, without changing both the absolute configurations and the chemical nature of the sugars.

The formation of the $\alpha\alpha\alpha\alpha$ atropisomer was not observed; however, thermal atropisomerization (at 80 °C) of the isolated three atropisomers, in a mix-

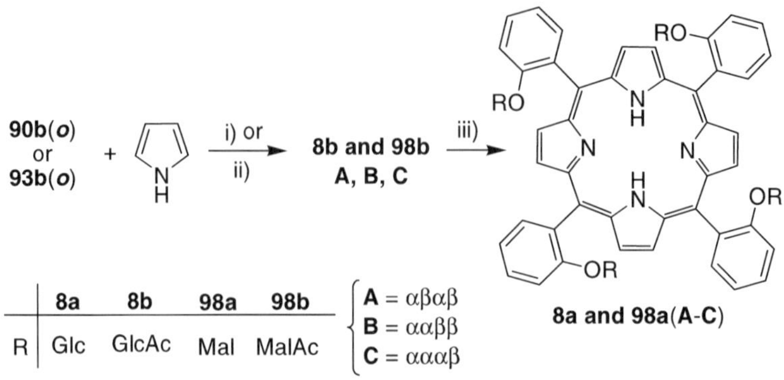

Scheme 18 i) AcOH, py, Δ; or ii) (a) $BF_3 \cdot OEt_2$, CH_2Cl_2, (b) *p*-chloranil, Δ; iii) NaOMe, MeOH, rt

ture of toluene/MeCN (10/3, v/v) and in the presence of silica gel, did afford the *αααα* isomer in 22% [116].

Protected porphyrins **8b(A–C)** are soluble in nonpolar solvents. The solubility of the three atropisomers **8a(A–C)** in water increases in a similar way as does the protection of the macrocycle faces (**8aC** < **8aB** < **8aA**); this might be presumably due to a difference in solvation.

The same approach was applied to prepare the octaglucosylated derivative **99a** (Fig. 12). It was obtained quantitatively, after deprotection of **99b** by sodium methoxide in dry methanol. The protected glucosylated derivative **99b** was synthesized in 32% yield by using Lindsey's method from pyrrole and bis(tetra-acetylglucopyranosyloxy)benzaldehyde. This aldehyde was obtained in 90% yield by condensation of 3,5-dihydroxybenzaldehyde with acetobromoglucose in the presence of Ag_2O [46].

The condensation of *ortho*- or *meta*-tetra-acetyl-*β*-D-glucosyloxybenzaldehyde **90b(o,m)**, *meta*-hepta-acetyl-*β*-D-maltosyloxybenzaldehyde **93b(m)**

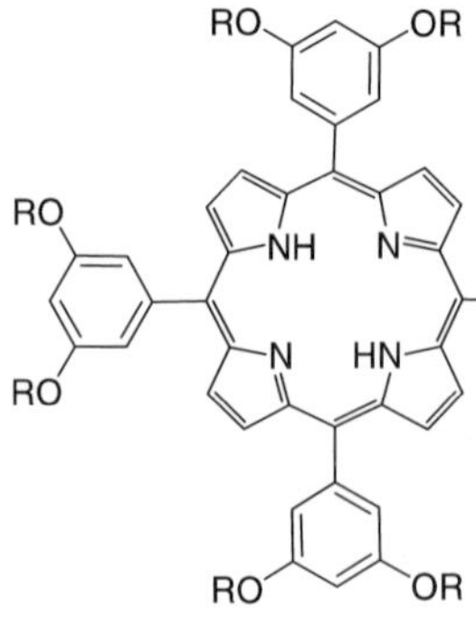

99a R = Glc
99b R = GlcAc

Fig. 12 Structure of octaglucosylated porphyrin **99**

or *meta*-hepta-acetyl-β-D-lactosyloxybenzaldehyde **94b(m)** with pyrrole, using Lindsey's methodology, gave the corresponding porphyrins **8b(o)**, **100b(m)**, **101b(m)** and **102b(m)**. These porphyrins have been metallated with manganese and iron ions (**1b–4b**) (Scheme 19), to be used as oxidative catalysts of certain substrates, such as 4-chlorostyrene, 1,2-dihydronaphthalene and cyclohexane, with H_2O_2 or PhIO as oxygen atom donors [58, 60].

	R
8b(o)	GlcAc
100b(m)	GlcAc
101b(o,m)	MalAc
102b(m)	LacAc

	1b(o,m)	**2b(o,m)**	**3b(o,m)**	**4b(m)**
M	Mn	Fe	Mn	Mn
R	GlcAc	GlcAc	MalAc	LacAc

Scheme 19 i) MCl_2, DMF, reflux, 1–6 h

Attachment of carbohydrates to metalloporphyrins can affect the regio- and stereo-selective oxidation features, since the glycosylated residues give a chiral environment for the catalytic site. However, these complexes have been demonstrated to be insufficiently robust against the oxidant action. To overcome this limitation, Momenteau et al. developed new glycosylated porphyrins whereby electron-withdrawing substituents are introduced at each *ortho*-aryl position. In fact, it is considered that the introduction of electron-withdrawing substituents on the four *meso*-aryl rings and on the β-pyrrolic positions induces a certain degree of stabilization of the metalloporphyrins towards any oxidative degradation during the catalytic process and also increases the reactivity by generating a more electrophylic metal-oxo intermediate [59].

meso-Tetraphenylporphyrin derivatives bearing acetylated β-D-glucosyl chiral moieties linked at the *ortho* position and one chlorine atom at the other *ortho* position of each *meso*-phenyl group have been synthesized [59].

The synthesis of 5,10,15,20-tetrakis[2-chloro-6-(2,3,4,6-tetra-acetyl-O-β-D-glucosyloxy)phenyl]porphyrins **104b(A–C)** (Scheme 20) involves, firstly, the preparation of the substituted benzaldehyde **103** bearing the β-Dacetylglucosyl group in one *ortho* position and a chlorine atom as an electron-withdrawing substituent at the other *ortho* position. The condensation of 2,3,4,6-tetra-*O*-acetyl-1-bromo-α-D-glucopyranose with 6-chloro-2-hydroxybenzaldehyde (it can be obtained by treating 2-fluoro-6-chlorobenzaldehyde with NaOH), did afford the glycosylated benzaldehyde **103** in 55% yield. Condensation of this chloroglucosylated benzaldehyde **103** with pyrrole, by the procedure described by Lindsey, afforded the corresponding glucosylated porphyrin atropisomer mixture **104b(A–C)** in 9% yield [59].

Scheme 20 i) NaOH (aq. sol. 5%), Bu₄NBr, CH₂Cl₂, 3 d, rt; ii) (a) Pyrrole, BF₃ · OEt₂, CH₂Cl₂, Argon, rt, overnight; (b) *p*-chloranil; iii) MCl₂, DMF, Argon, 140 °C, 5 h

Again, the formation of the αααα atropisomer was not observed. This fact was attributed to a strong steric hindrance brought about by both glucosyl and chlorine substituents.

The metallation yields were low for the αβαβ (24%) and ααββ (40%) atropisomers in which both macrocycle faces are more sterically hindered than those of the αααβ derivative (90%).

These *ortho*-glucosylated porphyrin metal complexes have been shown to be active catalysts for alkene epoxidation with certain asymmetric induction due to the presence of chiral sugar substituents in the vicinity of the metal center. However, as far as their medicinal applications are concerned, it was demonstrated that their deprotected, neutral derivatives exhibit neither toxicity nor phototoxicity against tumoral cells [117]. A possible explanation for that considers a certain degree of globular structure for the molecules, which prevents suitable cell penetration.

Thus, some research groups looked at the synthesis of other derivatives which do not have atropisomers formation. *meso*-Tetrakis(glycosylaryl)porphyrins, mixed *meso*-tetrakis(glycosylaryl, aryl)porphyrins, and mixed *meso*-tetrakis(glycosylaryl, alkyl)porphyrins having mono- or disaccharide moieties linked at the *para* positions of the *meso*-phenyl groups have been synthesized.

Momenteau et al. synthesized the symmetric *meso*-tetrakis(glycosylphenyl)porphyrins **105–108** (Fig. 13). This was performed by using the *para* glycosylated benzaldehydes. These porphyrins bearing protected monosaccharide substituents are very soluble in nonpolar solvents, however the corresponding unprotected tetra-monosaccharide porphyrins **105a–107a** are soluble in alcohols and weakly soluble in water. In contrast, the tetramaltosyl derivative **108a** is very soluble in aqueous solutions even at high concentrations [106].

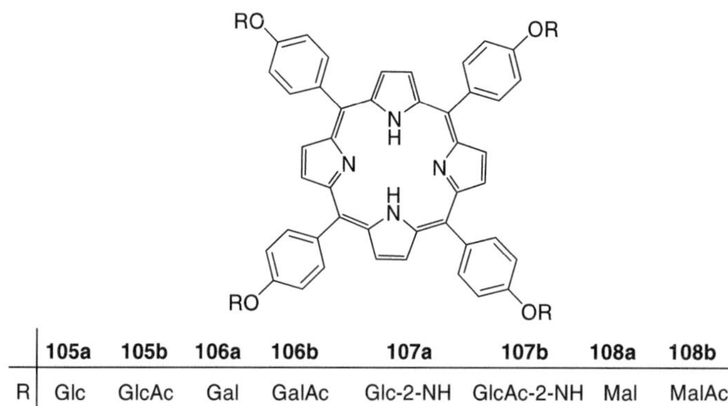

	105a	105b	106a	106b	107a	107b	108a	108b
R	Glc	GlcAc	Gal	GalAc	Glc-2-NH	GlcAc-2-NH	Mal	MalAc

Fig. 13 Symmetric *meso*-tetrakis(glycosylaryl)porphyrins

A similar methodology was used by Hombrecher to synthesize the derivative **59b**. This was carried out by reaction of di-*O*-isopropylidene-α-D-galactopyranosyl substituted benzaldehyde **109** with pyrrole (Scheme 21). The aldehyde **109** was synthesized by a CsF-catalyzed condensation reaction of 6-bromo-6-desoxy-1,2:3,4-di-*O*-isopropylidene-α-D-galactopyranose with 4-formylbenzoic acid.

The nature and the number of the saccharide substituents can rule the physical properties of such compounds; furthermore, the presence of hydrophobic *meso*-phenyl, *meso*-pentafluorophenyl or *meso*-alkyl substituents could increase the interactions of porphyrins with the lipid parts of cell membranes whereas the glycosyl moieties could become functional components involved in cell recognition [118, 119].

In order to obtain *meso*-mono, -bis, and -tris(glycosylaryl) asymmetric porphyrins **110–115**, pyrrole was condensed with a mixture of aldehydes and glycosylated benzaldehydes in relative proportions of 4 : 2 : 2 (Fig. 14) [106].

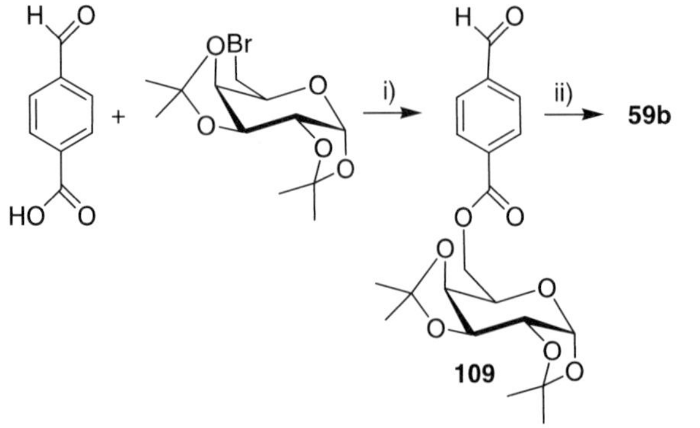

Scheme 21 i) CsF, DMF, 120 °C, 2 d; ii) pyrrole, dry CH_2Cl_2, $BF_3 \cdot OEt_2$, rt, 1 h, then *p*-chloranil, reflux, 1 h

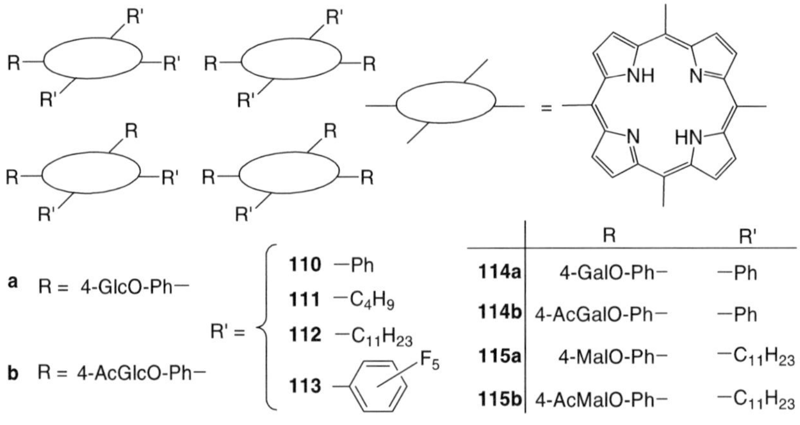

Fig. 14 Structure of asymmetric *meso*-glycosylarylporphyrins

In fact the pyrrole/aldehyde condensation methodology allows the use of a mixture of aldehydes, providing that the required product can be easily purified from the final statistical product-mixture. Using two aldehydes, that usually happens when their relative proportions are 3 : 1.

In all the glycoporphyrins presented above, a carbohydrate moiety is linked directly to an aryl substituent. However, it is possible to have *meso*-glycosylarylporphyrins with the carbohydrate moiety separated from the aryl group by a spacer. This has been achieved by Bolbach et al. either by direct glycosylation of the *ortho*- or *para*-hydroxyalkoxyarylporphyrins (Sect. 3.2.1) or by condensation of required aldehydes with pyrrole [120]. The latter procedure gave rise to the monoglycosylporphyrins **116(o,p)–119(o,p)** (Fig. 15).

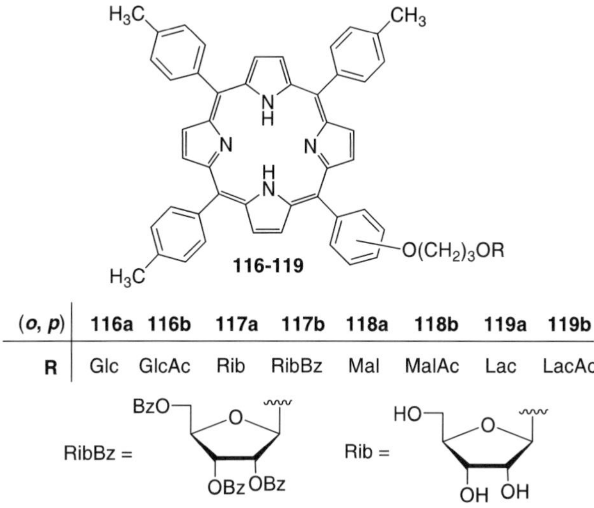

Fig. 15 Mono-glycosylated porphyrins obtained by condensation of pyrrole with mixtures of glycosylated and nonglycosylated aldehyde

Related compounds **120–124** have also been synthesized by the same group (Fig. 16) using the same procedure [120].

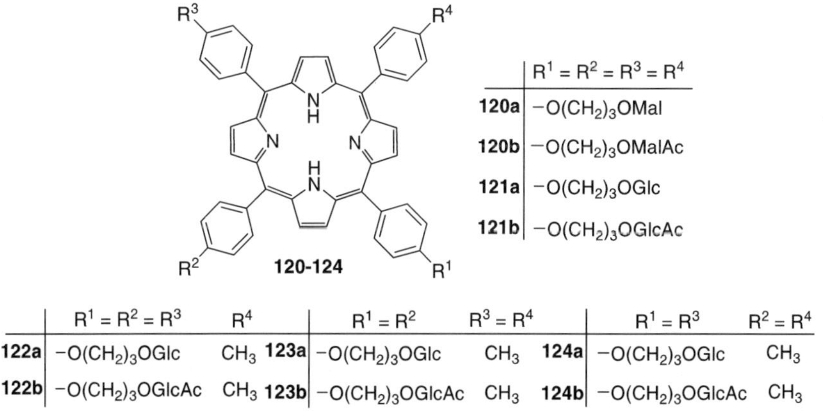

	R¹ = R² = R³	R⁴		R¹ = R²	R³ = R⁴		R¹ = R³	R² = R⁴
122a	−O(CH₂)₃OGlc	CH₃	123a	−O(CH₂)₃OGlc	CH₃	124a	−O(CH₂)₃OGlc	CH₃
122b	−O(CH₂)₃OGlcAc	CH₃	123b	−O(CH₂)₃OGlcAc	CH₃	124b	−O(CH₂)₃OGlcAc	CH₃

Fig. 16 Tetra-, tri- and di-glycosylated porphyrins obtained by condensation of pyrrole with adequate benzaldehydes

During the last two decades porphyrin dimers have been synthesized and evaluated for PDT cancer treatments [121, 122]. Porphyrin precursors leading to certain dimers can be obtained by following the pyrrole/aldehyde condensation methodology.

This has been followed in the synthesis of compounds **68(p)** and **125(o,p)**, which were precursors in the synthesis of glycoporphyrin dimers **128b(p)** and **129b(o,p)** (Scheme 22). Thus, *O*-glycosyl neutral porphyrin dimers with ether linkages **128a(p)** and **129a(o,p)** have been synthesized and evaluated for their biological activities [123, 124].

68(p) R = CH$_3$
125(o,p) R = OGlcAc

126(p) R = CH$_3$
127(o,p) R = OGlcAc

126(p) + **125(p)** $\xrightarrow{\text{ii)}}$ **128b(p)** $\xrightarrow{\text{iii)}}$ **128a(p)**

127(o,p) + **125(o,p)** $\xrightarrow{\text{ii)}}$ **129b(o,p)** $\xrightarrow{\text{iii)}}$ **129a(o,p)**

128, 129	R	R^1
128b(p)	CH$_3$	GlcAc
128a(p)	CH$_3$	Glc
129b(o,p)	OGlcAc	GlcAc
129a(o,p)	OGlc	Glc

Scheme 22 i) I(CH$_2$)$_3$I, K$_2$CO$_3$, DMF, 6 h; ii) K$_2$CO$_3$, DMF, 24 h; iii) NaOMe, MeOH/CH$_2$Cl$_2$ (8 : 2)

These dimers **128b(p)** and **129b(o,p)** were prepared from **68(p)** and **125 (o,p)** in two steps with acceptable yields; firstly by treating them with a tenfold excess of 1,3-diiodopropane, and secondly by reacting the products **126(p)** and **127(o,p)** with **125b(o,p)**. The deacetylated porphyrins **128a(p)** or **129a(o,p)** were achieved in, respectively, 85%, 67% and 77% yields, after treating **128b(p)** and **129b(o,p)** with sodium methoxide in methanol [123, 124].

Knowing that glycopeptides are involved in important biological recognition phenomena and transport processes, it can be anticipated that porphyrins containing hydrophilic moieties, linked to cellular recognition elements, can be considered as promising candidates in PDT of cancer cells.

In this way, several examples of coupling amino acids to porphyrins have been reported in the literature.

Two series of amino acid porphyrin derivatives have been synthesized by Krausz et al. The first involves the coupling of adequate glycoporphyrins with 9-fluorenylmethoxycarbonyl-L-alanine (Fmoc-L-alanine) to give tri-, di-, and mono-alanine glycoporphyrin derivatives **130–133**, after deprotection of the amino groups [125, 126]; the second group (**134**) involves a glucosylamino acid moiety instead of the alanine [125, 126]. Such examples of porphyrins are shown in Fig. 17. All the glycosylated porphyrinic precursors were prepared by pyrrole/aldehyde condensation methodology.

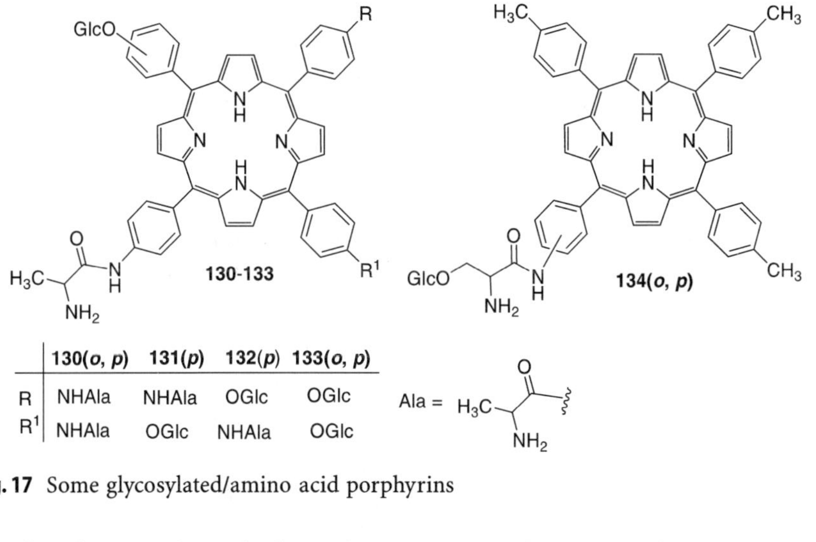

	130(*o, p*)	131(*p*)	132(*p*)	133(*o, p*)
R	NHAla	NHAla	OGlc	OGlc
R¹	NHAla	OGlc	NHAla	OGlc

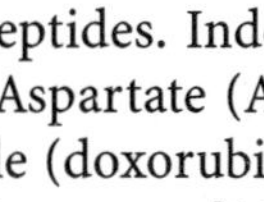

Fig. 17 Some glycosylated/amino acid porphyrins

Also glycoporphyrin biological properties can be enhanced in a compound of that type if it is coupled with certain peptides. Indeed, it was shown that certain tripeptides, like Arginine-Glycine-Aspartate (Arg-Gly-Asp; RGD motif), linked to a chemotherapeutic molecule (doxorubicin), helps the drug to reach the new blood vessels that nourish the tumor [125].

Glycoporphyrins bearing the RGD tripeptide might become promising candidates for medicinal applications [127]. Thus, several porphyrins bearing the RGD moiety were synthesized, of which **137a(o,p)** are two examples. Further studies have demonstrated that there was an increase in their plasmatic lifetimes, which presumably is due to the enhancement of the water solubility of the macrocycles.

The general procedure for the synthesis of porphyrins **137** consists of two steps: (i) the formation of porphyrins **135(o,p)**, each bearing a carboxylic acid function linked to the macrocycle through a spacer arm (Scheme 23), and (ii) the solid-phase connection of these porphyrins to the RGD peptide grafted on a Wang resin gave the corresponding precursors **136(o,p)**; these gave rise to porphyrins **137a(o,p)**, by treatment with TFA and then with sodium methoxide in methanol (Scheme 23) [127]. The porphyrins **135(o,p)** were achieved by treatment of porphyrin **125** with ethyl 4-bromobutyrate with K_2CO_3 in dry DMF followed by saponification.

Scheme 23 i) *N,N′*-diisopropylcarbodiimide, BuOH, DMF, 24 h, rt; ii) TFA, CH_2Cl_2/anisole, 4 h, rt; iii) NaOMe, MeOH/CH_2Cl_2, 2 h, rt

α,α'-Di-unsubstituted dipyrrylmethanes can be used in the synthesis of many carbohydrate-porphyrin derivatives. These dipyrrolic units can be substituted at the β-pyrrolic positions, at the *meso*-, or at both *meso*- and β-positions; also the aldehydes used to couple with two pyrromethane units can have several types of substituents. A wide variety of substituted end-products can then be obtained.

This synthetic methodology has been used in the condensation of *meso*(*p*-tolyl)pyrromethane with glycosylated aldehydes (Scheme 24), to give rise to porphyrins **124** [128–130].

Scheme 24 i) (**a**) BF₃ OEt₂, CH₂Cl₂; (**b**) *p*-chloranil; iii) NaOMe, MeOH/CH₂Cl₂

Hombrecher et al. also prepared β-substituted glycosylated di-arylporphyrins, to be used in photodynamic therapy studies, containing sugar and cholesterol moieties [131]. The synthesis was performed by coupling of a galactopyranosyl-substituted pyrromethane **138** with a cholesteryloxy-substituted pyrromethane **139**, in the presence of an imminium salt, followed by oxidation of the intermediate porphyrinogen (Scheme 25). The deprotection of the sugar moiety of compound **140b** was achieved by treatment with TFA/H₂O (9 : 1), giving **140a** in very good yield [108].

Using a similar methodology several *meso*-monoaryl-β-substituted derivatives **141-147** were prepared (Fig. 18) [107]. These glycosylated derivatives showed high binding constants with liposomes and LDL; this is an important property bringing high selectivity to cancer cells.

Lactosylated 5,15-diphenylporphyrinatoiron(III) chloride derivatives showed unique colorimetric responses to certain metal ions [79] (Sect. 1). The 5,15-bis(4-*per*-acetyl-β-lactosyloxyphenyl)porphyrin **149b** was synthesized by condensation, in dry dichloromethane and in the presence of TFA, of pyrromethane **148** with 4-*per*-acetyl-β-lactosyloxybenzaldehyde **94b(p)**, followed by *p*-chloranil oxidation of the intermediate porphyrinogen (Scheme 26). The corresponding iron(III) complex **150b** was obtained from **149b** after treatment with FeCl₂ in DMF at 60 °C.

Handling of **150b** with ammonia in methanol resulted in the deacetylation derivative 5,15-bis(4-β-lactosyloxyphenyl)porphyrinatoiron(III) **150a**, which dimerizes in basic solution affording the corresponding μ-oxo-bis[porphyrinatoiron(III)] **9** [67, 132].

Our research group [133, 134] has put forward a synthetic route to prepare 5,15-bis-aryl-10-monoglycoporphyrin derivatives **155** and **156** (Scheme 27).

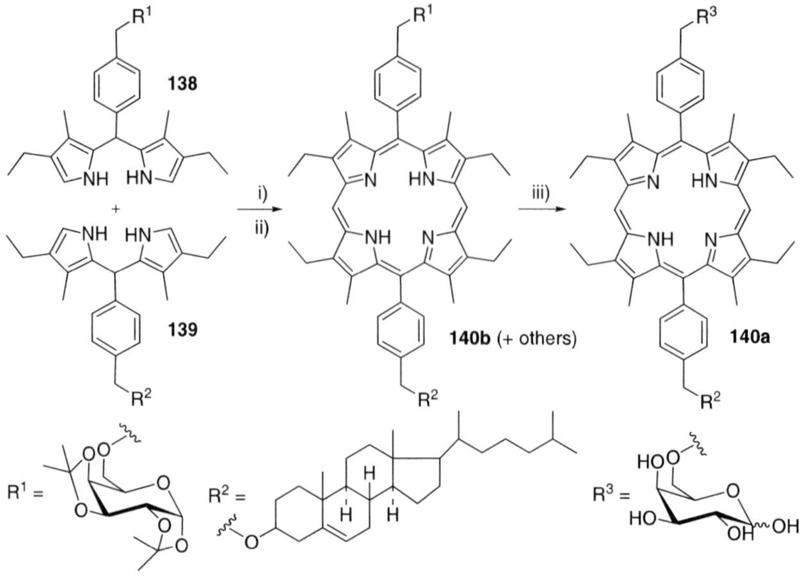

Scheme 25 i) $CH_2 = N(Et)_2Cl$, CH_2Cl_2/CH_3CN; ii) $K_3[Fe(CN)_6]$ or DDQ; iii) TFA/H_2O (9 : 1)

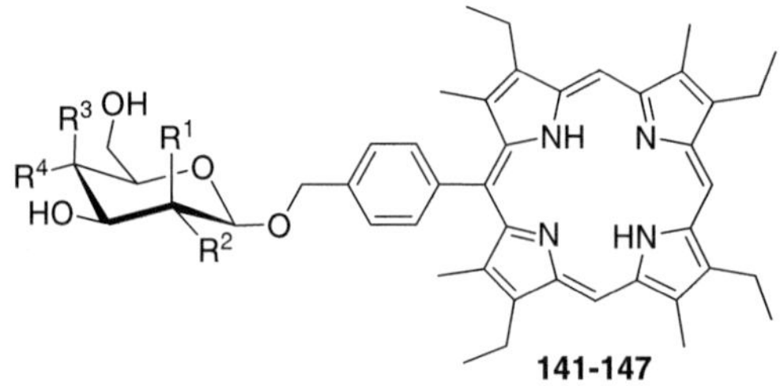

141-147

	141	142	143	144	145	146	147
	Man	Gal	Glc	Glc-2-NH	Mal	Lac	Cellobiose
R^1	OH	H	H	H	H	H	H
R^2	H	OH	OH	NHAc	OH	OH	OH
R^3	H	OH	H	H	H	H	H
R^4	OH	H	OH	OH	α(1,4)Glc	β(1,4)Gal	β(1,4)Glc

Fig. 18 Structures of several β-substituted *meso*-monoaryl carbohydrate-porphyrins

The work started with the synthesis of the free base 5,15-bis(3-metoxyphenyl)-porphyrin **151** by condensation of the pyrromethane **148** with 3-methoxy-benzaldehyde, followed by oxidation of the intermediate porphyrinogen. This porphyrin was metallated with nickel(II) and then, under Vilsmeier conditions (POCl$_3$/DMF), it was formylated in one of the two *meso*-free

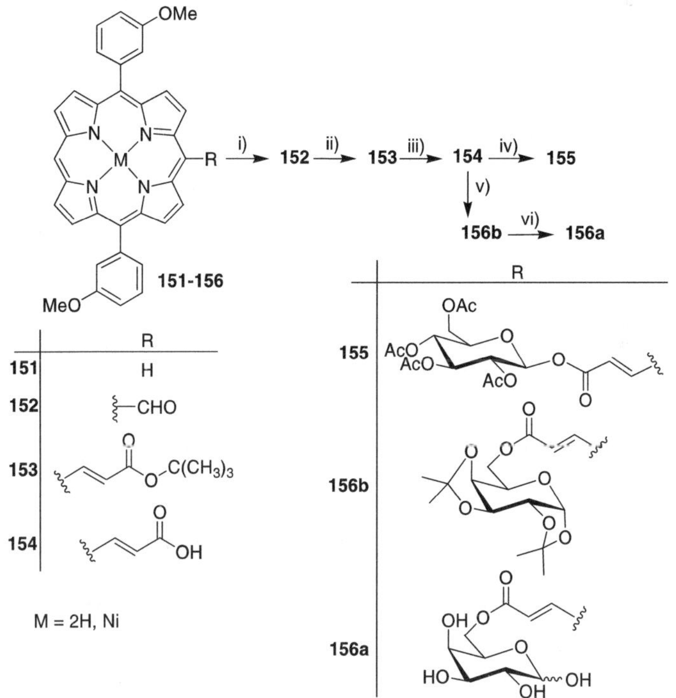

Scheme 26 i) (a) TFA, CH_2Cl_2, rt, (b) p-chloranil; ii) $FeCl_2$, DMF, 60 °C; iii) aqueous ammonia, MeOH, rt; iv) Tris-HCl buffer (pH > 8.5)

Scheme 27 i) $POCl_3$/DMF, dry $CHCl_3$, 40 °C, 8 h; ii) $Ph_3P = CHCO_2C(CH_3)_3$, CH_2Cl_2, reflux, 6.5–24 h; iii) HCO_2H, $CHCl_3$, rt, 7 h; iv) 2,3,4,6-tetra-O-acetyl-1-bromo-α-D-glucopyranose, DMF, K_2CO_3, rt, 4 h; v) 6-iodo-1,2:3,4-di-O-isopropylidene-α-D-galactopyranose, DMF, K_2CO_3, 80 °C, 24 h; vi) TFA/H_2O (9 : 1), rt, 30 min

positions giving **152** (86%). The appended formyl group was used to get the derivatives **153** in 97% yield, using a Wittig reaction with (*tert*-butoxycarbonylmethylene)triphenylphosphorane. Removal of the *tert*-butyl groups of **153** was performed with formic acid, at room temperature, giving the carboxylic acid **154** (96%). The reaction of **154** with 2,3,4,6-tetra-*O*-acetyl-1-bromo-α-D-glucopyranose and 6-iodo-1,2:3,4-di-*O*-isopropylidene-α-D-galactopyranose, in the presence of potassium carbonate, afforded the target compounds **155** and **156b** in very good yields (90–94%). Cleavage of the carbohydrate protection groups in **156b** was performed with aqueous TFA at room temperature, affording **156a** quantitatively [47, 133, 134].

Strapped glycoporphyrins have also received great interest in the past years in relation to their three-dimensional organization which confers to these structures stereoselective catalytic and biomimetic properties [135–137]. For example, such compounds have been designed as models to mimic the oxygen binding to hemoglobin and myoglobin and the oxidations that are catalyzed by cytochrome P_{450} enzymes.

The condensation of *meso*-(*p*-tolyl)pyrromethane with the glycosylated bisaldehyde 6,6′-di-*O*-(3-formyl)benzoyl-2,3,4,2′,3′,4′-hexa-*O*-trimethylsilyl-α,α'-trehalose **157**, using the Lindsey's method [138], gave compound **158b** (Scheme 28), an example of a strapped glycoporphyrin. The target compound **158a** was obtained by deprotection of the sugar hydroxyl groups [139].

Scheme 28 i) $BF_3 \cdot OEt_2$, *p*-chloranil, CH_2Cl_2; ii) Bu_4NF

3.3.2
Synthesis of C-Glycosylarylporphyrins

Both the yield and product-stability are crucial factors for the successful incorporation of carbohydrate moieties into porphyrins. A possible problem with *O*-glycoside derivatives is the possibility of their cleavage by glycohydrolases in biological systems. In this way, saccharide-drug conjugates using *C*- or *S*-glycoside linkages have been made, including several porphyrin derivatives; in some cases *C*- and *S*-glycosylated arylporphyrins have been obtained in high yields [109].

Franck et al. [109] suggested a strategy for the synthesis of *C*-glycosyl benzaldehyde derivatives to be used in porphyrin preparations. They started by condensing 2,3,4,6-tetra-*O*-acetyl-1-thio-β-D-glucopyranose with α,α'-dibromo-*p*-xylene to afford thioglucoside **159** (Scheme 29). The latter, by

Scheme 29 i) NaH, THF, rt, 2 h; ii) TBDMSOH, Ag(OTf), 2,6-di-*tert*-butylpyridine, CH$_2$Cl$_2$, rt, 3 h; iii) NaOMe, MeOH, rt, 3 h; iv) NaH, BnBr, Bu$_4$NI, THF/DMF, rt, 8 h, for 2 steps; v) monoperoxyphthalic acid, THF/EtOH/H$_2$O, 60 °C, 2 h; vi) CBr$_2$F$_2$, KOH 25% on alumina, CH$_2$Cl$_2$/*t*-BuOH, 0 °C to rt, 3 h; vii) H$_2$, 5% Pd on alumina, EtOAc, rt, 12 h; viii) Bu$_4$NF, THF, rt, 2 h; ix) Oxalyl chloride, DMSO, CH$_2$Cl$_2$, – 78 °C, then Et$_3$N, – 78 °C to rt, 1.5 h

treatment with *tert*-butyldimethylsilanol (TBDMSOH) in the presence of silver triflate and 2,6-di-*tert*-butylpyridine, was converted into the silylated derivative **160**; this compound was then successively treated with sodium methoxide and benzyl bromide, giving rise to thioglycoside **161**. The latter was then oxidized to the sulfone **162** with monoperoxyphthalic acid; the resulting sulfone was employed in the Ramberg–Bäcklund synthesis of the *exo*-glucal **163** under Chan's conditions [140]. The two isomers (*Z*)-**163** and (*E*)-**163** in the 8 : 2 ratio were isolated and identified by NOE measurements. In some cases the intermediate α-bromosulfone **164** was isolated from the reaction mixture and then converted to the *exo*-glucal **163** by treatment with base. Hydrogenation of **163** with 5% palladium on alumina afforded the β-*C*-glucosyl derivative **165**; this compound, after cleavage of the silyl ether, gave rise to the alcohol **166**, which afforded the desired aldehyde **167** by Swern oxidation.

The condensation of aldehyde **167**, or a mixture of **167** and benzaldehyde, with pyrrole, under Lindsey's conditions, gave porphyrins **168b** and **169b** in 53 and 15% yields, respectively (Scheme 30). The hydrogenolysis of the benzylic protecting groups yielded the porphyrins **168a** and **169a** [109].

Scheme 30 i) $BF_3 \cdot OEt_2$, NaCl, rt, 5 h, then DDQ, rt, 30 min; ii) H_2, 10% Pd on carbon, EtOAc/MeOH, rt, 16 h

3.3.3
Porphyrins Containing Glyco-Moieties Directly Linked to the Macrocycle

The interest in carbon-linked nucleoside models led Casiraghi et al. to study the direct *C*-heteroarylation of acyclic and cyclic sugars as new porphyrin derivatives with biological significance [141–143]. Glycoporphyrins with sugar moieties anchored to the *meso* carbon of the macrocycle by means of chemically and metabolically robust carbon-carbon bonds were obtained by condensation of the dipyrrylmethane units **170**, with 4-fluorobenzaldehyde or benzaldehyde, in dry CH_2Cl_2 in the dark and under an argon atmosphere in the presence of either trifluoroacetic acid or BF_3-etherate. The intermediate porphyrinogens were then oxidized with DDQ and the required porphyrins **171–177** were obtained in acceptable yields (Scheme 31) [142, 143].

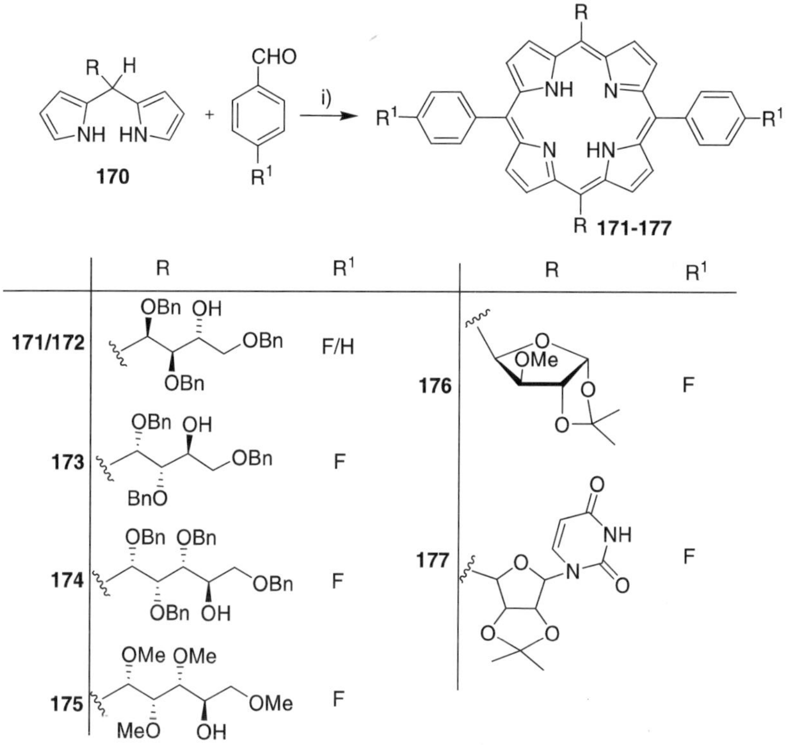

Scheme 31 i) (**a**) TFA or BF_3 OEt_2, dry CH_2Cl_2, rt, Argon; (**b**) DDQ

The required α,α'-free dipyrrylmethanes **170** were prepared by condensation of glyceraldehyde, furanosyl and pyranosyl aldose derivatives (arabinose, glucose, xylose) with pyrrole, in dichloromethane, at room temperature in the presence of $SnCl_4$ (Scheme 32).

meso-C-Tetraglycosylated porphyrins have also been prepared from the condensation of pyrrole and glycosylated aldehydes by the following known procedures [144–146]. When using 2,3-*O*-isopropylidene-D-glyceraldehyde **A** two major atropisomeric porphyrins $\alpha,\beta,\alpha,\beta$ and $\alpha,\alpha,\alpha,\beta$ of **179** were obtained [144]. The procedure was extended to the use of derivative **F**

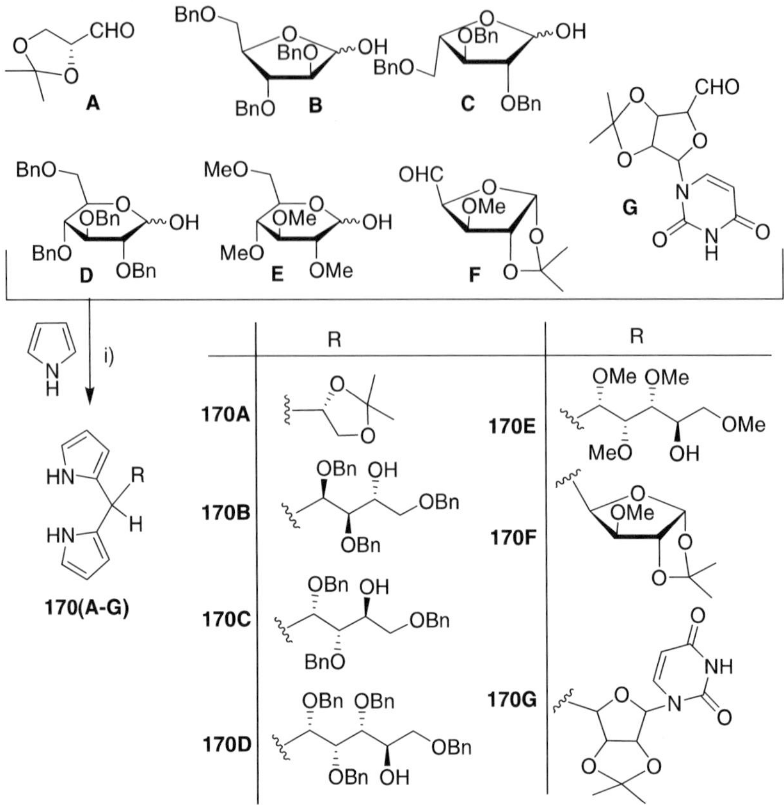

Scheme 32 i) CH_2Cl_2, $SnCl_4$, N_2, 1 h, rt

and to both arabinose enantiomer derivatives **178** and **ent-178**; however, in these cases, only the atropisomeric forms $\alpha,\beta,\alpha,\beta$ of **180–183** were detected. Moreover, when both arabinose derivatives **178** were used, monodeprotected porphyrin **181** and bis-deprotected porphyrin **182** were identified (Scheme 33).

The complete deprotection of all acetonide groups of **180, ent-180, 181** and **182** was attained by treatment with 50% TFA in CH_2Cl_2 at room temperature under ultrasonic irradiation. The resulting porphyrin derivatives became insoluble in most organic solvents but they were soluble in methanol or ethanol and also in water, and this might be a useful feature in the areas of chiral recognition and asymmetric catalysis [144].

Several metal complexes were prepared with the new products; such complexes were evaluated for their efficiencies as DNA cleavers upon exposure to visible light. The attained results enlighten the role played by the water-solubilizing carbohydrate moieties attached to the lipophilic porphyrin macrocycle during the cleavage of the DNA target [145, 146].

Scheme 33 i) BF$_3$·OEt$_2$, dry CH$_2$Cl$_2$, rt, Argon, 3 h, then DDQ, Et$_3$N, rt, 18 h

Maruyama et al. have also reported the synthesis of porphyrins with four sugar groups directly attached to the β-positions of the porphyrin ring [50]. For this, they synthesized the pyrrole carbohydrate derivatives described in Scheme 34. The glycoaldehyde **184** derived from galactose was converted into the corresponding nitroalkene **185** by condensation with nitroethane. Treatment of **185** with ethyl isocyanoacetate, in the presence of DBU at room temperature, gave pyrrole **186** in 63% yield (Scheme 34). Porphyrin **187b** was obtained in 18% yield after reduction of pyrrole ester **186** by LiAlH$_4$, followed by the acid-catalyzed tetramerization of the resultant pyrrole and oxidation of the formed porphyrinogen with p-chloranil.

Although the tetramerization of unsymmetrical pyrroles usually gives a mixture of four possible porphyrin isomers, in this case that was not observed presumably due to the steric effect brought by the sugar moieties.

The desired porphyrin water-soluble **187a** was obtained when the protecting groups of the carbohydrate moieties of **187b** were removed by treatment with cation exchange resin in refluxing water saturated with chloroform for 5 h.

Scheme 34 i) KOAc, HC(OCH$_3$)$_3$, NO$_2$CH$_2$CH$_3$, MeOH, reflux, 4 h; ii) CNCH$_2$CO$_2$Et, DBU, THF, rt, 8 h; iii) LiAlH$_4$, 0 °C to rt, 2 h; iv) p-TsOH, rt, 10 h; v) p-chloranil, rt, 3 h; vi) Amberlite IR-120, H$_2$O/CHCl$_3$, reflux, 5 h

3.4
Glycosylated Cationic Porphyrins

Application of porphyrins as PDT photosensitizers gave rise to a growing interest in water-soluble porphyrin derivatives. Thus, a number of water-soluble porphyrins have been studied so far, and most of them have ionic substituents. The combination of carbohydrate residues and positively charged groups allows not only water solubility, but also significant membrane permeability [11]. In addition, due to their enhanced affinity for DNA, they also have different binding modes with nucleic acids in comparison with the neutral porphyrin ones [147].

An easy way to prepare cationic porphyrins is concerned with the possibility to alkylate pyridyl substituent groups. To prepare pyridylglycoporphyrins, two different pathways can be followed: condensation of pyrrole with suitable pyridinecarboxaldehyde and glycobenzaldehyde derivatives or by coupling of a pyridyl porphyrin with an adequate sugar derivative.

In 1993, Bolbach et al. reported the synthesis of alkyl pyridinium-4-yl porphyrins bearing a carbohydrate moiety [148]. After that, other research groups have extended the range of structures of glycosyl and alkyl substituents.

A mixture of suitable glycosylated benzaldehyde and 4-pyridinecarbox-aldehyde was used to synthesize unsymmetrical porphyrin derivatives. For example, to prepare a porphyrin with three pyridyl and one protected mono- or disaccharide aryl group at the *meso*-positions, Bolbach at al. carried out the condensation of a mixture of pyrrole with glycosylacetylated benzaldehyde and 4-pyridinecarboxaldehyde in propionic acid with 10% acetic anhydride (which prevented the deacetylation of the glycosylated moiety). The yields of chromatographically pure porphyrins were about 6–7% (Scheme 35).

Scheme 35 i) $CH_3CH_2CO_2H/Ac_2O$ (10 : 1), reflux, 1.5 h; ii) R'-I, DMF; iii) $Et_3N/MeOH/H_2O$ (10 : 10 : 1)

Treatment of glycosylated pyridylporphyrins **188–190(o,p)** with an excess of methyl iodide in dry DMF gave the expected tri-cationic glycosylated *N*-methylpyridinium porphyrins **191–193(o,p)**. Treatment of **188b** with isopropyl iodide or n-octyl iodide, yielded the corresponding cationic porphyrins **194b** and **195b**. Compounds **188b–195b(o,p)** were then treated with $Et_3N/MeOH/H_2O$ (10 : 10 : 1) at 0 °C, giving rise to the unprotected-sugar porphyrins **188a–195a(o,p)** in nearly quantitative yields [147]. The deacetylated neutral and cationic porphyrin products are water soluble.

The second pathway to prepare pyridylglycoporphyrins considers the derivatization of pyridylporphyrins with carbohydrate derivatives (Scheme 36). Porphyrins should then have the required substituent groups for coupling with the sugar moieties. In this way the neutral and cationic acid series **198** and **199** were prepared from precursor **196**, previously activated to ester **197** and then reacted with the protected galactose derivative [134, 149].

Scheme 36 i) SOCl$_2$, pyridine, 50 °C, 30 min then *N*-hydroxysuccinimide, 50 °C, 3 h; ii) 1,2:3,4-di-*O*-isopropylidene-α-D-galactopyranose, NaH, toluene, rt, 90 min; iii) TFA/H$_2$O (9 : 1), rt, 30 min; iv) MeI, DMF, 40 °C, 3 h

As already described in Sect. 3.2.1 and in Sect. 3.2.2 the *p*-fluoro atoms of the *meso*-pentafluorophenyl groups can be substituted by nucleophilic compounds, such as thio derivatives [11]. The synthetic route leading to thioglycosylated ionic compounds was based on that methodology. Reaction between **200** and 2,3,4,6-tetra-*O*-acetyl-1-thio-β-D-glucopyranose, in DMSO at room temperature, gave porphyrin derivative **201** in 72% yield (Scheme 37). The alkylation of **201** with a range of alkyl iodides gave the cationic products **202b–205b**, as their iodide salts [11]. Finally, after deprotection of the

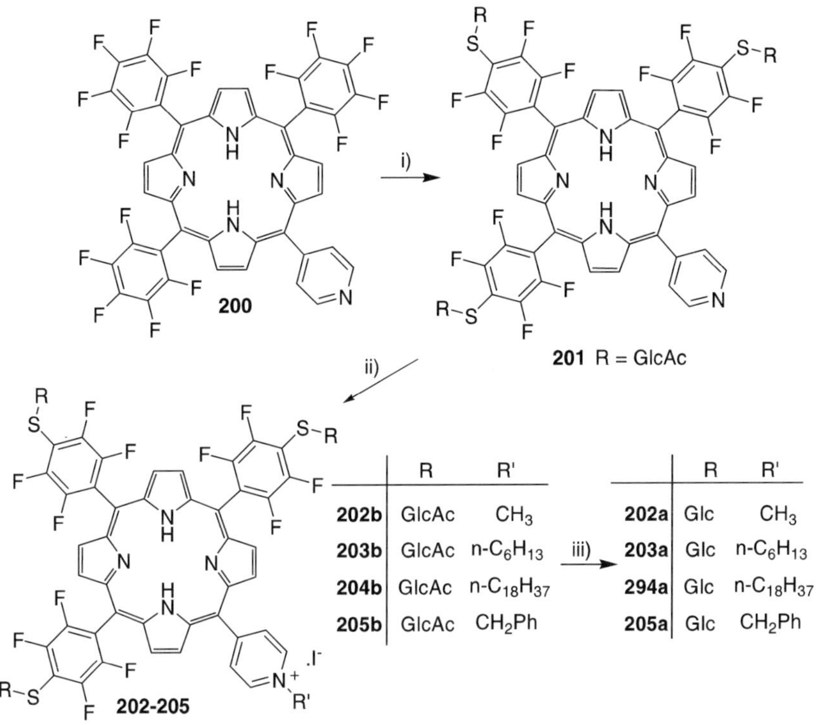

Scheme 37   i) 2,3,4,6-tetra-*O*-acetyl-1-thio-*β*-D-glucopyranose, DMF, rt, 16 h; ii) R'-I, DMF, rt, 16 h; iii) NaOMe, MeOH, rt, 1 h

sugar protecting groups using sodium methoxide in methanol, compounds **202a–205a** were obtained.

The alkylation of pyridyl groups can also be extended to the synthesis of oligomeric species. Using this strategy, Sol at al. prepared some glycosyl bis-porphyrin conjugates, based on the direct alkylation of the monopyridyl-triglycosylated porphyrin **206** with adequate iodo-compounds [123, 124]. The first example is concerned with the synthesis of dimer **207**. This was carried out as shown in Scheme 38. The glycosylderivative **206** (2 equiv.) reacted with 1,3-diiodopropane affording **207b**, which after deacetylation, under classical conditions, gave **207a** in high yield [123].

Using the same monopyridyl porphyrin **206**, Sol et al. prepared another two dimers [123]. In this case they used the derivative **206** and the suitable iodo-porphyrins **126b(p)** and **127b(p)** (Sect. 3.3.1.1) to prepare the asymmetric dimers **208b** and **209b** in 17 and 10% yields, respectively (Scheme 39); these dimers, by deacetylation, gave rise to the target glycosylated dimers **208a** and **209a**.

A variation of this *N*-alkylation procedure has allowed Kawakami et al. to synthesize tetracationic derivatives with the charge on each aryl group which

Scheme 38 i) $CH_3CH_2CO_2H$, reflux, 1 h; ii) $I(CH_2)_3I$, DMF, 20 h; iii) NaOMe, MeOH

Scheme 39 i) K_2CO_3, DMF, reflux, 48 h; ii) NaOMe, $MeOH/CH_2Cl_2$

also contains the sugar moiety. In this case, the cationization of 5,10,15,20-tetrakis(4-pyridyl)porphyrin **210** with 3-bromopropylamine hydrobromide afforded the derivative **211** (Scheme 40), which was then coupled to maltose [80] and lactose [81]. The resulting imines were reduced with sodium cyanoborohydride giving rise to the corresponding porphyrins **212** and **213**.

211 R = $-(CH_2)_3NH_2$

210

10, 11
212, 213

R =

212 M = 2H, R^1 = H, R^2 = OH
213 M = 2H, R^1 = OH, R^2 = H

10 M = Mn, R^1 = H, R^2 = OH
11 M = Mn, R^1 = OH, R^2 = H

Scheme 40 i) $Br(CH_2)_3NH_3Br$, DMF, reflux, 3 d, followed by precipitation with Et_4NCl; ii) (a) maltose monohydrate (**212**) or lactose monohydrate (**213**), sodium borate buffer (0.1 M, pH 8.5), 40 °C, 3 weeks for derivative **212** or 3 d for derivative **213**; (b) $NaBH_3CN$ 0.3 M, 40 °C, 4 d, iii) $Mn(OAc)_2 \cdot 4H_2O$, H_2O, 40 °C, 24 h

The corresponding manganese derivatives **10** and **11** were considered to mimic the SOD and cellular recognition.

Recently, another synthetic strategy has been considered for a different type of glycosylated cationic porphyrins. This involved the pyridyl groups cationization with iodocarbohydrate derivatives [149, 150]. In this case cationic porphyrin derivatives having a galactose or a bis(isopropylidene)-galactose unit directly linked to a pyridyl group have been synthesized. The groups containing the charges can be located at the *meso*-positions, as in porphyrins **214** and **215**, or at β-positions, as in porphyrins **216** (Scheme 41) [149, 150].

The new glycopyridinium porphyrins **214b** and **215b** were synthesized by coupling 5-(4-pyridyl)-10,15,20-triphenylporphyrin and 5,10,15,20-tetrakis(4-pyridyl)porphyrin with 6-iodo-1,2:3,4-di-*O*-isopropylidene-α-D-galactopyranose. The cationic porphyrin **216b** was synthesized in a one-step procedure involving an aldol condensation of the 2-formyl-5,10,15,20-tetraphenylporphyrin with *p*-methyl-*N*-galactopyridinium chloride [150, 151].

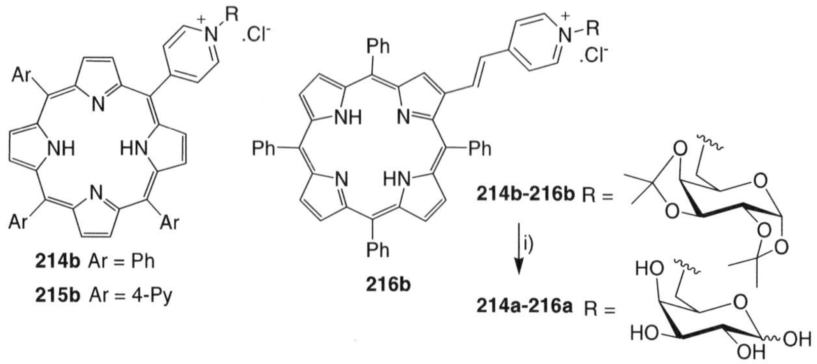

Scheme 41 i) TFA/H$_2$O (9 : 1), 20 min followed by washing with brine solution

The isopropylidene protecting groups of the sugar units were easily removed by acid treatment, leading to porphyrins **214a–216a**.

Cavaleiro et al. also proposed a different way to obtain cationized glycoporphyrins **217**, by promoting the alkylation of *p*-aminophenylTPP with 6-iodo-1,2:3,4-di-*O*-isopropylidene-α-D-galactopyranose followed by *N,N*-dimethylation (Fig. 19). In this case glyco-quaternized ammonium porphyrin salts **217b** and **217a** [150] were obtained.

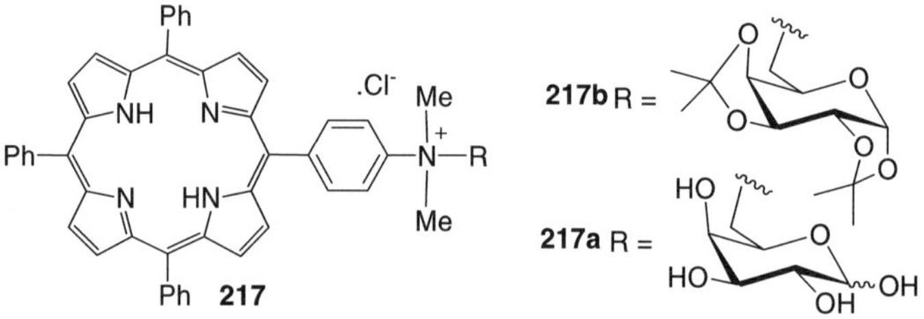

Fig. 19 Quaternary ammonium glycoporphyrin derivatives

4
Glycochlorins as new Dihydroglycoporphyrin Derivatives

While Photofrin® has proven to be effective in the treatment of various types of cancers, there are features that could be improved by new photosensitizers. The development of PDT agents that absorb light above 630 nm (where tissues are more transparent) is desirable because it would allow a greater depth of penetration and potentially greater destruction of malignant tissues. With such targets in mind a second generation of photosensitizers, like chlorins and related compounds, have been developed and evaluated as putative PDT agents in recent years [7, 8, 36].

4.1
Synthesis of Glycoconjugated Chlorins

Several research groups probably inspired by both the attractive features of PDT and the commercialization of Photofrin® and Visudyne®, have put considerable effort into the development and study of the so-called second-generation photosensitizers [152].

Momenteau et al. described the preparation of amphiphilic glycoconjugated *meso*-monoarylbenzochlorins from *meso*-monoarylporphyrins [153]. The metalloporphyrin **218** when treated with 3-(dimethylamino)acrolein under Vilsmeier conditions led to two isomeric compounds **219** and **220** in 85% total yield and 2 : 3 ratio; treatment of isomer **219** with TFA afforded the benzochlorin **221** in 58% yield and only one of the two possible benzochlorin isomers was obtained (Scheme 42). The dealkylation of the methoxy group

221 $\xrightarrow{\text{iii)}}$ **222** $\xrightarrow{\text{iv)}}$ **223** $\xrightarrow{\text{v)}}$ **224b** $\xrightarrow{\text{vi)}}$ **224a**

	M	R
221	Ni	CH_3
222	Ni	H
223	2H	H
224b	2H	AcMal
224a	2H	Mal

Scheme 42 i) 3-(dimethylamino)acrolein/POCl$_3$; ii) TFA, Argon; iii) BBr$_3$, dry CH$_2$Cl$_2$; iv) H$_2$SO$_4$ conc.; v) 1-bromoethoxy-*per*-acetyl-maltose, K$_2$CO$_3$, DMF, 60 °C, vi) NaOMe, MeOH

by boron tribromide gave compound **222**, which was demetallated in concentrated sulfuric acid to afford the metal free benzochlorin **223**.

The glycosylation of the benzochlorin **223** was performed, in 95% yield, with 1-bromoethoxy-*per*-acetyl-maltose, in DMF and in the presence of potassium carbonate, giving compound **224b**, from which the maltosylchlorin **224a** was obtained after deacetylation.

On the basis of the fact that glycoconjugation of the porphyrin macrocycle appears to create a balance between hydrophobicity and hydrophilicity, Blais et al. prepared tri- and tetra-*meta*-glucosylated chlorin derivatives [154]. The aim was to assess how the sugar units linked to chlorin derivatives affect the photoactivity, cell internalization and subcellular localization in HT29 human adenocarcinoma cells. This was compared with the action due to *meso*-tetrakis(*m*-hydroxyphenyl)chlorin, a compound being formulated as a photosensitizer (Foscan®) for palliative treatment of head and neck cancers.

The chlorin derivatives **226**, **227** and **228** were prepared from the corresponding glucoconjugate porphyrin derivatives **100b(m)** and **225** [106], by the diimide reduction procedure [155] to achieve the reduced macrocycles (Scheme 43).

While the reduction of the tetra-glucoconjugated porphyrin **100b(m)** reached the compound **226b** in 89% yield, the triglucosylated porphyrin **225**, when submitted to the same reduction process gave, in 72% yield, a mixture of the two isomeric chlorins **227b** and **228b** in a 1 : 1 ratio. The gluco-unprotected derivatives **226a–228a** were obtained in quantitative yield by treatment of chlorins **226b–228b** with sodium methoxide in methanol. The presence of the sugar moieties in the chlorin structures brought a change in the octanol-water partition coefficient. Also the biological evaluation carried out with these compounds revealed that the tetra-glucoconjugated chlorin **226a** was poorly internalized and weakly photoactive but the asymmetrical triglucosylated chlorins **227a** and **228a** exhibit a high in vitro photoactivity and a preferential mitochondrial affinity. In spite of the lower levels of cell uptake of **227a** and **228a** compared to Foscan®, the phototoxicity of the triglucosylated compounds was higher. The cellular uptake and the cells localization seem to be correlated with the amphiphilicity of the compounds. These results have encouraged the evaluation of the photodynamic activity of the glucosylated chlorins in vivo and to study their metabolism in vivo or in cell-based assays. The in vivo metabolism is an important issue when considering the use of glycoconjugated compounds in PDT treatment, since cleavage of the glycoside bond by glycosidases will result in modifications of amphiphilic properties, biodistribution, blood clearance and drug-cell interaction [156]. There is indication that the glycosidase activity is higher in certain tissues especially in tumoral tissues when compared to normal ones. Glycosidase-promoted metabolism of glycoconjugated porphyrins and chlorins can thus occur after in vivo administration of these compounds. However, a certain specificity and stability can be achieved. In fact, studies carried out on the in

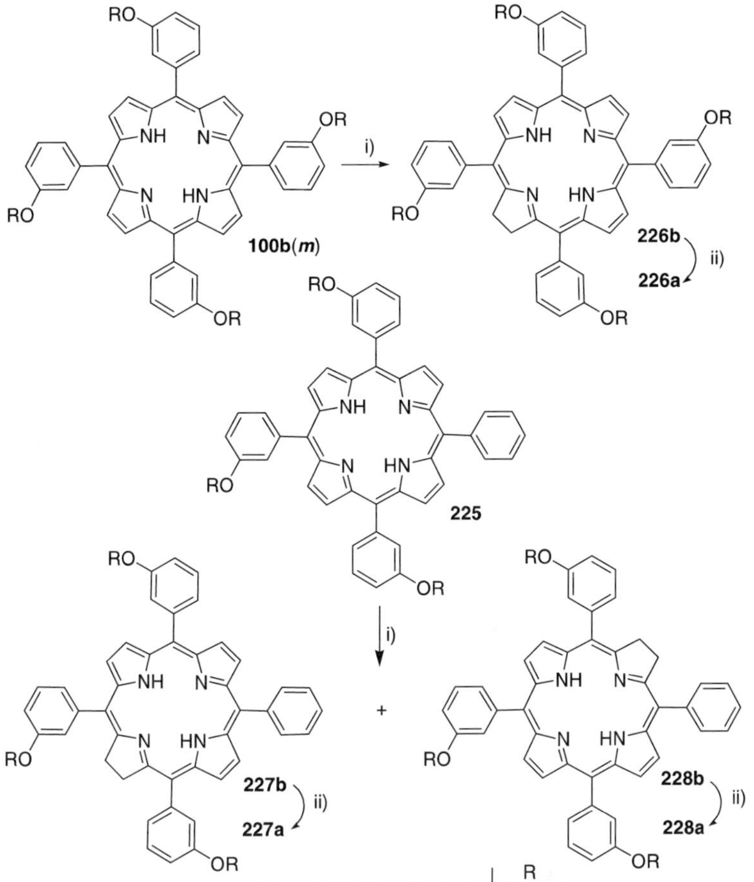

	R
100b(*m*), 225, 226b-228b	GlcAc
226a-228a	Glc

Scheme 43 i) (a) *p*-toluenesulfonylhydrazide, K_2CO_3, dry pyridine, 100 °C, argon, (b) ethyl acetate, *o*-chloranil; ii) NaOMe, MeOH

vitro metabolism of compounds **227a** and **228a** have demonstrated that the three sugar moieties in chlorins **227a** and **228a** undergo sequential hydrolysis and also there is dehydrogenation of the macrocycles to the corresponding porphyrins [156].

Following the same methodology Yano et al. prepared several *meta* and *para* tetra-glycoconjugate chlorin derivatives [157]. Starting with *meta* and *para* tetra-*β*-glycopyranosylporphyrin conjugates **106b(p)**, **229b–231b**, the chlorin derivatives **232b–234b** were obtained after treatment with *p*-toluenesulfonylhydrazide (Scheme 44) [155]. These, after deacetylation, gave the chlorins **232a–234a**.

106(*p*),229-231

(*m*)	(*p*)	(*m,p*)	R
229b	106b		GalAc
229a	106a		Gal
		230b	XylAc
		230a	Xyl
		231b	AraAc
		231a	Ara

232b-234b / **232a-234a**

(*m,p*)	R
232b	GalAc
232a	Gal
233b	XylAc
233a	Xyl
234b	AraAc
234a	Ara

Scheme 44 i) (a) *p*-toluenesulfonylhydrazide, K_2CO_3, dry pyridine, $100\,°C$, N_2; (b) benzene, *o*-chloranil, rt; ii) NaOMe, MeOH/CHCl$_3$

The deacetylated glycopyranosyl conjugate derivatives were biologically evaluated in vitro with HeLa cells and the results compared with those due to tetrasulfonated porphyrin (TPPS$_4$) [157, 158]. The glycochlorins **232a–234a(m,p)** showed higher uptake by HeLa cells than TPPS$_4$ and cellular uptake similar or greater than the corresponding porphyrins **106a, 229a–231a**.

Cavaleiro et al. developed methodologies based on cycloaddition transformations of porphyrins leading to the synthesis of compounds like chlorins and bacteriochlorins which have strong absorptions in the visible region near or above 650 nm. The reactive diene and dipolarophile species have been *o*-quinodimethanes [159], azomethine ylides [160] and nitrones [161]. The 1,3-dipolar cycloaddition approach led to the synthesis of glycoderivatives of the chlorin and bacteriochlorin types (Scheme 45) [162].

Heating *meso*-tetrakis(pentafluorophenyl)porphyrin **48** with an excess of sugar nitrones **235(A–D)** in toluene has afforded chlorins **236(A–D)** and, in some cases, bacteriochlorins **237** and **238** (Fig. 20) by *endo*-additions. The galactochlorin **236A** and two galactobacteriochlorins **237A** and **238A** (four diasteriomeric bacteriochlorins are possible under the *endo*-addition) were formed when galactosyl nitrone **235A** was used. Moreover, it was possible to

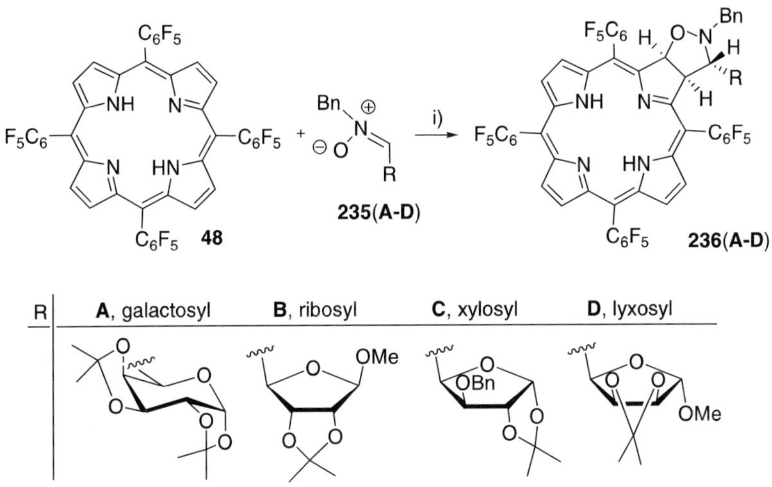

R	A, galactosyl	B, ribosyl	C, xylosyl	D, lyxosyl

Scheme 45 i) Toluene, N_2, 60–100 °C, 2–5 d

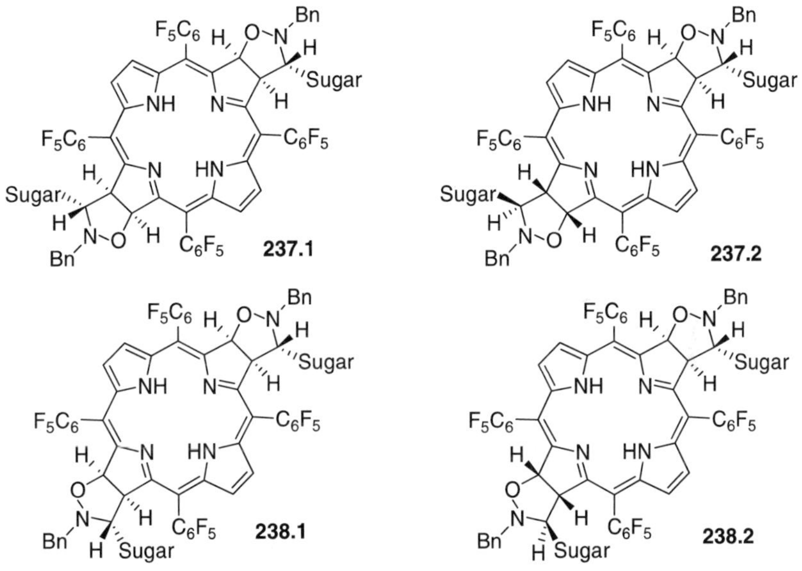

Fig. 20 Four possible diasteriomeric bacteriochlorin products, with two *endo*-1,3-dipolar cycloadditions

obtain the bacteriochlorins **237A** and **238A** by treatment of chlorin **236A** with an additional amount of sugar nitrone **235A**.

With the lyxosyl nitrone **235D**, a single bacteriochlorin **237D** as the major product and chlorin **236D** as the minor one have been formed; the use of ribosyl nitrone was quite stereoselective giving rise to the chlorin **236B** while xylosyl nitrone afforded the chlorin **236C** and trace amounts of the corresponding bacteriochlorin.

Cavaleiro et al. extended this cycloaddition approach to the synthesis of pyrrolidinochlorin glycoconjugates via sugar azomethine ylides [163]. For that purpose *C*- and *N*-glycoconjugated pyrrolidine-fused chlorins **240** and **243** were prepared. The reaction of *meso*-tetrakis(pentafluorophenyl)-porphyrin **48** with azomethine ylide **239**, generated from the galactosyl aldehyde **184** and *N*-methylglycine in refluxing toluene, afforded the two *C*-glycoconjugated pyrrolidine-fused chlorins **240** in 51% total yield (Scheme 46). The two diasteriomeric chlorins **240.1** and **240.2** were formed by an *endo*-cycloaddition process to each diastereoface of the azomethine ylide **239**; however, this facial selectivity was rather low as the two products were isolated in 32% and 19% yield, respectively. To avoid the problem of diastereoisomer formation, porphyrin **48** was reacted with the *N*-substituted symmetrical azomethine ylide **242** generated in situ by condensation of the galactose-substituted glycine derivative **241** with formaldehyde. The chlorin **243** was isolated in 19% yield [163].

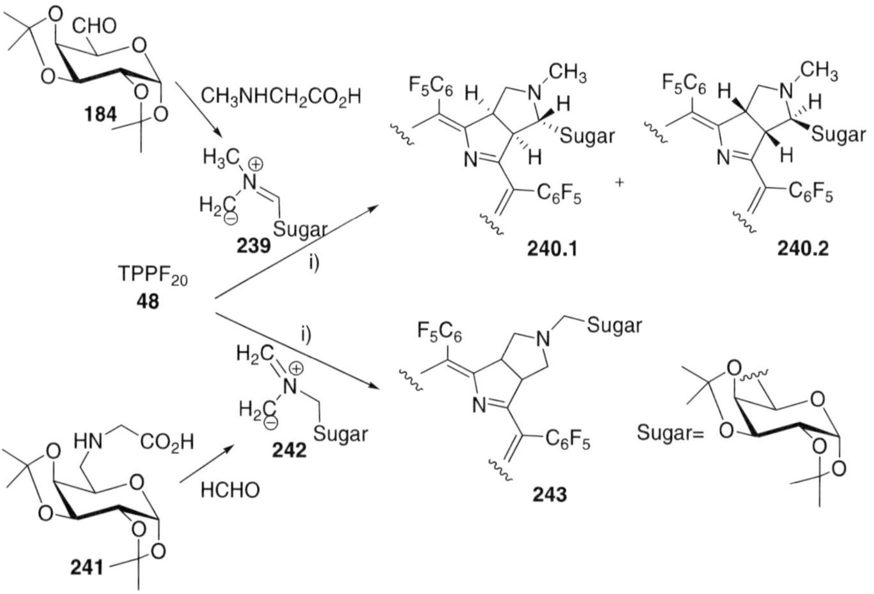

Scheme 46 i) Toluene, N_2, reflux, 2–4 h

Pandey et al. developed a synthesis leading to β-substituted glycobenzochlorins with variable lipophilicity [48].

The reactions between benzochlorin **244** and acetylated galactosamine, glucosamine and lactosamine were performed in dry dichloromethane in the presence of triethylamine and the coupling reagent BOP. The corresponding benzochlorin-carbohydrate conjugates **245b**, **246b** and **247b** were obtained in 40%, 97% and 43% yields, respectively. The sodium methoxide treatment gave the corresponding hydroxyl derivatives **245a–247a** in good yields (Scheme 47).

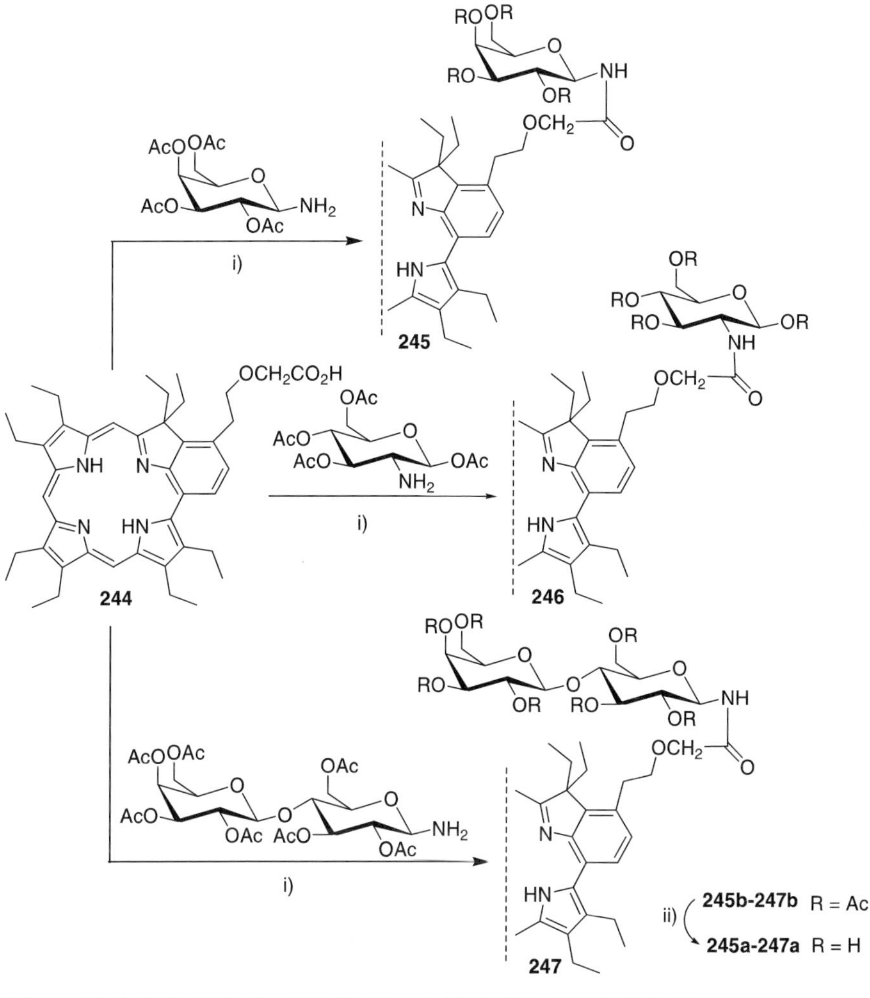

Scheme 47 i) BOP, Et$_3$N, dry CH$_2$Cl$_2$, N$_2$, rt; ii) NaOMe, MeOH/CH$_2$Cl$_2$

The same research group has also prepared benzochlorin-*C*-galactose conjugates to investigate their PDT efficacy.

The benzochlorin-*C*-galactose conjugate **249b** was obtained in high yield by condensation of benzochlorin **244** with the galactopyranose derivative **248** (see Scheme 49) using the DCC method. Its Zn(II) complex **250b** was hydrogenolized to afford compound **250a** in 75% yield. Treatment of **250a** with

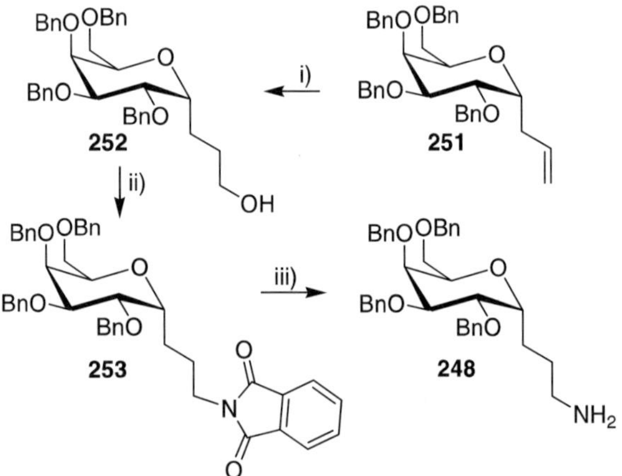

Scheme 48   i) DCC/DMAP; ii) (a) DMAP, CH_2Cl_2, N_2, rt; (b) $Zn(CH_3CO_2)_2$, MeOH; iii) EtOH, Pd/C; N_2, rt; iv) TFA

Scheme 49   i) 9-BBN in THF, 65 °C, 16 h followed by NaOH (3N), EtOH, 30% H_2O_2, 55 °C, 1 h; ii) Ph_3P/phthalimide/diethyl azodicarboxylate, 0 °C to rt, N_2, 20 h; iii) EtOH, $NH_2NH_2 \cdot H_2O$, reflux

TFA gave rise to the free-base benzochlorin-galactose conjugate **249a** in 97% yield (Scheme 48) [48].

The galactopyranose derivative **248** required for coupling with the chlorin macrocycle was synthesized by following the reaction sequences described in Scheme 49 and using compounds **251–253** [48].

Another approach to prepare other substituted glycochlorins involves the transformation of natural derived macrocycles. For instance, with mesopurpurin-18 methyl ester **254**, and following a methodology developed by Pandey et al. [164], the β-galactopyranosyl conjugate purpurinimides were prepared. The reaction of **254** with propargylamine, at reflux in benzene, produces the corresponding propargylimide derivative **255** in 80% yield. This derivative was used as the alkyne component for an enyne cross metathesis [165]. The alkene component, the acetylated allyl β-D-galactopyranoside **256bA**, was prepared from galactopyranose penta-acetate by BF$_3$-etherate-induced glycosylation [166]. The enyne cross metathesis occurs between compounds **255** and **256bA**, in CH$_2$Cl$_2$ and in the presence of Grubbs ruthenium catalyst [Cl$_2$(PCy$_3$)$_2$RuNCHPh (Cy = cyclohexyl)] at room temperature [167]. The galactopyranosyl-chlorin derivatives **257bA** were obtained as an *E/Z* mixture in 40% yield. A similar methodology was followed to prepare the lactose derivative **257bB**. Treatments of the galactopyranosyl-chlorin **257bA** derivative and the one containing a lactose moiety **257bB** with sodium methoxide in methanol/dichloromethane led to the expected conjugates **257aA** and **257aB**, respectively (Scheme 50) [165].

The Diels–Alder cycloaddition of galactopyranosyl-chlorin conjugate **257bA** with dimethyl acetylenedicarboxylate, in refluxing toluene, afforded the corresponding adduct **258** (Scheme 51). Cleavage of the acetyl groups in **258** gave rise to the required glycosylated chlorin derivate **259** as the major product in 44% yield [165].

Furthermore, some *O*- and *S*-glycosylated pyropheophorbide *a* derivatives have been prepared [168]. The *O*-glycosylated pyropheophorbide derivative **261** was obtained by Lewis acid catalysis *per*-acetate carbohydrate glycosylation of derivative **260A** (Fig. 21). Pyropheophorbide methyl ester **260A** was prepared from the acid derivative **260B**, in order to avoid the formation of an ester linkage between the propionic acid group and the sugar residues. The derivative **260B** was prepared by treating pyropheophorbide *a* **260D** with 50% HBr in acetic acid, followed by hydrolysis of the bromide intermediate **260C**.

The reagent proportions and the reaction times have been shown to rule the reaction yield. The best results were found when the reagents ratio was 2 : 1 : 2 (glycoside : pyropheophorbide : Lewis acid) and the reaction time was 12–15 h. Under these conditions the reaction yield was 36% with 83% of anomeric purity. The glycosylated derivatives **261** are diastereomeric mixtures.

For the preparation of *S*-glycosylated pyropheophorbide derivatives **262–265** a different approach was followed. In this case the pyropheophorbide *a*

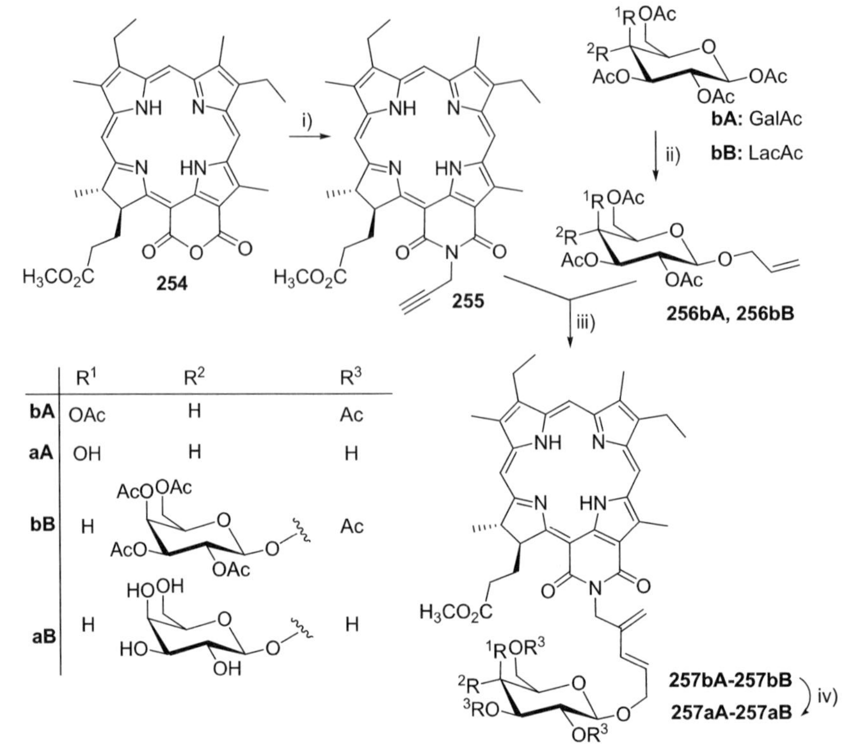

Scheme 50 i) $NH_2CH_2C \equiv CH$ in benzene, reflux, 12 h; ii) $BF_3 \cdot OEt_2$, allyl alcohol in CH_2Cl_2, 0 °C to rt; iii) Grubbs's catalyst, CH_2Cl_2, Argon, 48 h; iv) NaOMe, MeOH/CH_2Cl_2

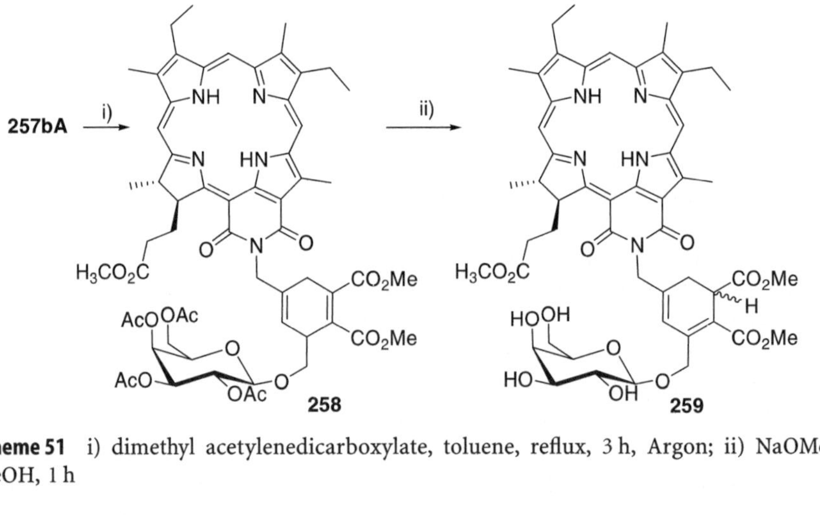

Scheme 51 i) dimethyl acetylenedicarboxylate, toluene, reflux, 3 h, Argon; ii) NaOMe, MeOH, 1 h

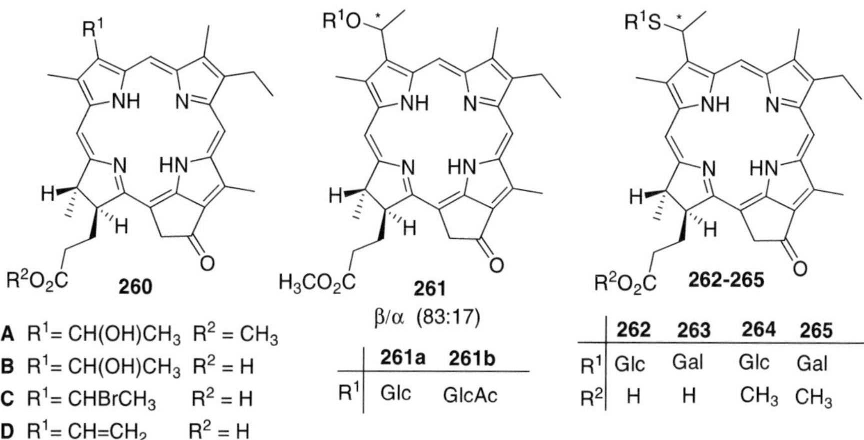

A R^1= CH(OH)CH$_3$ R^2 = CH$_3$
B R^1= CH(OH)CH$_3$ R^2 = H
C R^1= CHBrCH$_3$ R^2 = H
D R^1= CH=CH$_2$ R^2 = H

β/α (83:17)

	261a	261b
R^1	Glc	GlcAc

	262	263	264	265
R^1	Glc	Gal	Glc	Gal
R^2	H	H	CH$_3$	CH$_3$

Fig. 21 Pyropheophorbide *a* and *O*- and *S*-glycosylated pyropheophorbide *a* derivatives

derivative **260C** was gradually added to sodium β-D-glucopyranosylmercaptide or sodium β-D-galactopyranosylmercaptide. The thioglucoside derivative **262** was obtained in 54% yield, and the related thiogalactoside derivative **263** was obtained in 44% yield.

In order to compare the photophysical and biological properties of all the pyropheophorbide sugar derivatives the methyl ester derivatives **264** and **265** were also prepared by treatment with the corresponding acids with 3% sulfuric acid in methanol. The initial results obtained with these amphiphilic conjugates **262–265** showed a tendency to aggregate in aqueous and aqueous/ethanolic solutions [169].

5
Porphyrin-Glycodendrimers

It is known that proteins can incorporate porphyrin units by embedded them into their peptidic frameworks [170]. With such systems, the polypeptidic shells modulate the response of the porphyrin units to the external environment. Non-natural counterparts of these biologically occurring systems can be generated by embedding one or more porphyrin systems inside dendritic frameworks. Indeed, several dendrimers incorporating porphyrin cores have been designed and synthesized already [171, 172].

In order to explore the influence of glycodendrimer branches on the photophysical properties of photoactive cores, Stoddart at al. presented two carbohydrate shells, in their protected **267b** and deprotected **267a** forms, containing a porphyrin core. Dendrimer **267b** was prepared in 16% yield from β-D-glucopyranose building block **266** [173] and *meso*-tetrafunctionalized por-

Scheme 52 i) Et$_3$N, CH$_2$Cl$_2$; ii) NaOMe, MeOH/THF

phyrin **62** (Scheme 52). Removal of all acetyl protecting groups with sodium methoxide in a mixture of methanol and THF, has afforded **267a** quantitatively [174].

References

1. Dolphin D (1978) The Porphyrins. Academic Press, New York
2. Sheldon RA (1994) Metalloporphyrins in Catalytic Oxidations. Marcel Dekker, New York
3. Ortiz de Montellano PR (1995) Cytochrome P450 Structure, Mechanism, and Biochemistry. Plenum Press, New York
4. Meunier B (2000) Biomimetic Oxidations Catalyzed by Transition Metal Complexes. Imperial College Press, London
5. Wang PG, Bertozzi CR (2001) Glycochemistry: Principles, Synthesis, and Applications. Marcel Dekker, New York
6. Nicolaou KC, Mitchell HJ (2001) Angew Chem Int Ed 40:1577
7. Bonnett R (2000) Chemical Aspects of Photodynamic Therapy. Gordon and Breach Science, Amsterdam
8. Kadish KM, Smith KM, Guilard R (2000) The Porphyrin Handbook. Academic Press, San Diego
9. Konan YN, Gurny R, Allemann E (2002) J Photochem Photobiol B 66:89
10. Chen X, Hui L, Foster DA, Drain CM (2004) Biochemistry 43:10918
11. Ahmed S, Davoust E, Savoie H, Boa AN, Boyle RW (2004) Tetrahedron Lett 45:6045
12. Papazoglou TG (1995) J Photochem Photobiol B 28:3
13. D'Hallewin MA, Bezdetnaya L, Guillemin F (2002) Eur Urol 42:417
14. Miyoshi N, Ogasawara T, Nakano K, Tachihara R, Kaneko S, Sano K, Fukuda M, Hisazumi H (2004) Appl Spectrosc Rev 39:437
15. Ando A, Kumadaki I (1996) Heterocycles 42:885
16. Bradshaw JE, Gillogly KA, Wilson LJ, Kumar K, Wan XM, Tweedle MF, Hernandez G, Bryant RG (1998) Inorg Chim Act 276:106
17. Kumadaki I, Ando A, Omote M (2001) J Fluor Chem 109:67
18. Zheng G, Chen Y, Intes X, Chance B, Glickson JD (2004) J Porphyrins Phthalocyanines 8:1106
19. Montanari F, Casella L (1994) Metalloporphyrins Catalyzed Oxidations. Kluwer Academic, Dordrecht
20. Feiters MC, Rowan AE, Nolte RJM (2000) Chem Soc Rev 29:375
21. Kropf M, van Loyen D, Schwarz O, Durr H (1998) J Phys Chem A 102:5499
22. Choi MS, Yamazaki T, Yamazaki I, Aida T (2004) Angew Chem Int Ed 43:150
23. James TD, Shinkai S (2002) Host-Guest Chemistry. Top Curr Chem 218:159
24. Wada K, Mizutani T, Matsuoka H, Kitagawa S (2003) Chem Eur J 9:2368
25. Maiya BG (2004) J Porphyrins Phthalocyanines 8:1118
26. Guldi DM (2002) Chem Soc Rev 31:22
27. Guldi DM, Rahman GMA, Prato M, Jux N, Qin S, Ford W (2005) Angew Chem Int Ed 44:2015
28. Tagmatarchis N, Prato M, Guldi DM (2005) Physica E 29:546
29. Tome JPC, Pereira AMVM, Alonso CMA, Neves MGPMS, Tome AC, Silva AMS, Cavaleiro JAS, Martínez-Díaz MV, Torres T, Rahman GMA, Ramey J, Guldi DM (2005) Eur J Org Chem 257
30. Guldi DM, Rahman GMA, Jux N, Tagmatarchis N, Prato M (2004) Angew Chem Int Ed 43:5526
31. Ehli C, Aminur Rahman GM, Jux N, Balbinot D, Guldi DM, Paolucci F, Marcaccio M, Paolucci D, Melle-Franco M, Zerbetto F, Campidelli S, Prato M (2006) J Am Chem Soc 128:11222
32. Henderson BW, Dougherty TJ (1992) Photochem Photobiol 55:145

33. Bonnett R (1995) Chem Soc Rev 24:19
34. Dougherty TJ, Gomer CJ, Henderson BW, Jori G, Kessel D, Korbelik M, Moan J, Peng Q (1998) J Natl Cancer Inst 90:889
35. Hilmey DG, Abe M, Nelen MI, Stilts CE, Baker GA, Baker SN, Bright FV, Davies SR, Gollnick SO, Oseroff AR, Gibson SL, Hilf R, Detty MR (2002) J Med Chem 45:449
36. Nyman ES, Hynninen PH (2004) J Photochem Photobiol B 73:1
37. Ceburkov O, Gollnick H (2000) Eur J Dermatol 10:568
38. Maisch T, Szeimies RM, Jori G, Abels C (2004) Photochem Photobiol Sci 3:907
39. Babilas P, Landthaler M, Szeimies RM (2006) Eur J Dermatol 16:340
40. Scott LJ, Goa KL (2000) Drugs Aging 16:139
41. Keam SJ, Scott LJ, Curran MP (2003) Drugs 63:2521
42. Fenton C, Perry CM (2006) Drugs Aging 23:421
43. Derycke ASL, de Witte PAM (2004) Adv Drug Deliv Rev 56:17
44. Lang K, Mosinger J, Wagnerova DM (2004) Coord Chem Rev 248:321
45. Bautista-Sanchez A, Kasselouri A, Desroches MC, Blais J, Maillard P, de Oliveira DM, Tedesco AC, Prognon P, Delaire J (2005) J Photochem Photobiol B 81:154
46. Mikata Y, Onchi Y, Tabata K, Ogura SI, Okura I, Ono H, Yano S (1998) Tetrahedron Lett 39:4505
47. Aksenova AA, Sebyakin YL, Mironov AF (2003) Russ J Bioorg Chem 29:201
48. Li G, Pandey SK, Graham A, Dobhal MP, Mehta R, Chen Y, Gryshuk A, Rittenhouse-Olson K, Oseroff A, Pandey RK (2004) J Org Chem 69:158
49. Laville I, Pigaglio S, Blais JC, Doz F, Loock B, Maillard P, Grierson DS, Blais J (2006) J Med Chem 49:2558
50. Ono N, Bougauchi M, Maruyama K (1992) Tetrahedron Lett 33:1629
51. Sylvain I, Zerrouki R, Granet R, Huang YM, Lagorce JF, Guilloton M, Blais JC, Krausz P (2002) Bioorg Med Chem 10:57
52. Sol V, Branland P, Granet R, Kaldapa C, Verneuil B, Krausz P (1998) Bioorg Med Chem Lett 8:3007
53. Gabor F, Szocs K, Maillard P, Csik G (2001) Radiat Environ Biophys 40:145
54. Tome JPC, Neves MGPMS, Tome AC, Cavaleiro JAS, Mendonça AF, Pegado IN, Duarte R, Valdeira ML (2005) Bioorg Med Chem 13:3878
55. Meunier B (1992) Chem Rev 92:1411
56. Meunier B, Adam W (2000) Metal-oxo and Metal-peroxo Species in Catalytic Oxidations. Springer, Berlin Heidelberg New York
57. Maillard P, Guerquin-Kern JL, Momenteau M (1991) Tetrahedron Lett 32:4901
58. Vilain-Deshayes S, Robert A, Maillard P, Meunier B, Momenteau M (1996) J Mol Catal A 113:23
59. Vilain-Deshayes S, Maillard P, Momenteau M (1996) J Mol Catal A 113:201
60. Zhang XB, Guo CC, Xu JB, Yu RQ (2000) J Mol Catal A 154:31
61. Guo CC, Zhang XB, Song JX, Li HP (2000) Act Chim Sinica 58:332
62. Vilain S, Maillard P, Momenteau M (1994) J Chem Soc Chem Comm p 1697
63. James TD, Sandanayake K, Shinkai S (1996) Angew Chem Int Ed Eng 35:1911
64. Ogoshi H, Mizutani T (1999) Curr Opin Chem Biol 3:736
65. Bell TW, Hext NM (2004) Chem Soc Rev 33:589
66. Murakami H, Nagasaki T, Hamachi I, Shinkai S (1993) Tetrahedron Lett 34:6273
67. Takeuchi M, Chin Y, Imada T, Shinkai S (1996) Chem Commun p 1867
68. Arimori S, Takeuchi M, Shinkai S (1996) Chem Lett p 77
69. Arimori S, Takeuchi M, Shinkai S (1996) J Am Chem Soc 118:245
70. Takeuchi M, Imada T, Shinkai S (1996) J Am Chem Soc 118:10658
71. Takeuchi M, Kijima H, Hamachi I, Shinkai S (1997) Bull Chem Soc Jpn 70:699

72. Takeuchi M, Imada T, Shinkai S (1998) Bull Chem Soc Jpn 71:1117
73. Arimori S, Takeuchi M, Shinkai S (1998) Supramolec Sci 5:1
74. Sugasaki A, Ikeda M, Takeuchi M, Koumoto K, Shinkai S (2000) Tetrahedron 56:4717
75. Ikeda M, Shinkai S, Osuka A (2000) Chem Commun p 1047
76. Takeuchi M, Yoda S, Chin Y, Shinkai S (1999) Tetrahedron Lett 40:3745
77. Gong FC, Zhang XB, Guo CC, Shen GL, Yu RQ (2003) Sensors 3:91
78. Yang DW, Gong FC, Cao Z (2006) Sensor Actuat B 114:152
79. Hasegawa T, Numata M, Asai M, Takeuchi M, Shinkai S (2005) Tetrahedron 61:7783
80. Asayama S, Mori T, Nagaoka S, Kawakami H (2003) J Biomater Sci Polym Ed 14:1169
81. Asayama S, Mizushima K, Nagaoka S, Kawakami H (2004) Bioconjugate Chem 15:1360
82. Prinsep MR, Caplan FR, Moore RE, Patterson GML, Smith CD (1992) J Am Chem Soc 114:385
83. Prinsep MR, Patterson GML, Larsen LK, Smith CD (1995) Tetrahedron 51:10523
84. Morliere P, Maziere JC, Santus R, Smith CD, Prinsep MR, Stobbe CC, Fenning MC, Golberg JL, Chapman JD (1998) Cancer Res 58:3571
85. Minehan TG, Cook-Blumberg L, Kishi Y, Prinsep MR, Moore RE (1999) Angew Chem Int Ed 38:926
86. Minehan TG, Kishi Y (1999) Angew Chem Int Ed 38:923
87. Wang W, Kishi Y (1999) Org Lett 1:1129
88. Prinsep MR, Patterson GML, Larsen LK, Smith CD (1998) J Nat Prod 61:1133
89. Garrido JL, Otero J, Maestro MA, Zapata M (2000) J Phycol 36:497
90. Smith KM (1975) Porphyrins and Metalloporphyrins. Elsevier, New York
91. Fuhrhop J-H, Demoulin C, Boettcher C, Köning J, Siggel U (1992) J Am Chem Soc 114:4159
92. Fulling G, Schroder D, Frank B (1989) Angew Chem Int Ed Engl 101:1550
93. Matsuo T, Nagai H, Hisaeda Y, Hayashi T (2006) Chem Commun p 3131
94. Maillard P, Guerquin-Kern JL, Momenteau M, Gaspard S (1989) J Am Chem Soc 111:9125
95. Hombrecher HK, Ohm S, Koll D (1996) Tetrahedron 52:5441
96. Lloyd-Williams P, Albericio F, Giralt E (1997) Chemical Approaches to the Synthesis of Peptides and Proteins. CRC Press, Boca Raton, Fl
97. Tamaru SI, Nakamura M, Takeuchi M, Shinkai S (2001) Org Lett 3:3631
98. Tamaru SI, Takeuchi M, Sano M, Shinkai S (2002) Angew Chem Int Ed 41:853
99. Kawano S-I, Tamaru S-I, Fujita N, Shinkai S (2004) Chem Eur J 10:343
100. Sol V, Charmot A, Krausz P, Trombotto S, Queneau Y (2006) J Carbohydrate Chem 25:345
101. Redl FX, Lutz M, Daub J (2001) Chem Eur J 7:5350
102. Hasegawa T, Umeda M, Numata M, Li C, Bae AH, Fujisawa T, Haraguchi S, Sakurai K, Shinkai S (2006) Carbohydrate Res 341:35
103. Holzer W, Penzkofer A, Redl FX, Lutz M, Daub J (2002) Chem Phys 282:89
104. Lee LV, Mitchell ML, Huang SJ, Fokin VV, Sharpless KB, Wong CH (2003) J Am Chem Soc 125:9588
105. Defaye J, Driguez H, Ohleyer E, Orgeret C, Viet C (1984) Carbohydrate Res 130:317
106. Oulmi D, Maillard P, Guerquin-Kern J-L, Huel C, Momenteau M (1995) J Org Chem 60:1554
107. Schell C, Hombrecher HK (1999) Chem Eur J 5:587
108. Schell C, Hombrecher HK (1999) Bioorg Med Chem 7:1857
109. Pasetto P, Chen X, Drain CM, Franck RW (2001) Chem Commun p 81
110. Halazy S, Berges V, Ehrhard A, Danzin C (1990) Bioorg Chem 18:330

111. Maillard P, Guerquin-Kern J-L, Huel C, Momenteau M (1993) J Org Chem 58:2774
112. Treibs A, Haberle N (1968) Liebigs Ann Chem 718:183
113. Lindsey JS, Hsu HC, Schreiman IC (1986) Tetrahedron Lett 27:4969
114. Lindsey JS, Schreiman IC, Hsu HC, Kearney PC, Marguerettaz AM (1987) J Org Chem 52:827
115. Kohata K, Higashio H, Yamaguchi Y, Koketsu M, Odashima T (1994) Bull Chem Soc Jpn 67:668
116. Maillard P, Vilain S, Huel C, Momenteau M (1994) J Org Chem 59:2887
117. Momenteau M, Oulmi D, Maillard P, Croisy A (1995) In: Brault D, Jori G, Moan J, Ehrenberg B (eds) Proc SPIE - The Int Soc for Optical Engineering (Photodynamic Therapy of Cancer II), vol 2325. Soc Photo-Optical Instrumentation Engineers, Lille, France, p 13
118. Lee RT, Lee YC (2000) Glycoconjugate J 17:543
119. Fujimoto K, Miyata T, Aoyama Y (2000) J Am Chem Soc 122:3558
120. Gaud O, Granet R, Kaouadji M, Krausz P, Biais JC, Bolbach G (1996) Can J Chem 74:481
121. Faustino MAF, Neves M, Vicente MGH, Cavaleiro JAS, Neumann M, Brauer HD, Jori G (1997) Photochem Photobiol 66:405
122. Faustino MAF, Neves M, Cavaleiro JAS, Neumann M, Brauer HD, Jori G (2000) Photochem Photobiol 72:217
123. Kaldapa C, Blais JC, Carre V, Granet R, Sol V, Guilloton M, Spiro M, Krausz P (2000) Tetrahedron Lett 41:331
124. Sol V, Lamarche F, Enache M, Garcia G, Granet R, Guilloton M, Blais JC, Krausz P (2006) Bioorg Med Chem 14:1364
125. Sol V, Blais JC, Carre V, Granet R, Guilloton M, Spiro M, Krausz P (1999) J Org Chem 64:4431
126. Sol V, Blais JC, Bolbach G, Carre V, Granet R, Guilloton M, Spiro M, Krausz P (1997) Tetrahedron Lett 38:6391
127. Chaleix V, Sol V, Huang YM, Guilloton M, Granet R, Blais JC, Krausz P (2003) Eur J Org Chem 1486
128. Lee CH, Lindsey JS (1994) Tetrahedron 50:11427
129. Gaud O, Granet R, Kaouadji M, Krausz P, Biais JC, Bolbach G (1996) Can J Chem 74:481
130. Carre V, Gaud O, Sylvain I, Bourdon O, Spiro M, Blais J, Granet R, Krausz P, Guilloton M (1999) J Photochem Photobiol B 48:57
131. Hombrecher HK, Schell C, Thiem J (1996) Bioorg Med Chem Lett 6:1199
132. Takeuchi M, Imada T, Shinkai S (1996) J Am Chem Soc 118:10658
133. Tome JPC, Neves MGPMS, Tome AC, Cavaleiro JAS (2000) First Int Conf on Porphyrins and Phthalocyanines (ICPP-1), Dijon, France, p 587
134. Tomé JPC (2001) Synthesis and Biological Evaluation of Amphiphilic Porphyrins. PhD Thesis, University of Aveiro
135. Traylor TG (1981) Accounts Chem Res 14:102
136. Momenteau M (1986) Pure Appl Chem 58:1493
137. Groves JT, Viski P (1990) J Org Chem 55:3628
138. Wagner RW, Johnson TE, Lindsey JS (1997) Tetrahedron 53:6755
139. Davoust E, Granet R, Krausz P, Carre V, Guilloton M (1999) Tetrahedron Lett 40:2513
140. Chan TL, Fong S, Li Y, Man TO, Poon CD (1994) J Chem Soc Chem Commun p 1771
141. Casiraghi G, Cornia M, Rassu G, Del Sante C, Spanu P (1992) Tetrahedron 48:5619
142. Casiraghi G, Cornia JM, Zanardi JF, Rassu G, Ragg E, Bortolini R (1994) J Org Chem 59:1801

143. Cornia M, Binacchi S, Del Soldato T, Zanardi F, Casiraghi G (1995) J Org Chem 60:4964
144. Cornia M, Casiraghi G, Binacchi S, Zanardi F, Rassu G (1994) J Org Chem 59:1226
145. Cornia M, Valenti C, Capacchi S, Cozzini P (1998) Tetrahedron 54:8091
146. Cornia M, Menozzi M, Ragg E, Mazzini S, Scarafoni A, Zanardi F, Casiraghi G (2000) Tetrahedron 56:3977
147. Driaf K, Granet R, Krausz P, Kaouadji M, Thomasson F, Chulia AJ, Verneuil B, Spiro M, Blais JC, Bolbach G (1996) Can J Chem 74:1550
148. Krausz DK, Verneuil B, Spiro M, Blais JC, Bolbach G (1993) Tetrahedron Lett 34:1027
149. Tomé JPC, Silva EMP, Pereira AMVM, Alonso CMA, Faustino MAF, Neves MGPMS, Tomé AC, Cavaleiro JAS, Tavares SAP, Duarte RR, Caeiro MF, Valdeira ML (2007) Bioorg Med Chem: in press
150. Silva EMP, Serra VV, Ribeiro AO, Tomé JPC, Domingues P, Faustino MAF, Neves MGPMS, Tomé AC, Cavaleiro JAS, Ferrer-Correia AJ, Iamamoto Y, Domingues MRM (2006) Rapid Commun Mass Spectrom 20:3605
151. Silva EMP, Giuntini F, Faustino MAF, Tome JPC, Neves MGPMS, Tome AC, Silva AMS, Santana-Marques MG, Ferrer-Correia AJ, Cavaleiro JAS, Caeiro MF, Duarte RR, Tavares SAP, Pegado IN, D'Almeida B, De Matos APA, Valdeira ML (2005) Bioorg Med Chem Lett 15:3333
152. Pandey RK, Zheng G (2000) Porphyrins as Photosensitizers in Photodynamic Therapy. Academic Press, San Diego
153. Maillard P, Hery C, Momenteau M (1997) Tetrahedron Lett 38:3731
154. Laville I, Figueiredo T, Loock B, Pigaglio S, Maillard P, Grierson DS, Carrez D, Croisy A, Blais J (2003) Bioorg Med Chem 11:1643
155. Whitlock HW, Hanauer R, Oester MY, Bower BK (1969) J Am Chem Soc 91:7485
156. Laville I, Pigaglio S, Blais JC, Loock B, Maillard P, Grierson DS, Blais J (2004) Bioorg Med Chem 12:3673
157. Hirohara S, Obata M, Ogata S, Ohtsuki C, Higashida S, Ogura S, Okura I, Takenaka M, Ono H, Sugai Y, Mikata Y, Tanihara M, Yano S (2005) J Photochem Photobiol B 78:7
158. Hirohara S, Obata M, Saito A, Ogata S, Ohtsuki C, Higashida S, Ogura S, Okura I, Sugai Y, Mikata Y, Tanihara M, Yano S (2004) Photochem Photobiol 80:301
159. Tome AC, Lacerda PSS, Neves M, Cavaleiro JAS (1997) Chem Commun p 1199
160. Silva AMG, Tome AC, Neves M, Silva AMS, Cavaleiro JAS (1999) Chem Commun p 1767
161. Tome AC, Lacerda PSS, Silva AMG, Neves M, Cavaleiro JAS (2000) J Porphyrins Phthalocyanines 4:532
162. Silva AMG, Tome AC, Neves MGPMS, Silva AMS, Cavaleiro JAS, Perrone D, Dondoni A (2002) Tetrahedron Lett 43:603
163. Silva AMG, Tome AC, Neves MGPMS, Cavaleiro JAS, Perrone D, Dondoni A (2005) Synlett p 857
164. Zheng G, Potter WR, Camacho SH, Missert JR, Wang G, Bellnier DA, Henderson BW, Rodgers MAJ, Dougherty TJ, Pandey RK (2001) J Med Chem 44:1540
165. Zheng G, Graham A, Shibata M, Missert JR, Oseroff AR, Dougherty TJ, Pandey RK (2001) J Org Chem 66:8709
166. Dahmen J, Frejd T, Magnusson G, Noori G (1983) Carbohydrate Res 114:328
167. Grubbs RH, Chang S (1998) Tetrahedron 54:4413
168. Aksenova AA, Sebyakin YL, Mironov AF (2001) Russ J Bioorg Chem 27:124
169. Aksenova AA, Sebyakin YL, Mironov AF (2001) Bioorganicheskaya Khimiya 27:150
170. Moore MR, Pettigrew GW (1990) Cytochrome C: Evolution and Physicochemical Aspects. Springer, Berlin Heidelberg New York

171. Choi MS, Yamazaki T, Yamazaki I, Aida T (2003) Angew Chem Int Ed 43:150
172. Choi MS, Aida T, Luo H, Araki Y, Ito O (2003) Angew Chem Int Ed 42:4060
173. Ashton PR, Boyd SE, Brown CL, Jayaraman N, Nepogodiev SA, Stoddart JF (1996) Chem Eur J 2:1115
174. Ballardini R, Colonna B, Gandolfi MT, Kalovidouris SA, Orzel L, Raymo FM, Stoddart JF (2003) Eur J Org Chem 288

Top Heterocycl Chem (2007) 7: 249–285
DOI 10.1007/7081_2007_051
© Springer-Verlag Berlin Heidelberg
Published online: 15 March 2007

New Developments in the Synthesis of Anisomycin and Its Analogues

A. El Nemr[1] · E. S. H. El Ashry[2] (✉)

[1]Environmental Division, National Institute of Oceanography and Fisheries, Kayet Bey,
El Anfoushy, Alexandria, Egypt

[2]International Center for Chemical and Biological Science,
HEJ Research Institute for Chemistry, Karachi University, Karachi, Pakistan
Eelashry60@hotmail.com

Abstract Anisomycin is a natural product that received much attention due to its interesting biological activity against certain pathogenic protozoa, strains of fungi, and in the treatment of certain diseases such as amoebic dysentery and tricomonas vaginitis. This review is focused on the synthesis of natural anisomycin and its analogues. Different approaches to the synthesis of naturally occurring anisomycin and its unnatural analogues are described and detailed experimental conditions are given. It is intended in this review to give enough information to enable a chemist to design a multi-step synthesis.

Keywords Anisomycin · Antibiotic · Streptomyces grieolus ·
Streptomyces roseochromogenes · Sugars

Abbreviations

Ac	acetyl
AIBN	azobis(isobutironitrile)
All	allyl
CAN	ceric ammonium nitrate
Cbz	benzyloxycarbonyl
CDI	carbonyldiimidazolide

DBU   1,8-diazabicyclo[5.4.0]undec-7-ene
DHQ-CLB   Sharpless asymmetric dihydroxylation reagent
DMAP   4-dimethylaminopyridine
DMSO   dimethyl sulphoxide
DME   1,2-dimethoxyethane
DMF   dimethylformamide
DMP   2,2-dimethoxypropane
Im   imidazole
m-CPBA   *m*-chloroperoxybenzoic acid
MS   molecular sieves
NBS   *N*-bromosuccinimide
NMO   *N*-methylmorpholine *N*-oxide
PCC   pyridinium chlorochromate
Ph   phenyl
PMB   *p*-methoxybenzyl
PPTs   pyridinium *p*-toluenesulfonate
Py   pyridine
TBAF   tetra-butyl ammonium fluoride
TBDMS   *t*-butyl dimethylsilyl
Tf   trifluoromethylsulphonyl
TFA   trifluoroacetic acid
THF   tetrahydrofuran
TIBSCl   triisobutylsilylchloride
TMS   trimethylsilyl
TPAP   tetra-*n*-propyl ammonium perruthenate

1
Introduction

The chemistry of the antibiotic (–)-anisomycin (**1**) (Fig. 1) has been included amongst other naturally occurring nitrogen heterocycles in a book by El Ashry and El Nemr [1]. (–)-Anisomycin (**1**) was first isolated from *Streptomyces grieolus* and *Streptomyces roseochromogenes* in 1954 [2]. It was also found in two related strains, *Streptomyces* sp. No. 638 [3] and *Streptomyces SA* 3097 [4]. Anisomycin has a broad activity against certain pathogenic protozoa, strains of fungi [1–12]. It is effective in the treatment of amoebic dysentery and tricomonas vaginitis [13, 14] and in plants as a fungicide [15]. Deacetyl anisomycin (**2**) has also been used as a fungicide in the eradication of bean mildew and for the inhibition of other pathogenic fungi in plants. It has high antitumor activity in vitro, with IC_{50} values in the nanomolar range [3, 4]. Anisomycin may be used in a synergistic fashion with a cyclin-dependent protein kinase inhibitor to kill carcinoma cells [19–21]. It has a widespread use as a tool in molecular biology, where it inhibits protein synthesis [22, 23] and activates JNK and p38 kinases [24]. The anisomycin alkaloid blocks the aminoacyl-SRNA transfer reaction in protein biosynthesis

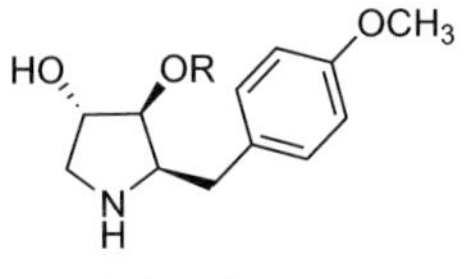

1 R = Ac
2 R = H

Fig. 1 Structure of anisomycin (**1**) and deacetyl anisomycin (**2**)

and exhibits a remarkably selective inhibition of peptide chain elongation on 60S eukaryotic ribosomes [16–18].

The structure and relative stereochemistry of anisomycin were firstly investigated chemically [10] and then determined by X-ray crystallographic analysis [25]. The absolute stereochemistry was established as ($2R, 3S, 4S$) by chemical correlation with L-tyrosine [10–12]. The activity and structural features of anisomycin have attracted the attention of several synthetic organic chemists to produce synthetic routes to anisomycin and its analogues. Few routes have been achieved with good stereoselectivity, while others have an inherent problem in separating unwanted isomers [26–90]. Herein, we report a new development in the synthesis of anisomycin and its analogues.

2
Synthetic Approaches

Various synthetic approaches have been developed for the synthesis of anisomycin and its analogues. The structural features of anisomycin, particularly the presence of three asymmetric centers could lead via a retro synthetic analysis to the required precursors which may have two hydroxyl groups with the assigned stereochemistry. This has reasoned the advantage of using carbohydrate as precursors for the synthesis of these compounds. Moreover, as a consequence of their biological significance, the required precursors have been also approached from non-carbohydrate starting materials.

2.1
Synthesis from D-Galactose

Anisomycin (**1**) and deacetylanisomycin (**2**) have been synthesized [46] from D-galactose by conversion firstly to ethyl 2,3-di-O-benzyl-β-D-galactofuranoside (**3**) [46–49] in 70% overall yield (Scheme 1). Oxidation of **3** using periodate gave the corresponding aldehyde, which was treated with (4-methoxyphenyl)magnesium bromide to afford a mixture of epimeric alcohols **4** in 87% yield. Ionic deoxygenation of **4** with triethylsilane in the presence of trifluoroacetic acid was followed by overnight-boiling with 80% acetic

Scheme 1 **a** [47–49]; **b** *1*. 1 : 1 EtOH – H$_2$O, NaIO$_4$, rt, 3 hr, 95%; *2*. 4-CH$_3$OC$_6$H$_4$MgBr, ether, reflux, 2 hr, 87%; **c** *1*. Et$_3$SiH, CH$_2$Cl$_2$, TFA, rt, 48 hr, 76%; *2*. 80% acetic acid, reflux, overnight, 97%; **d** HONH$_2$·HCl, Py, EtOH, reflux, 5 hr, 86%; **e** Py, MsCl, rt, 1 hr, then 60–70 °C, 2 hr, 71%; **f** 1 M BH$_3$ in THF, reflux, 3 hr, then 2 M HCl, reflux, 20 min, 65%; **g** 10% Pd *on* C, H$_2$, EtOH, HCO$_2$H, sonication, 90 min, 77%; **h** CH$_2$Cl$_2$, Na$_2$CO$_3$, CbzCl, 2.5 hr, 72%; **i** DMF, imidazole, TBDMSCl, rt, 1 hr, 80%; **j** Ac$_2$O, Py, rt, 3 days, 96%; **k** *1*. 0 °C, THF, 1 M TBAF, 30 min, 85%; *2*. EtOH, 10% Pd *on* C, H$_2$, 15 min, 95%

acid to produce the anomeric mixture **5**. The latter epimeric mixture was condensed with hydroxylamine hydrochloride to afford the *E*- and *Z*-oximes **6**. Treatment of **6** with methanesulfonyl chloride in pyridine caused dehydration and simultaneous *O*-mesylation to afford the nitrile mesylate **7**.

Nitrile reduction with BH$_3$ proceeded with cyclization to give **8**. This underwent catalytic hydrogenation in the presence of formic acid to afford the (–)-deacetylanisomycin (**2**). Compound **2** was subjected to a five-step sequence to produce **1** [2, 26, 50–52]. Thus, *N*-carbamate protection of **2** afforded carbamate **9**, which was selectively sililated to give **10**. This was acetylated to produce **11** that on treatment with TBAF followed by catalytic hydrogenation furnished (–)-anisomycin (**1**).

2.2
Synthesis from L-Arabinose

(–)-Anisomycin (**1**) and deacetylanisomycin (**2**) were prepared from 2,3,5-tri-*O*-benzyl-*β*-L-arabinofuranose (**12**) (Scheme 2) [53]. This was converted to

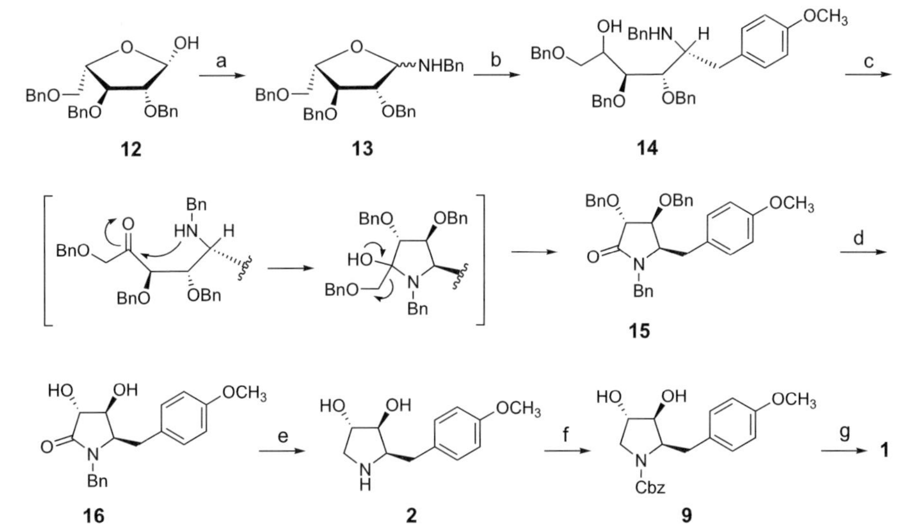

Scheme 2 **a** BnNH$_2$, CH$_2$Cl$_2$, MS 4 Å, quant.; **b** 4-CH$_3$OC$_6$H$_4$CH$_2$MgCl, – 78 °C to 0 °C, THF, 78%; **c** PCC, MS 4 Å, CH$_2$Cl$_2$, 63%; **d** Pd black, HCO$_2$H, CH$_3$OH, 99%; **e** *1*. LiAlH$_4$, THF, 91%; *2*. Pd-black, HCO$_2$H, CH$_3$OH; **f** CbzCl, NaHCO$_3$, CH$_3$OH, 77% for two steps; **g** *1*. DMF, imidazole, TBDMSCl, rt, 1 hr, 80%; *2*. Ac$_2$O, Py, rt, 3 days, 96%; *3*. 0 °C, THF, 1 M TBAF, 30 min, 85%; *4*. EtOH, 10% Pd *on* C, H$_2$, 15 min, 95%

furanosylamine **13** by amination with benzylamine which was smoothly reacted with *p*-methoxybenzylmagnesium chloride at low temperature to provide the adduct **14** in high yield [54]. Oxidative degradation of **14** with PCC in the presence of 4 Å molecular sieves afforded the *cis*-functionalized lactam **15**. Removal of the *O*-benzyl groups from **15** by 4% HCOOH – MeOH on Pd(black) followed by reduction of the resulting lactam **16** with lithium aluminum hydride and subsequent hydrogenation afforded deacetylanisomycin (**2**), which was treated with CbzCl in the presence of NaHCO$_3$ to give the carbamate **9**. Conversion of **9** into the (–)-anisomycin (**1**) has been achieved as mentioned in the former scheme [52–54].

2.3
Synthesis from D-Ribose

D-Ribose was used for the synthesis of (–)-anisomycin (**1**) and its phenylanalogue **34** (Scheme 3) [55, 56]. Thus, 2,3-*O*-isopropylidene-D-ribose (**17**) [57] was treated with Grignard reagents, benzyl- and *p*-methoxybenzyl-magnesium bromide to give the corresponding triols **18** (77%) and **19** (70%); only the D-*allo* stereoisomer could be isolated in each case. Periodate oxidation of **18** and **19** gave the corresponding hemiacetals **20** and **21**, respectively. Reaction of **20** and **21** with hydroxylamine hydrochloride in pyridine gave the corresponding oximes **22** and **23** which upon treatment with methane-

Scheme 3 **a** ArCH$_2$MgCl, THF, rt, 1 hr, **18** (77%), **19** (70%); **b** *1.* NaIO$_4$, EtOH, H$_2$O, 30 min, **20** (93%), **21** (76%); **c** Py, NH$_2$OH·HCl, 3 hr, **22** (99%), **23** (94%); **d** Py, MsCl, 1 hr, 60 °C, **24** (91%), **25** (88%); **e** ether, LiAlH$_4$, 2 hr; then EtOAc, 1 hr, **26** (54%), **25** (51%); **f** *1.* CH$_3$OH, 1 M HCl, reflux, 3 hr; *2.* Glacial AcOH, HBr, 50 °C, 1 hr; **g** CH$_3$OH, H$_2$O, KOH, 10 min, **32** (68%), **33** (70%); **h** Allyl alcohol, CHCl$_3$, perchloric acid (70%), 60 °C, 36–48 hr, **34** (63%), **35** (67%); **i** *1.* CHCl$_3$, BnBr, Et$_3$N, 2 hr; then Ac$_2$O, 60 °C, 5 hr; *2.* CH$_3$OH, 2 M HCl, 10% Pd *on* C, reflux, 48 hr, then 3 hr under H$_2$, **36** (66%), **1** (68%)

sulphonyl chloride in pyridine gave the corresponding nitriles **24** and **25**. Lithium aluminium hydride reduction of **24** and **25** led to the respective pyrrolidines **26** and **27** in 48 and 42% overall yield, respectively, from the hemiacetals **20** and **21**.

Acid hydrolysis of the acetal group in **26** and **27**, followed by treatment with glacial acetic acid containing HBr gave the bromo-acetates (**28**, **30**) and (**29**, **31**), which were treated with aqueous methanolic potassium hydroxide to give the epoxides **32** and **33**. The epoxides **32** and **33** were reacted with allyl alcohol in the presence of perchloric acid to produce the allyl ethers (**34** and **35**), which were *N*-benzylated, *O*-acetylated, and finally hydrogenated to produce **36** and **1** in an overall yield of 10 and 7%, respectively, from **17**.

2.4
Synthesis from L-Threose

A chiral total synthesis of optically pure (–)-anisomycin (**1**) has been achieved from 2,3-*O*-bis(methoxymethyl)-L-*threo*-furanose (**37**) (Scheme 4) [52]. The L-*threo*-furanose **37**, obtained from diethyl L-tartarate, was treated with (4-methoxybenzyl)magnesium chloride to furnish a mixture of the two diastereomers, *xylo* **38** and *lyxo* **39**, in a ratio of 79 : 21. The primary hydroxyl group of the mixture was selectively benzylated by treatment with 1 equiv of benzyl chloride in the presence of phase-transfer catalyst *n*-Bu$_4$NBr, followed by oxidation of the remaining free secondary hydroxyl group to give the ketone **40**. Reduction of **40** with Zn(BH$_4$)$_2$ afforded only the desired *lyxo* isomer **41** in 40% yield from **37**. Removal of the benzyl group from **41** furnished **39** that underwent mesylation followed by S$_N$2 displacement of the primary mesy-

Scheme 4   **a** 4-CH$_3$OC$_6$H$_4$CH$_2$MgCl, THF, rt, 14 hr, 69%, **36 : 37**, 79 : 21; **b** *1.* *n*-Bu$_4$NBr, 6 N NaOH, CH$_2$Cl$_2$, BnCl, CH$_2$Cl$_2$, rt to 60 °C, 24 hr, 69%; *2.* (COCl)$_2$, DMSO, CH$_2$Cl$_2$, – 78 °C; then Et$_3$N, 93%; **c** Zn(BH$_4$)$_2$, ether, 0 °C, 10 min to rt, 50 min 91%; **d** *1.* CH$_3$OH, 10% Pd *on* C, H$_2$, 1 hr, 100%; *2.* Et$_3$N, CH$_2$Cl$_2$, MsCl, 10 min, 87%; *3.* NaN$_3$, DMF, 80 °C, 0.5 hr, 45%; **e** CH$_3$OH, 10% Pd *on* C, H$_2$, 95%; **f** *1.* CH$_3$OH, HCl, H$_2$O (2 : 1 : 1), reflux, 20 hr, 81%; *2.* CbzCl, CH$_2$Cl$_2$, Na$_2$CO$_3$, 2.5 hr, 72%; **g** *1.* DMF, imidazole, TBDMSCl, rt, 1 hr, 80%; *2.* Ac$_2$O, Py, rt, 3 days, 96%; **h** *1.* 0 °C, THF, 1 M TBAF, 30 min, 85%; *2.* EtOH, 10% Pd *on* C, H$_2$, 15 min, 95%

late group with sodium azide to produce the azide **42** whose hydrogenation gave **43**.

Removal of the methoxymethyl groups from **43** with HCl, followed by protection of the secondary amine with benzyl chloroformate afforded the carbamate **9**. Selective silylation of the 4-hydroxy group of **9** by treatment with 1.2 equiv of *tert*-butyldimethylsilyl chloride and 2.4 equiv of imidazole in DMF at room temperature for 1 h afforded the corresponding silyl ether as a single isomer in 80% yield that was subjected to acetylation at the *C*-3 hydroxyl group to afford the carbamate **11**, which upon removal of the Cbz and TBDMS groups furnished **1**.

2.5
Synthesis from L-Threitol

The L-threitol was used for the syntheses of deacetylanisomycin (**2**) and its derivatives using a highly selective addition of Grignard reagents or organolithium to the L-threose imine **47** (Scheme 5) [57]. Compound **46** was prepared [58] in two steps from the commercially available 2-*O*-benzyl-L-threitol (**44**) through the intermediate **45**. Coupling of **46** with benzylamine gave quantitatively the *N*-benzyl imine **47**, which on treatment with three equivalents of organolithium compound afforded the aminotriols **48** and **49** in a ratio of 5 : 95. The diastereomer **49** was transformed into 2-substituted

Scheme 5 **a** [58]; **b** CrO_3, Py., 74%; **c** Al_2O_3 (63–200 μm), $BnNH_2$, 1 hr, rt, quantitative; **d** *1*. RLi, ether, – 78 °C, 15 min, *or* RMgX, ether, 0 °C to rt, 2–5 hr; aqueous NH_4Cl; *2*. HCl, H_2O, dioxane (1 : 1), 62%; **e** PPh_3, DEAD, Py, 0 °C, 1.5 hr; H_2O, LiOH, dioxane, 80 °C, *or* PPh_3, CCl_4, NEt_3, DMF, rt, CH_3OH; **f** H_2 (4 bar), Pd *on* C, CH_3OH, HCl, rt, 3 days *or* H_2 (4 bar), $Pd(OH)_2$ *on* C, CH_3OH, rt, 2 days; HBr, rt, 1 day, 99%

trans-dihydroxypyrrolidine **50** in 77% yield, either by an intramolecular Mitsunobu reaction or by cyclization using the Appel reaction [59, 60]. Hydrogenation of **50** afforded the deacetylanisomycin hydrobromide **2** (34%), which could be converted to anisomycin (**1**) in 45% yield [53].

Intramolecular iodoamidation of pentenetriol derivative **52**, obtained from L-threitol derivative **51**, was used to prepare (–)-anisomycin in 17 ~ 20% overall yield (Scheme 6) [61]. Oxidation of **51** [62], followed by methylenation, acid hydrolysis, and protection with CCl₃CN gave the olefin **52**, whose reaction with iodine in the presence of sodium bicarbonate gave a 4.5 : 1 mixture of **53** and **54**. Hydrolysis of the mixture, followed by *N*-protection with Boc₂O afforded a 12 : 1 mixture of **55** and **56**. On the other hand, when **52** was reacted with iodine monobromide in the presence of potassium carbonate, compound **53** only was obtained that could be transformed into a 37 : 1 mix-

Scheme 6 a *1.* (COCl)₂, DMSO, CH₂Cl₂, – 78 °C; then Et₃N; *2.* Ph₃P⁺CH₃I⁻, *n*-BuLi, THF; *3.* 2 M HCl, THF, 20 °C, 78% for three steps; *4.* CCl₃CN, DBU, CH₃CN, – 30 °C; **b** I₂, NaHCO₃, CH₃CN, 0 °C, 4.5 : 1 mixture of **53** : **54** *or* IBr, K₂CO₃, CH₃CH₂CN, – 60 °C; **c** *1.* 6 M HCl, CH₃OH, 20 °C; *2.* Boc₂O, NaHCO₃, CH₃OH, 0 °C, 75% for four steps; **d** *1.* acetone, TFA, 20 °C; *2.* LDA, THF, – 20 °C, 77% for two steps; **e** *1.* 4-CH₃OC₆H₄MgBr, CuBr, DMS, toluene, – 30 °C; *2.* TFA, 20 °C; *3.* Boc₂O, NaHCO₃, CH₃OH, 0 °C; 81% for three steps; **f** DEAD, Ph₃P, PPTs, THF, 0 °C; **g** *1.* TBDMSCl, imidazole, DMF, 20 °C; *2.* Ac₂O, DMAP, Et₃N, CH₂Cl₂, 20 °C, 56% from **59**; **h** 6 M HCl, CH₃OH, 20 °C, 88%

ture of **55** and **56**. Isopropylidenation of **55**, followed by treatment with LDA afforded a 3 : 1 mixture of aziridines **57** and **58**. This mixture was treated with 4-methoxyphenylmagnesium bromide in the presence of copper(I) bromide dimethyl sulfide complex in toluene followed by acid hydrolysis and *N*-protection to give **59**. The pyrrolidine ring was formed from **59** by treatment with diethyl azodicarboxylate (DEAD) and triphenyl phosphine in the presence of PPTs to produce **60**. Silylation of **60** followed by acetylation afforded **61**, which underwent acidic hydrolysis to give (–)-anisomycin (**1**).

A total synthesis of (+)-*N*-methylanisomycin (**87**) from 4-*O*-benzyl-2,3-*O*-bis-(methoxymethyl)-L-threose (**75**) has been reported (Schemes 7 and 8) [30]. Methoxymethylation of L-diethyl tartrate (**62**) with dimethoxy-

Scheme 7 **a** *1*. Chloromethyl methyl ether, P_2O_5; *2*. $LiAlH_4$, THF; **b** BnBr, 4 N NaOH, Bu_4NBr, 50 °C; **c** *1*. Et_3N, Ac_2O, rt, 8 h, 98%; *2*. H_2, 5% Pd/C, AcOEt, 94%; **d** *1*. $(COCl)_2$, CH_2Cl_2, DMSO, – 78 °C, 1 h, then Et_3N, 15 min, 100%; *2*. (4-methoxybenzyl)triphenylphosphonium bromide, THF, NaH, rt, overnight, 78%; **e** $(COCl)_2$, CH_2Cl_2, DMSO, – 78 °C, 1 h, then Et_3N, 15 min, 100%; **f** $MeNH_2$·HCl, MeOH, then sodium cyanoborohydride, rt, overnight, then 5% NaOH, 84%; **g** Et_3N, Ac_2O, rt, 8 h, 58%; **h** 10% KOH, MeOH, rt, 1 h, 100%; **i** H_2, 10% Pd/C then $(COCl)_2$, CH_2Cl_2, DMSO, – 78 °C, 1 h, then Et_3N, 15 min, 100%; **j** Wittig reaction with 4-methoxyphenylmethylenetriphenylphosphorane; **k** $MeNH_2$·HCl, MeOH, N_2, 1 h, then sodium cyanoborohydride, rt, overnight, 88%

75 R = H; (*E*)
76 R = H; (*Z*)

77

78

79

80

81

Scheme 8 **a** BuLi, THF, – 78 °C, 10 min, then HMPA, 5 min; then transferred to anode chamber containing 0.25 M LiClO$_4$, THF, HMPA and electrolyzed at a constant current, – 10 °C, 53%; **b** MeOH, conc. HCl, reflux, 15 min, 82%; **c** DMF, imidazole, TBDMSCl, rt, 8 h, 85%; **d** Ac$_2$O, Et$_3$N, rt, 8 h, 90%; **e** THF, 1.0 M Bu$_4$NF, 0 °C, 30 min, then rt, 8 h, 75%

methane, followed by reduction with LiAlH$_4$, gave diol **63** in 94% yield. Monobenzylation of the diol **63** with 1 equiv. of benzyl bromide in the presence of a phase-transfer catalyst gave monobenzyl ether **64** in 85% yield. Acetylation of **64** followed by removal of the benzyl group gave the corresponding alcohol **65** in 92% yield. Swern oxidation of alcohol **65** gave the corresponding aldehyde, which underwent a Wittig reaction to give a 80 : 20 mixture of (*E*)-(**66**) and (*Z*)-alkenes (**67**) in 78% yield from **65**. The hydrolysis of the mixture of **66** and **67** with KOH in methanol gave a mixture of the corresponding isomeric alcohols after workup, the recrystallization of which gave a pure crystalline (*E*)-alcohol and an oily (*Z*)-alcohol containing 18% of (*E*)-alcohol. Removal of the acetyl group from **66** followed by Swern oxidation gave **72** in 93% yield. Reductive amination of **72** with methylamine gave **75** in 88% yield. (*Z*)-Isomer **76** was similarly obtained from **67**.

On the other hand, Swern oxidation of **64** gave **68** in quantitative yield. Reductive amination of aldehyde **68** with methylamine in the presence of sodium cyanoborohydride gave *N*-methylamine **69** in 83% yield. Acetylation of amine **69** followed by removal of the benzyl group of the resulting *N*-acetate **70** by hydrogenolysis over Pd/C and subsequently by Swern oxidation of the resulting alcohol, gave **71** in 87% yield. Wittig reaction of **71** gave olefin **74** in 36% yield. Deacetylation of **74** with a base under various conditions failed to give an acceptable yield of methylamine **75** and **76**.

Treatment of **75** with butyllithium at – 78 °C, followed by anodic oxidation of the resulting lithium amide gave the pyrrolidine **77** as a single stereoisomer in 53% yield (Scheme 8). A similar anodic oxidation of the lithium amide of (*Z*)-isomer **76** also gave the pyrrolidine **77**, which was identical to that obtained from (*E*)-isomer **75**. Removal of the methoxymethyl groups of **77** with hydrochloric acid in methanol gave diol **78** in 82% yield. The treatment of **78** with *tert*-butyldimethyisilyl chloride and imidazole gave monosilyl ether **79** in 85% yield. Acetylation of the free hydroxy group of **79** followed by treatment of the resulting acetate **80** with tetrabutylammonium fluoride gave (+)-*N*-methylanisomycin (**81**) in 77% yield.

(−)-Deacetylanisomycin (**2**) has been synthesized via the stereocontrolled reductive alkylation of a protected trihydroxynitrile **84** derived from L-threonic acid derivative **82** (Scheme 9) [63]. Transformation of **82** to the primary amide **83** by reaction with ammonia in ethanol, followed by dehydration with trifluoroacetic anhydride gave the nitrile **84**. Condensation of **84** with 4-methoxybenzylmagnesium chloride afforded a metalloimine intermediate which upon treatment with *N*-benzylamine and subsequently reduced with sodium borohydride afforded the *syn* and *anti* diastereomeric *N*-benzylamines **85** and **86** in 19 : 81 ratio and 80% overall yield from **84**. Compound **86** was desilylated with tetrabutylammonium fluoride to give **87**. This was regioselectively mesylated on the hydroxyl group, deisopropylidenated, and then intramolecularly cyclized to furnish *N*-benzyl deacetylanisomycin **88**. Catalytic hydrogenation of **88** gave (−)-deacetylanisomycin (**2**).

Scheme 9 **a** NH_3, EtOH, 90%; **b** TFAA, Py, CH_2Cl_2, 87%; **c** *1*. $4\text{-}CH_3OC_6H_4CH_2MgCl$, ether; *2*. $BnNH_2$, CH_3OH; *3*. $NaBH_4$, 80% for three steps; **d** TBAF, THF, 80%; **e** *1*. MsCl, Et_3N, cat. DMAP, CH_2Cl_2; *2*. 10% HCl, THF; then $NaHCO_3$, 80% for two steps; **f** H_2, 10% Pd *on* C, AcOEt, 90%

A nitrone-based approach to the enantioselective total synthesis of (−)-anisomycin (**1**) has been developed from $(3S, 4S)$-1-benzylpyrrolidine-3,4-diol (**89**), which was obtained from L-tartaric acid (Scheme 10) [29]. L-Tartaric acid was converted to pyrrolidine **89** [79] by condensation with benzylamine and then reduction with $NaBH_4/BF_3{\cdot}Et_2O$.

Methoxymethyl ethers protection of the hydroxyl groups was carried out by reaction of **89** with dimethoxymethane in the presence of P_2O_5. Deprotection of the *N*-benzyl group from **90** by hydrogenation in the presence

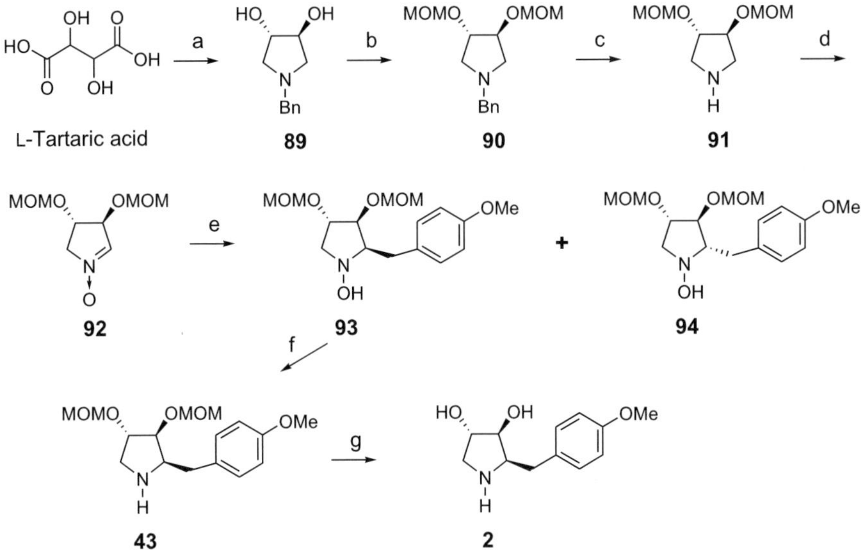

Scheme 10 a [79], **b** dimethoxymethane was cooled to 0 °C, P_2O_5, then rt, 48 h; then 20% methanolic KOH, 75%; **c** MeOH, 20% $Pd(OH)_2/C$, H_2, rt, 3 h, 85%; **d** SeO_2, acetone, 0 °C, 30% H_2O_2, 60%; **e** CH_2Cl_2, $MgBr_2 \cdot Et_2O$, rt, 15 min, 4-methoxybenzylmagnesium chloride, 0 °C, 30 min, **93** (42%), **94** (18%); **f** MeOH, H_2, Raney Ni, rt, 6 h, 85%; **g** MeOH, 6 N HCl, reflux, 24 h, 75%

of 20% $Pd(OH)_2$ on carbon afforded the amine **91** in 45% yield from L-tartaric acid. The conversion of pyrrolidine **91** into the desired nitrone **92** was carried out by oxidation with 30% H_2O_2 in the presence of SeO_2 as the catalyst. Reaction of (4-methoxybenzyl)magnesium chloride with **92** at 0 °C in THF produced a chromatographically separable mixture of diastereomers **93** and **94** in a ratio of 2 : 3 and 60% yield from **91**. Conducting the reaction in the presence of 1 equivalent of $MgBr_2 \cdot Et_2O$ in CH_2Cl_2 effected a change in the diastereoselectivity. Since the $MgBr_2$ is capable of coordinating with the nitrone oxygen, the magnesium atom of the Grignard reagent would possibly be forced to interact with the acetal oxygen in position 3. This coordination allows the attack to the double bond through a cyclic five-membered ring transition state giving preferentially a *cis* addition. Catalytic hydrogenation of **93** in the presence of Raney Ni produced pyrrolidine **43** in 85% yield. Removal of the methoxymethyl groups gave deacetylanisomycin (**2**) in 12% overall yield from (+)-tartaric acid. Conversion of **2** to (–)-anisomycin (**1**) can be achieved as shown above in Scheme 6.

Another approach for the total synthesis of anisomycin derivatives from (+)-tartaric acid has been reported via the *N*-benzyl tartarimide (**95**) (Scheme 11) [89]. Thus, (+)-tartaric acid was refluxed with benzyl amine in a xylene solution to give **95**, which was subjected to reaction with the Grignard reagent of anisyl chloride in THF to give the keto-amide **96** in 55% yield.

Scheme 11 **a** BnNH$_2$, xylene, reflux, 3 h with continuous water removal; **b** Mg, anisyl chloride, THF, rt, 30 min; then added N-benzyl tartarimide, rt, overnight, 55%; **c** LiAlH$_4$, THF, reflux, 6 h; **d** Ac$_2$O, 20 min; **e** EtOH, HCl, 5% Pd – C, H$_2$, 3 h; **f** P$_2$O$_5$, CHCl$_3$, 10 min, evaporation; then added MeOH, 20% KOH in EtOH, 30 min; **g** glacial acetic acid, reflux, 90 min, evaporation; then 1 N NaOH, reflux, 2 h

Treatment of **96** with lithium aluminum hydride afforded via the intermediate I three compounds **97**, **98a**, and **98b** which were isolated in 2, 12, and 3% yield, respectively. Acetylation of **98a** with acetic anhydride gave the hydroxy acetate **99**. The acetate **99** was treated with phosphorus pentachloride in dry chloroform at room temperature followed by hydrolysis in alkaline alcoholic solution to give the epoxide **100**. Refluxing **100** in glacial acetic acid gave a mixture of hydroxy acetates which were hydrolyzed by aqueous sodium hydroxide solution to two glycols **98** and **101** in equal amounts in total 56% yield from **100**. Debenzylation of the diol **101** in dilute ethanolic hydrochloric acid solution with 5% Pd/C under hydrogenation conditions gave the deacetylanisomycin hydrochloride quantitatively. Liberation of the free base by aqueous sodium hydroxide gave (+)-deacetylanisornycin (**102**). Similarly, debenzylation of **98b** as above gave quantitatively deacetyl anisomycin hydrochloride which on treatment with aqueous base gave (–)-deacetylanisomycin (**2**).

N-Benzylhomo-(–)-anisomycin has been synthesized also from tartarimide via amidoalkylation followed by introduction of a *p*-methoxyphenyl group (Schemes 12 and 13) [32, 33]. The allylic amide **104** was obtained from tartarimide **103** via *cis*-amidoalkylation [84]. The amide **104** was subjected

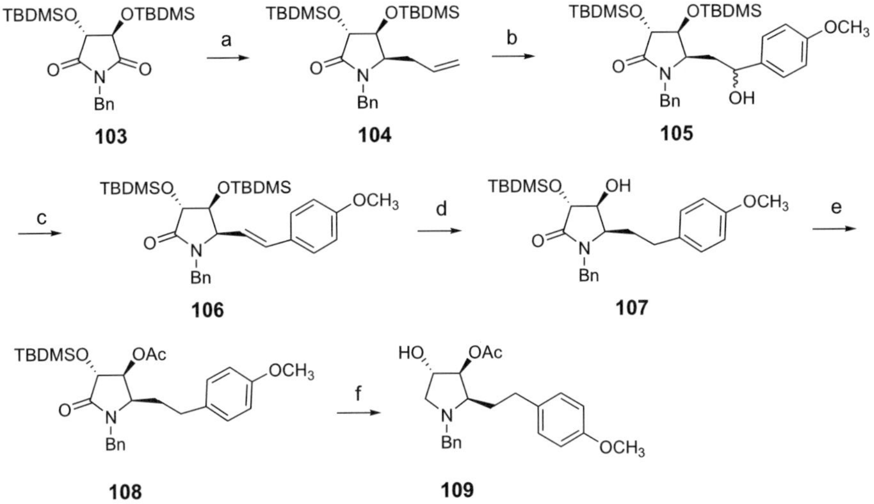

Scheme 12 a [84]; **b** *1*. O_3, CH_2CH_2, MeOH; *2*. Methyl sulfide; *3*. (*p*-methoxyphenyl)magnesium bromide, THF; **c** Et_3SiH, CH_2Cl_2, TFA; **d** *1*. 5% Pd–C, H_2; *2*. TBAF, THF; *3*. TBDMSCl, imidazole, DMF; **e** Ac_2O, Py; **f** *1*. TBAF, THF; *2*. BH_3–DMS, THF, rt

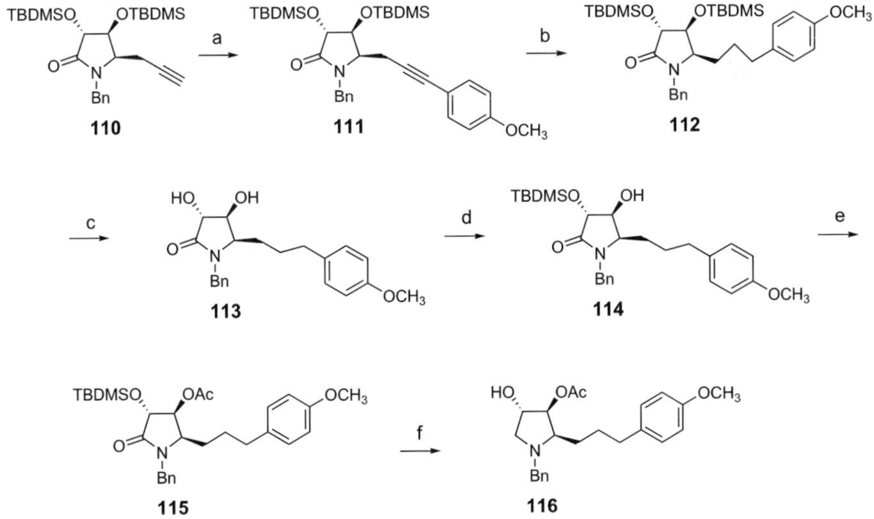

Scheme 13 a $4\text{-}CH_3OC_6H_4I$, DMF, CuI, PPh_3, K_2CO_3, 76%; **b** 5% Pd–C, H_2, MeOH; **c** TBAF, THF; **d** TBDMSCl, imidazole, CH_2Cl_2; **e** Ac_2O, Py; **f** *1*. TBAF, THF; *2*. BH_3–DMS, THF

to ozonolysis followed by reductive work-up using methyl sulfide to give the corresponding aldehyde. It was treated with (*p*-methoxyphenyl)magnesium bromide in THF to yield an epimeric mixture of benzylic alcohols **105** in 59% overall yield. To remove the alcohol group of **105**, various reductive conditions have been applied. The treatment of **105** with triethylsilane in the presence of trifluoroacetic acid in THF followed by β-elimination and then hydrogenation in the presence of Pd-catalyst, proved to be the best condition, to give **107**.

The lactam ring functional groups of **106** holding the desired side group then needs transformation to the desired structure. As the stereochemistry in **106** was positioned as required; and the acetate functionality at the OH – 3 position could be installed by using the steric hindrance resulting from the side chain. Thus, deprotection of TBDMS groups of **106** by TBAF provided a diol, and the sterically less hindered α-hydroxy group of the diol was specifically silylated when treated with 1.2 equivalents of *tert*-butyldimethylsilyl chloride in DMF to give **107**. Acetylation of the 3-hydroxyl group with acetic anhydride in pyridine yielded the acetate **108** in an overall yield of 52%. Finally, removal of the protecting silyl group with TBAF was followed by reduction of the amide group with borane-dimethyl sulfide complex to afford *N*-benzylhomo-(–)-anisomycin **109** with 34% yield in 2 steps.

For the preparation of other related compounds which have two more carbons extending the side chain, compound **110** was selected as a proper convenient intermediate (Scheme 13). The terminal acetylene functional group would be suitable for the attachment of various aromatic functionalities. The conventional conditions of the terminal acetylene coupling reaction, using a palladium reagent ($PdCl_2(Ph_3P)_2$, CuI, Et_3N) afforded **111** in 76% yield. Compound **111** was reduced to **112** using palladium on charcoal (5%) under H_2 in 63% yield. The same sequence of transformations as for the homoanisomycin derivative **109** was applied. TBAF treatment of compound **112** in THF provided diol **112** in 82% yield. Selective acetylation of 3-β-hydroxyl group of the diol was achieved by treatment of **113** with 1.2 equivalents of *tert*-butyldimethylsilyl chloride in CH_2Cl_2 to provide **114** in 38% yield, and treatment with acetic anhydride in pyridine yielded **115** in 68% yield. Deprotection of the TBDMS group in **115** with TBAF followed by reduction of the amide group gave **116** in 28% yield.

A total synthesis of (–)-anisomycin (**1**) from malimide has been achieved by a highly regio and *trans* stereoselective reductive alkylation of (*S*)-*N,O*-dibenzyl malimide **117** (Scheme 14) [85]. Reductive alkylation of (*S*)-*N,O*-dibenzyl malimide **117**, prepared from (*S*)-malic acid [86], with *p*-methoxybenzylmagnesium chloride gave the α-hydroxylactam **118** as a diastereomeric mixture. Hydroxylactam **118** in the presence of 3 equivalents of boron trifluoride etherate was reduced with excess of triethylsilane to yield predominantly *trans*-**119** in 94.8% yield. Catalytic hydrogenation of **119** afforded **120** in quantitative yield which was reduced to pyrrolidine **121** in 90%

Scheme 14 **a** p-MeOC$_6$H$_4$CH$_2$MgCl, THF, $-15\,°$C $\sim-10\,°$C, 1 h, 94.5%; **b** BF$_3$.OEt$_2$, Et$_3$SiH, $-78\,°$C, 6 h, then rt, 6 h, 94.8%; **c** H$_2$, 1 atm, 10% Pd $-$ C, EtOH, rt, 7 days, 100%; **d** LiAlH$_4$, THF, rt, 2 h, then 50 °C, 4 h, 90%; **e** H$_2$, 1 atm, 20% Pd(OH)$_2$/C, EtOH, rt, 15 h, 93.7%; **f** CbzCl, Et$_3$N, CH$_2$Cl$_2$, rt, 4 h, 82.8%; **g** NaH, THF, imidazole, 65 °C; CS$_2$, 60 °C, 2 h; MeI, 60 °C, 72%; **h** heat 190 $\sim$ 200 °C, under reduced pressure < 10 mm Hg, 2 h, 87.3%; **i** [66]

yield. *N*-Debenzylation in the presence of Pearlman's catalyst followed by selective *N*-benzyloxycarbonylation gave *N*-protected pyrrolidine **123** in an overall yield of 77%. Xanthation of alcohol **123** provided **124** in 72% yield, that on thermolysis of **124** at 190–200 °C under reduced pressure gave **125**. This was previously converted to the natural enantiomer of (–)-anisomycin (**1**) [66].

2.6
Synthesis from D-Mannitol

An enantioconvergent synthesis [64] of (–)-anisomycin (**1**) has been established starting from both (*R*)-(**126**) and (*S*)-(**134**) enantiomers of epichlorohydrin [65] that could be obtained from D-mannitol (Schemes 15 and 16). The (*R*)-epichlorohydrin (**126**) was first transformed to (*R*)-*O*-benzylglycidol (**128**) in 60% overall yield by treatment with benzyl alcohol in the presence

Scheme 15 **a** BF$_3$·OEt$_2$, BnOH, 50 °C; **b** NaOH, H$_2$O, Et$_2$O; **c** 4-bromoanisole, n-BuLi, CuCN, THF, -78 °C; **d** 1. H$_2$, Pd(OH)$_2$, CH$_3$OH; 2. PhCHO, p-TsOH, benzene, reflux; **e** NBS, CCl$_4$; **f** K$_2$CO$_3$, CH$_3$OH

Scheme 16 **a** Lithium acetylide ethylenediamine complex, DMSO, rt; **b** H$_2$, Pd, CaCO$_3$, AcOEt; **c** phthalimide, diisopropyl azodicarboxylate, PPh$_3$, THF, -20 °C; **d** H$_2$NNH$_2$, EtOH, reflux; **e** BzCl, Et$_3$N, CH$_2$Cl$_2$; **f** I$_2$, H$_2$O, CH$_3$CN; **g** 1. CbzCl, Et$_3$N, CH$_2$Cl$_2$; 2. K$_2$CO$_3$, CH$_3$OH; 3. CS$_2$, NaOH, n-Bu$_4$NHSO$_4$; then CH$_3$I, benzene, 87%; **h** 1. ODB, reflux, 70%; 2. NaOH, (CH$_2$OH)$_2$, 120 °C, 89%

of boron trifluoride etherate followed by cyclization of the resulted chlorohydrin **127**. Then, **128** was treated with 4-methoxyphenyllithium to give **129** (98%), which on catalytic debenzylation followed by benzylidenation gave the benzylidene acetal **130**. The latter was treated with *N*-bromosuccinimide to give the bromobenzoate **131** whose methanolysis in the presence of potassium carbonate afforded the epoxide **132** in 67% overall yield. Alternatively, treatment of (*S*)-epichlorohydrin (**134**) with 4-methoxyphenyllithium in the presence of copper(I) cyanide afforded the chlorohydrin **133** which was immediately exposed to methanolic potassium carbonate to give (*S*)-(4-methoxybenzyl)oxirane (**132**) in 74% overall yield.

Treatment of **132** with lithium acetylide ethylenediamine complex afforded the acetylene derivative **135** (85%), which was transformed into the vinyl alcohol **136**, by partial hydrogenation using Lindlar catalyst. Employing the Mitsunobu reaction, compound **136** was transformed into the phthalimide **137**, which was converted into the benzamide **139** (64%) via the amine **138** by sequential deacylation and benzoylation. When the amide **139** was exposed to three equivalents of iodine in aqueous acetonitrile, 2-(4-methoxybenzyl)-4-benzoyloxypyrrolidine **140** was obtained in 90% yield in a single step as a 2 : 1 mixture of epimers at the C_4-center. The mixture of **140** was *N*-protected, debenzoylated, and the resulting hydroxyl group was converted into the xanthate **141** in 87% overall yield. Thermolysis of **141** in *o*-dichlorobenzene followed by alkaline hydrolysis furnished the amine **142** in 89% yield; a precursor for the natural (–)-anisomycin (**1**) [26, 66].

An efficient approach to enantiomerically pure (+)-deacetylanisomycin (*anti*-**2**) (**102**) and a formal synthesis of (+)-anisomycin (*anti*-**1**) (**153**) have been achieved from 1,2 : 3,4 : 5,6-tri-*O*-isopropylidene-D-mannitol (**143**); Grignard reaction and intramolecular cyclization have been used as key steps (Scheme 17) [67]. The fully protected D-mannitol **143** was treated with H_5IO_6 to give aldehyde **144** followed by immediate reduction with $NaBH_4$ to give the corresponding arabinitol derivative, which was treated with benzyl bromide and NaH to give benzyl ether derivative **145**. Selective deisopropylidenation of the terminal acetal **145** with 50% aqueous AcOH gave the corresponding diol followed by regioselective tosylation to produce **146**, which on further treatment with K_2CO_3/MeOH yielded epoxide **147**. Grignard reaction of **147** with *p*-methoxyphenylmagnesium bromide gave **148**. Treatment of **148** with MsCl/Et_3N, followed by treatment with NaN_3 yielded the azido derivative **151**. Reduction of the azido group with LiAlH$_4$/THF followed by protection of the resulting amine by in situ treatment with (Boc)$_2$O/Et_3N afforded compound **150**. Removal of the benzyl group in **150** by using Li/liq. NH$_3$ and subsequent mesylation gave the corresponding mesyl derivative **151**, which without purification was treated with TFA followed by Et_3N to give (+)-deacetylanisomycin (*anti*-**2**) (**102**). This was treated with CbzCl to give (+)-*N*-benzyloxycarbonyl deacetylanisomycin (**152**), which can be converted to (+)-anisomycin (*anti*-**1**) (**153**) as mentioned above [66].

A synthesis of the intermediate **39** used in Scheme 4 for the synthesis of anisomycin (**1**) has been developed from D-mannitol (Scheme 18) [76, 77] by its conversion to the commercially available (*E*)-unsaturated ester (4*S*)-

143 → 144 → 145 → 146 → 147 → 148 → 149 → 150 → 151 → 102 → 152 → 153

Scheme 17 **a** H_5IO_6, [68]; **b** *1*. $NaBH_4$, MeOH, rt, 3 h, 67% for two steps; *2*. NaH, BnBr, DMF, 0 °C to rt, overnight, 84%; **c** *1*. 50% aq. AcOH, rt, overnight, 84%; *2*. *p*-TsCl, Et_3N, DCM, 0 °C, 16 h; **d** K_2CO_3, MeOH, 30 min, 71% for two steps; **e** 4-Bromo anisole, Mg, I_2/CuI (cat.), 0 °C to rt, overnight, 86%; **f** *1*. MsCl, Et_3N, DCM, rt, 1 h; *2*. NaN_3, 18-Crown-6, DMSO, 65 °C, 24 h, 83% for two steps; **g** *1*. $LiAlH_4$, THF, 0 °C to rt, overnight; *2*. $(Boc)_2O$, Et_3N, THF, rt, 6 h, 81% for two steps; **h** *1*. Li/liq. NH_3, – 78 °C, 30 min, 83%; *2*. MsCl, Et_3N, DCM, rt, 1 h; **i** *1*. TFA, DCM, 0 °C to rt, 10 h; *2*. Et_3N, MeOH, 0 °C to rt, 5 h; **j** CbzCl, Na_2CO_3, THF, 2 h, 69% for four steps; **k** [66]

154 [78] via treatment with 80% aqueous AcOH at 80 °C to afford the diol (4*S*)-**155** in quantitative yield. Bromination of (4*S*)-**155** with CBr_4 and triphenylphosphine provided a mixture of bromohydrins (4*S*)-**156** and (4*R*)-**157**. This mixture was subjected to silylation to give (4*S*)-**158** and (4*R*)-**159** in 32% and 15% overall yield from (4*S*)-**154**, respectively. The bromohydrin (4*S*)-**158** was converted to (4*S*)-4,5-epoxy-2(*E*)-pentenoate **159** in 85% yield by treatment with K_2CO_3 in MeOH. Conversion of (4*S*)-**159** into (4*R*)-**163** without loss of optical purity was carried out by reaction of (4*S*)-**159** with benzyl alcohol in the presence of $BF_3 \cdot Et_2O$ to give (4*R*)-**160** in 55% yield followed by bromination of (4*R*)-**160** to provide (4*R*)-**161** in 95% yield, which underwent deprotection of the benzyl group to afford (4*R*)-**162** in 87% yield.

Scheme 18 diagram showing the synthetic route from D-mannitol to compound 1 via intermediates (4S)-154, (4S)-155, (4S)-156, (4R)-157, (4R)-158, (4S)-159, (4R)-160, (4R)-161, (4R)-162, (4R)-163, (4R)-164, (4R)-165, (4S)-166, (4S)-167, (4S)-168, (4S)-169, and (4S)-39.

Scheme 18 **a** [78]; **b** 80% aq. AcOH, 80 °C, 30 min. quantitative yield; **c** CH_2Cl_2, Ph_3P, CBr_4, reflux, 1 h; **d** TBDMSCl, imidazole, DMF, 0 °C, 2 h; **e** molecular sieves 3 Å, K_2CO_3, MeOH, 0 °C, 2.5 h, 85%; **f** BnOH, $BF_3 \cdot Et_2O$, CH_2Cl_2, – 20 °C to 0 °C, 1.5 h, 56%; **g** Ph_3P, CH_2Cl_2, CBr_4, 0 °C to rt 1 h, 95%; **h** $AlCl_3$, CH_2Cl_2, m-xylene, – 20 °C, 30 min, 87%; **i** anisole, $BF_3 \cdot Et_2O$, CH_2Cl_2, – 20 °C, 1 h, 22%; **j** Ph_3P, CH_2Cl_2, CBr_4, 0 °C to rt 1 h, 92%; **k** CH_3NO_2, molecular sieves 3 Å, $AgNO_3$, rt, 24 h, 91%, **l** Zn-dust, CH_3COONH_4, MeOH, 0 °C, 1 h, 87%; **m** N-methylmorpholine N-oxide, 2% aq. OsO_4, acetone, 0 °C, 2 h, 78%; **n** N,N-diisopropylethylamine, MeCN, CH_3OCH_2Cl, 0 °C to rt, 24 h, 78%; **o** 1 M diisobutylaluminum hydride, benzene, 0 °C, 1 h, 81%; **p** see Scheme 4, [52]

Alkaline treatment of (4R)-**162** yielded (4R)-**163** in 85% yield. The reaction of (4R)-**163** and anisole in the presence of $BF_3 \cdot Et_2O$ followed by enzymatic separation gave (4R)-**164** and analogue in 47% and 14% yield, respectively. The former (4R)-**164** was rearranged into the bromide (4R)-**165** in 92% yield, which was treated with $AgNO_3$ in $MeNO_2$ to furnish the nitrate (4S)-**166** in 91% yield followed by subsequent conversion into (4S)-**167** in 87% yield.

Osmium tetroxide-catalyzed dihydroxylation followed by treatment with *N*-methylmorpholine *N*-oxide gave the lactone (4*S*)-**168** in 78% yield. Treatment of (4*S*)-**168** with chloromethyl methyl ether furnished the di-MOM ether (4*S*)-**169** in 78% yield. Reduction of (4*S*)-**169** with Dibal-H gave the (4*S*)-diol **39** in 81% yield, which can be converted into (–)-anisomycin (**1**) [52].

2.7
Synthesis from Divinylcarbinol

The functionalized epoxide **171**, readily available from divinyl carbinol **170**, has been used for the synthesis of (–)-anisomycin (**1**) and 3097-B1 (**181**) (Scheme 19) [31]. The (3*R*, 4*S*)-trimethylsilyloxy oxirane **171** was obtained in 65% yield from divinylcarbinol (**170**) using Sharpless asymmetric epoxidation [69–71]. Regioselective ring opening of **171** with *p*-anisyl magnesium bromide gave **172** in 85% yield, which was mesylated in 94% yield then treated with sodium azide in DMF followed by removal of the silyl group with HCl to produce the azide **173** in 75% yield. Sharpless asymmetric dihydroxylation of **173** using DHQ-CLB as the chiral ligand and $K_3Fe(CN)_6$ as the cooxidant gave the corresponding triol which was subjected to regioselective silylation of the primary alcohol with TBDMSCl in DMF and imidazole at 0 °C to give the diol **174**. Addition of 1.1 equivalent of acetyl chloride or

Scheme 19 **a** *1*. D-(–)-DIPT, TBHP, Ti(OPr)$_4$, CH$_2$Cl$_2$, – 20 °C; *2*. trimethylsilyl chloride, Et$_3$N, DMAP, CH$_2$Cl$_2$; **b** *p*-anisyl magnesium bromide, CuI (10%), THF, – 10 °C; **c** *1*. MsCl, Py, CH$_2$Cl$_2$, rt; *2*. NaN$_3$, DMF, 80 °C, 7 hr then 2 N aq. HCl, 30 min; **d** *1*. OsO$_4$ (cat), K$_3$Fe(CN)$_6$, K$_2$CO$_3$, DHQ-CLB, *t*-BuOH/water, rt; *2*. TBDMSCl, imidazole, DMF, 0 °C; **e** AcCl (1.1 eq) or propionyl chloride, Py, CH$_2$Cl$_2$, rt, 2 h, then MsCl (1.5 eq), 4 h; **f** 1.0 M TBAF, THF, 0 °C, 1 h; **g** LiBr (10 eq), AcOH, THF, rt, 24 h; **h** H$_2$, 10% Pd/C, rt, 1 atm, 2 h, then NaOAc (1.5 eq), MeOH, reflux, 10 hr

propionyl chloride in pyridine to the diol **174** followed by the addition of 1.5 equivalent of methanesulfonyl chloride produced the mesylate **175** or **176** in 78% and 85% yield, respectively. Removal of the silyl group from **177** or **178** was achieved by treatment with 1 M n-Bu$_4$NF. The resultant alkoxide anion, acting as the nucleophile, displaced spontaneously the mesyloxy group (S_N2) to produce the oxirane **179** or **178** in 91% and 98% yield, respectively. The stereochemistry of C-2 was *threo* with C-3 as desired in the synthesis of anisomycin (**1**) and 3097-B1 (**132**). Hydrogenation of the oxiranes **177** and **178** did not form the desired (–) anisomycin (**1**) or 3097-B1 (**181**) due to the poor nucleophilic activity of the resulting amine in opening the oxirane. However, regioselective cleavage of the oxirane with lithium bromide gave the bromohydrin **179** and **180** in 90% and 89% yield, respectively. Finally, the bromohydrin was hydrogenated with 10% Pd/C followed by cyclization upon refluxing with NaOAc in MeOH to produce the (–)-anisomycin (**1**) and 3097-B1 (**181**) in 58% and 54% yield, respectively.

2.8
Synthesis from Tyrosine

Chiral intermediates for the synthesis of (–)-anisomycin (**1**) and (+)-anisomycin (*anti*-1) (**153**), (R)-2-(p-methoxyphenyl)methyl-2,5-dihydro-pyrrole (**142**) and its (S)-isomer (+)-**187**, have been efficiently synthesized from D-tyrosine and L-tyrosine, respectively (Scheme 20) [28]. D-tyrosine was converted to O-methyl D-tyrosine methyl ester (**182**) [72–75] which was treated with di-*tert*-butyl dicarbonate to protect the amino group. Subsequent reduction of the ester group with sodium borohydride in the presence of lithium chloride furnished the alcohol **183**. Swern oxidation of **183** followed by chain extension with the anion derived from bis(2,2,2-trifluoroethyl)(ethoxycarbonylmethyl)-phosphonate afforded (Z)-α,β-unsaturated ester (**184**), which was used immediately without purification to avoid or minimize any possible racemization of the chiral center. Reduction of the ester group of **184** with diisobutylaluminium hydride afforded the alcohol **185** which after mesylation followed by intramolecular cyclization gave the desired 2,5-dihydropyrrole derivative **186**. Removal of the *tert*-butyloxycarbonyl group was achieved by treatment with trifuoroacetic acid to give (–)-**142** in 62% overall yield from **182**. The (S)-2,5-dihydropyrrole (+)-**187** was also prepared in the same manner starting from L-tyrosine. Since (+)-**187** had been transformed into (+)-anisomycin (*anti*-1) (**153**), (–)-**142** could also be transformed to the (–)-anisomycin (**1**) [26, 66].

A total synthesis of enantiomerically pure (–)-deacetylanisomycin (**2**) from D-tyrosine, which also constitutes a formal total synthesis of **1**, has been reported (Scheme 21) [88]. Compound **188** was obtained in 70% yield from D-tyrosine by protecting the amino functionality using di-*tert*-butyl dicarbonate, followed by methylation of the acid and phenolic hydroxyl group

Scheme 20 a [72–75]; **b** *1*. (Boc)$_2$O, Na$_2$CO$_3$, H$_2$O-dioxane (1 : 1), RT, overnight, 95%; *2.* NaBH$_4$: LiCl (1 : 1; 5 eq), EtOH, THF (4 : 3), RT, 95%; **c** *1*. (COCl)$_2$, DMSO, CH$_2$Cl$_2$, Et$_3$N, – 78 °C; *2.* (CF$_3$CH$_2$O)$_2$P(O)CH$_2$CO$_2$Et, 18-crown-6, (Me$_3$Si)$_2$NK, THF, – 78 °C, 80% for two steps; **d** DIBAH, CH$_2$Cl$_2$, – 78 °C, 98%; **e** *1.* MsCl, CH$_2$Cl$_2$, Et$_3$N, – 10 °C, 20 min, 93%; *2.* NaH, CH$_2$Cl$_2$, rt, overnight, 94%; **f** TFA, CH$_2$Cl$_2$, rt, 4 h, 100%; **g** [26, 66]

with MeI and K$_2$CO$_3$. Reduction of ester **188** with lithium borohydride gave the alcohol **183** (72%), which was subjected to a Swern oxidation, followed immediately, without purification, by reaction with Ph$_3$P = CHCO$_2$Et in CH$_2$Cl$_2$ to produce the *trans* α,β-unsaturated ester **184** in 73% yield. Upon reacting the *N*-Boc-protected γ-amino-α,β-unsaturated ester **184** with the modified Sharpless asymmetric dihydroxylation conditions using AD-mix-α, the (2*R*,3*S*,4*R*) ester **189** and **190** were obtained in 60% yield with high stereoselectivity. Then diol **190** was subjected to acetonation with 2,2-dimethoxypropane (2,2-DMP) in the presence of a catalytic amount of camphorsulfonic acid (CSA) in anhydrous CH$_2$Cl$_2$ to afford the acetonide **191** in 72% yield, which was reduced with LiBH$_4$ to the corresponding alcohol **192** in 68% yield. The alcohol **192** was converted to its bromide **193** in 65% yield using LiBr, via the tosylate intermediate. The resulting Boc-amine **193** was subjected to acidolysis using TFA and the resulting amine was in situ treated with triethylamine to facilitate cyclization to the desired (–)-deacetylanisomycin (**2**) in 70% yield.

A highly stereo-divergent methodology for the synthesis of a range of stereoisomers of deacetylanisomycin has been demonstrated starting from a single amino acid with excellent selectivity (Schemes 22, 23, and 24) [80]. Syntheses of β-iodoacetates **199** and **210** were achieved by the following reaction sequence. The aminoaldehyde **194** was synthesized from homochiral

Scheme 21 a *1.* 1 M NaOH, Dioxane, H_2O, Boc_2O, rt, 30 min; *2.* MeI, acetone, K_2CO_3, reflux, 8 h 70%; **b** $LiBH_4$, Et_2O, MeOH, reflux, 5 h, 72%; **c** *1.* $(COCl)_2$, DMSO, DCM, Et_3N, −78 °C; *2.* $Ph_3P = CHCO_2Et$, CH_2Cl_2, rt, 73%; **d** AD-mix-α, $(DHQ)_2PHAL$, potassium osmate, *t*-BuOH, H_2O, 0 °C, then rt, 18 h, 60%; **e** 2,2-DMP, CSA, CH_2Cl_2, rt, 2–3 h under N_2, 72%; **f** $LiBH_4$, Et_2O, MeOH, reflux, 5 h, 68%; **g** *1.* TsCl, CH_2Cl_2, Py, rt, 10 h; *2.* LiBr, DMF, 80 °C, 10 h, 65%; **h** *1.* TFA, H_2O, CH_2Cl_2, rt, 10 h; *2.* Et_3N, MeOH, rt, 5 h, 70%

amino acid, D-tyrosine [81, 82] in 65% yield. The 9-phenyl-9-fluorenyl (Pf) group has been chosen for protection of the amine to inhibit deprotonation at the α-position of the α-amino aldehyde and to provide an electron-donating effect to the nitrogen atom, and accelerate iodoamidation. Treatment of **194** with vinylmagnesium bromide yielded a separable mixture of the *syn* and *anti* isomers of allylic alcohol **195**, in a 1 : 1 ratio, in 93% total yield. The conversion of the mixture of the allylic alcohol **195** to a single diastereoisomer of either **197** or **208** was achieved by Swern oxidation of **195** to afford the unsaturated ketone **196** followed by a highly regio- and stereoselective reduction at the carbonyl moiety by complex reducing agents to give both epimers **197** (Scheme 22) and **208** (Scheme 24) in high diastereoselectivity in accordance with the polar Felkin model, where the most electron-withdrawing substituent occupies the position usually adopted by the largest group in the classical Felkin–Ahn-type transition state. The nature of this group being perpendicular to the carbonyl adds extra delocalization in the transition state, allowing electron density from the nascent bond to overlap with the best acceptor orbital. The electrophilic BH_3SMe_2 was found to be the best reducing agent to give the Felkin product. On the other hand, **197** was formed in

Scheme 22 **a** [81, 82]; **b** 1 M vinylmagnesium bromide, THF, – 40 °C, 30 °C then saturated NaHCO$_3$; **c** oxalyl chloride, DMSO, Et$_3$N, – 78 °C, CH$_2$Cl$_2$ then saturated NaHCO$_3$, 91%; **d** 3 equiv. (*S*)-BINAL, THF, – 78 °C, 1 h, 89%; **e** Et$_3$N, CH$_2$Cl$_2$, DMAP, Ac$_2$O, rt, 2 h, 98%; **f** sat NaHCO$_3$, THF, Et$_2$O, iodine, rt, 36 h, 90%; **g** silver benzoate, toluene, 12 h, 92%; **h** LiAlH$_4$, THF, 0 °C, 30 min, 94%; **i** H$_2$, Pd/C, EtOAc, rt, 6 h, then MeOH, Dowex 50W-X8, then added 3 N NH$_4$OH, 89%

an excellent enantioselective manner by the use of (*S*)-BINAL, while the use of (*R*)-BINAL favored **208**. Subsequent reaction with Ac$_2$O furnished acetate **198** in 98% yields. Treatment of **198** with I$_2$ led to a stereoselective formation of an epiiodonium ion in situ that was followed by subsequent highly regioselective ring closure to give **199** in 90% yield. Formation of **202** could be achieved by treatment of **199** with dry AgOBz giving the corresponding protected diester **200** being obtained in excellent yield and with a high regioselectivity. Cleavage of both ester protecting groups in **200** with LiAlH$_4$ gave **201**, which was easily deprotected to the (2*R*,3*R*,4*R*)-deacetyl-anisomycin **202** (Scheme 22).

Treatment of **199** with wet AgBF$_4$ gave a 1 : 1 mixture of easily separable hydroxy acetates **203** and **204**. Proof that the two products differed only in the position of the acetate group was given by cleavage of a 1 : 1 mixture of acetates **203** and **204** to the corresponding single diol derivative (Scheme 23). Subsequent hydrogenolysis yielded (2*R*, 3*R*, 4*S*)-deacetylanisomycin (**207**). On the other hand, compound **199** was treated with silver triflate, using Prevost conditions. Trapping of the cyclic acetoxonium ion by trifluoroacetate

Scheme 23 **a** AgBF$_4$, toluene, H$_2$O, rt, 12 h, then MeOH; **b** *1.* LiAlH$_4$, THF, 0 °C, H$_2$O, 73%; *2.* H$_2$, Pd/C, EtOAc, rt, 6 h; then the residue was dissolved in MeOH, Dowex 50W-X8, 93%; **c** AgBF$_4$, toluene, reflux 6 h, then saturated NaHCO$_3$, 84%; **d** H$_2$, Pd/C, EtOAc, rt, 6 h, 91%

Scheme 24 **a** 2 M BH$_3$(CH$_3$)$_2$S, toluene, – 78 °C, 30 min, then saturated NaHCO$_3$, 91%; **b** Et$_3$N, CH$_2$Cl$_2$, DMAP, Ac$_2$O, rt, 2 h, 96%; **c** sat NaHCO$_3$, THF, Et$_2$O, iodine, rt, 36 h, 90%; **d** AgBF$_4$, toluene, H$_2$O, rt, 12 h, then MeOH; **e** *1.* LiAlH$_4$, THF, 0 °C, H$_2$O, 70%; *2.* H$_2$, Pd/C, EtOAc, rt, 6 h; then the residue was dissolved in MeOH, Dowex 50W-X8, 92%

formed the *trans* diol derivative, with a labile protecting group installed regioselectively on the C-4 position. Deprotection was easily effected during chromatography on silica gel to yield *N*-protected anisomycin **205** in 84% yield. Deprotection by hydrogenolysis furnished (2*R*, 3*R*, 4*R*)-anisomycin **206** in 91% yield. Thus, the synthesis of anisomycin was accomplished from **199** in two steps in 76% overall yield.

However, in an analogous manner to that documented above, **210** was treated with AgBF$_4$ in aqueous conditions, and as observed previously, a 1 : 1 mixture of separable acetates **211** and **212** was generated (Scheme 24). Cleavage of the acetyl group from the mixture **211/212** using LiAlH$_4$ gave the corresponding diol as a single product in 70% yield whose deprotection furnished (2R, 3S, 4R)-deacetylanisomycin (**213**).

The synthesis of C(4)Me **220** and C(4)H **229** analogues of anisomycin was achieved from oxazolidinone **214** and aldehyde **215** (Schemes 25 and 26) [90].

Scheme 25 a *1.* Bu$_2$BOTf, NEt$_3$, CH$_2$Cl$_2$, 0 °C, 15 mm; *2.* **215**, – 78 °C to 0 °C, 5 h (87%); **b** LiBH$_4$, MeOH, THF, 0 °C to rt, 18 h (82%); **c** *1.* p-TsCl, DMAP. CH$_2$Cl$_2$, 0 °C, 18 h; *2.* Dowex Cl$^-$ (83%); **d** H$_2$, 5% Pd/C, K$_2$CO$_3$, MeOH, rt, 20 min (93%); **e** *1.* Ac$_2$O, NEt$_3$, CH$_2$Cl$_2$, rt, 18 h (82‰); *2.* H$_2$, 20% Pd(OH)$_2$/C, 1 M HCl – Et$_2$O, MeOH, rt, 18 h (100%)

Scheme 26 a LiOH, THF – H$_2$O, reflux, 18 h (95%); **b** *1.* CDI, THF, rt, 2.5 h; *2.* CH$_2$ = C(OLi)OEt, – 78 °C, 2 h (82%); **c** NaCNBH$_3$, AcOH, Et$_2$O, MeOH, 0 °C to rt, 18 h (83%); **d** LiAlH$_4$, THF, – 78 °C, 6 h (95%); **e** *1.* TIBSCl, DMAP, CH$_2$Cl$_2$, 0 °C, 18 h; *2.* Dowex Cl$^-$ (85%); **f** H$_2$, 5% Pd/C, K$_2$CO$_3$, MeOH, rt, 20 min (80%); **g** Ac$_2$O, NEt$_3$, CH$_2$Cl$_2$, rt, 18 h (87%); **h** H$_2$, 20% Pd(OH)$_2$/C, 1 M HCl – Et$_2$O, MeOH, rt, 20 h (100%)

The propionate derivative of oxazolidinone **214** was allowed to condensate with the aldehyde **215** to furnish **216** in 87% yield. Borohydride reduction of **216** gave diol **217**. Selective tosylation of the primary alcohol resulted in spontaneous cyclization to give the pyrrolidinium tosylate salt, which was converted to its chloride salt **218** in 83% yield. Selective mono-deprotection of the salt afforded the free hydroxyl **219**, which was acetylated and then subjected to hydrogenolysis in the presence 1M hydrochloric acid to allow the isolation of the C(4)Me analogue **220** as its hydrochloride salt in quantitative yield.

The synthesis of the C(4)H analogue **229** of anisomycin has been obtained from tyrosine derivative **221** (Scheme 26) [90]. Hydrolysis of methyl ester **221** gave **222**, which was converted to the more reactive acylimidazole intermediate **223**. The *syn* 3-hydroxy ester **225** was obtained after sodium cyanoborohydride reduction of ester **224**. Subsequent lithium aluminum hydride reduction of the ester **225** to give the diol precursor to the pyrrolidine ring proceeded in 95% yield. Treatment with Dowex Cl exchange resin facilitated purification of the pyrrolidinium salt **226** (81% from ester **225**). Monodebenzylation of **226** gave alcohol **227** followed by acylation to give **228** and finally benzyl deprotection were achieved as in the C(4)Me series allowing the isolation of the C(4)H analogue **229**.

2.9
Synthesis from Pyrrolealdehyde

Total synthesis of (±)-anisomycin and its analogues **36** and **247** from 2-benzylpyrroles **232**, **233**, and **234**, which could be prepared by arylation-reduction of pyrrole-2-carboxaldehyde (**230**) (Scheme 27) has been reported [26, 83]. The *syn* epoxides **32**, **33**, and **241** were stereospecifically prepared by reduction of 2-(substitutedbenzyl)pyrroles **232–234** with zinc in HCl to give the corresponding 2-(substitutedbenzyl)-3-pyrrolines, which were then protected with benzyl chloroformat to give *N*-((benzyloxy)carbonyl)-2-(substitutedbenzyl)-3-pyrrolines **235–237**. Subsequent treatment with *N*-iodosuccinimide in perchloric acid and THF at 0 °C gave only one isomer of the iodohydrins, *N*-((benzyloxy)carbonyl)-4β-hydroxy-3α-iodo-2-(substitutedbenzyl)pyrrolidines **238–240**. This regioselective and stereospecific reaction reflected the dominant influence of the bulky (substitutedbenzyl) group at the C-2 position. *N*-iodosuccinimide rather than *N*-bromoacetamide had to be used to prepare the halohydrins **238** (R = OCH$_3$) and **239** (R = CH$_3$) to avoid halogenation of the activated aromatic ring; however, for the preparation of halohydrin **240** (R = H), this side reaction was not a problem and *N*-bromoacetamide could be used. The halohydrins **230–240** were then converted to the corresponding *N*-protected *syn* epoxides in 5–10% KOH (alcohol). The *N*-((benzyloxy)carbonyl) protecting group was then removed by catalytic hydrogenation to yield 2-(*p*-methoxybenzyl)-

Scheme 27 **a** RC_6H_4Li, Et_2O; **b** $Li-NH_3$, NH_4Cl; **c** *1*. Zn, aqueous HCl, EtOH or MeOH; *2*. $C_6H_5CH_2OCOCl$, toluene, 2 N NaOH *or* $NaHCO_3$ (MeOH); **d** NIS or NBA, aqueous $HClO4$, THF; **e** *1*. 5–10% KOH (EtOH or MeOH); *2*. H_2, 10% Pd–C, MeOH or EtOH; **f** *1*. CF_3CO_2H, CF_3CO_2Na, 120 °C; *2*. 10% Na_2CO_3, THF, $C_6H_5CH_2OCOCl$; **g** *1*. Cl_3CCH_2OCOCl, Py, CH_2Cl_2; *2*. Ac_2O, Py; **h** *1*. Zn, AcOH, THF; then 2 N NaOH; *2*. H_2, 10% Pd/C, MeOH

$3,4\beta$-oxidopyrrolidine (**33**) quantitatively. The *syn* epoxide **33** was cleaved in refluxing trifluoroacetic acid containing sodium trifluoroacetate and then the acid removed in vacuo to produce *N*-((benzyloxy)carbonyl)-$3\beta,4\alpha$-dihydroxy-2β-(*p*-methoxybenzyl)pyrrolidine (**9**) in 97% yield. The less hindered 4α-hydroxy group was selectively protected by the slow addition of an excess of 2,2,2-trichloroethyl chloroforrnate. The anisomycin derivative **244** was then sequentially deprotected with zinc in HOAc-THF and catalytic hydrogenation (H_2/Pd – C). The overall isolated yield of (±)-anisomycin (**1**) from the *syn* epoxide **33** was 75%. The isolated yields of the anisomycin analogues **36** and **247** from the corresponding *syn* epoxides **32** and **241** were 75% and 71%, respectively.

Comparison of the crystalline synthetic (±)-anisomycin product with authentic natural (–)-anisomycin indicated total identity with ^{1}H NMR, ^{13}C NMR, IR, UV, and mass spectroscopy. In addition, in vitro tube-dilution biological testing against protozoa (*Trichomonas vaginalis*, *T. foetus*, and *Entai'noeba hystolytica*), yeasts (*Candida albicans* and *Saccharomyces cerevisiae*), and dermatophytes (*Trichophyton mentogrophytes* and *Epidermophyton floccosum*) indicated that the synthetic (±)-anisomycin possess half of the activity of the natural (–)-anisomycin. The new synthetic analogues **244** and **36** were also effective antibiotics against these microorganisms, with the general trend of (±)-**1** > (±)-**244** > (±)-**36**.

Total syntheses of (–)-anisomycin (**1**) and the analogues **244**, **36**, and **258–261** have been achieved from furfuraldehyde (Scheme 28) [25]. *N*-Protection was necessary during the generation of the halohydrin **251** and during the selective acylation of the *trans*-diol **255**, and for these purposes the *N*-Cbz group performed this task. However, when the *N*-Cbz protecting group was employed in the d–g series, the protecting group could not be removed by catalytic hydrogenation from the *N*-Cbz derivatives of the *syn*-2-benzyl-3-pyrrolidine epoxides **252d–g**. Medium-pressure (45 psi) catalytic hydrogenation or HBr/HOAc conditions destroyed the epoxide. A procedure of limited success was to reflux the *N*-Cbz derivatives (R″ = Cbz) of the *syn*-epoxides

230 **248a–g**

249a–g R″ = H
250a–c R″ = Cbz
250d–g R″ = TCE

251a,b X = I, R″ = Cbz
251c X = Br, R″ = Cbz
251d–g X = I, R″ = TCE

252a–c R″ = Cbz
252d–g R″ = TCE
253a–g R″ = H

254a–g R″ = H
255a–c R″ = Cbz
255d–g R″ = Fmoc

256a–c R″ = Cbz
257d–g R″ = Fmoc

1   R = *p*-OCH$_3$, R' = H
247 R = *p*-CH$_3$, R' = H
36 R = H, R = H
258 R = *m*-OCH$_3$, R' = H
259 R = *o*-OCH$_3$, R' = H
260 R = *p*-OCH$_3$, R' = Me
261 R = *p*-OCH$_3$, R' = Ph

Scheme 28 **a** *1.* Et$_2$O; *2.* Li – NH$_3$, NH$_4$Cl; **b** *1.* Zn, aqueous HCl, EtOH; *2.* C$_6$H$_5$CH$_2$OCOCl *or* CCl$_3$CH$_2$OCOCl in toluene, 2N NaOH; **c** NIS *or* NBA, aqueous HClO$_4$, THF; **d** *1.* 5–10% KOH in EtOH; *2.* H$_2$, 10% Pd – C, MeOH *or* Zn, HOAc, THF, then 2 N, NaOH; **e** *1.* CF$_3$CO$_2$H, CF$_3$CO$_2$Na, 120 °C; *2.* C$_6$H$_5$CH$_2$OCOCl *or* 9-fluorenylmethyl 4,6-dimethyl-2-pyrimidylthiocarbonate in 10% Na$_2$CO$_3$, THF; **f** *1.* CCl$_3$CH$_2$OCOCl, Py, CH$_2$Cl$_2$; *2.* Ac$_2$O, Py; **g** *1.* Zn, AcOH, THF, then 2N NaOH; *2.* H$_2$, 10% Pd – C, MeOH *or* DBU, THF

252e and **252g** in triethylsilane containing a catalytic amount of PdCl$_2$ and Et$_3$N to yield the corresponding deprotected *syn*-epoxides **252e** and **252g** in 70% yields. The procedure failed for the *N*-Cbz derivatives (R″ = Cbz) of the *syn*-epoxides **252d** and **252f**. The best solution to this problem was to change to 2,2,2-trichloroethoxycarbonyl (TCE) as the *N*-protecting group, which could be removed efficiently with zinc in THF-HOAc to yield the *syn*-epoxides **252d–g** in 85% yield.

The second major problem which developed in the d-g series was neither the *N*-Cbz group nor the *N*-TCE group could be removed from the protected anisomycin analogues of **257** (R = Cbz or TCE) at the end of the syntheses. The *N*-protection was required during the previous step to perform selective sequential acylation of the *trans*-diols **254a–g**. The protecting groups, allyloxycarbonyl and 9-fluorenylmethoxycarbonyl (Fmoc) proved satisfactory and could, in turn, be removed under nonhydrolytic conditions. The bulkier Fmoc group was selected, since its presence on the secondary amine of the *trans*-diols **255d–g** made the subsequent acylation of the α-hydroxy group at C-4 with 2,2,2-trichloroethyl chloroformate more selective. After acetylation of the β-hydroxy group at C-3, the TCE group was removed with zinc (THF-HOAc) and then the *N*-Fmoc group with 1,8-diazabicyclo[5.4.0]undec-7-ene (DBU) in THF to complete the syntheses of the anisomycin analogues **258-261d–g**.

2.10
Synthesis from 1,1,2-Trichloroethene

The 1,1,2-trichloroethene and (*S*)-1-(2,4,6-triisopropylphenyl)ethanol (**262**) have been also used for the synthesis of the enantiopure (–)-anisomycin (**1**) (Scheme 29) [34]. Thus, the alcohol **262** was transformed with potassium hydride and trichloroethylene to the *E* dichloro enol ether **263** in 82% yield. The corresponding lithium acetylide, formed from **263** with *n*-butyllithium, was then treated with 4-methoxybenzyl bromide in THF-HMPA under carefully optimized conditions to yield the unstable alkylated acetylenic ether, which was converted without purification into the *Z* enol ether **264** in 81% yield for the two steps. In the presence of dichloroketene, generated in situ from trichloroacetyl chloride and zinc-copper couple, enol ether **264** underwent cycloaddition to produce in high yield the desired dichlorocyclobutanone **265**, contaminated with only 7% of diastereomeric material. Since it was anticipated that removal of this contaminant would be more readily achieved at the pyrrolidinone stage, the crude cycloadduct was converted into the corresponding dichloropyrrolidinone, which was dechlorinated with zinc-copper couple in methanol saturated with ammonium chloride to provide the highly crystalline pyrrolidinone **266**. Recrystallization of this material from pentane-dichloromethane afforded in 68% overall yield from enol ether **263** the stereochemically pure pyrrolidinone **266**. The resulting pyrrolidine **267**

Scheme 29 **a** 35% KH in mineral oil, $Cl_2C=CHCl$, 82%; **b** _1._ n-C_4H_9Li, p-$CH_3OC_6H_4$ CH_2Br; _2._ H_2, $Pd-BaSO_4$, C_5H_5N, 81%; **c** Cl_3CCOCl, $Zn-Cu$; **d** _1._ MSH, Al_2O_3; _2._ $Zn-Cu$, NH_4Cl, 68% for 3 steps; **e** _1._ $CBZ-Cl$, Et_3N, _2._ $LiEt_3BH$, Et_3SiH, BF_3, 61%; **f** CF_3COOH, 99%; **g** [38]

on brief treatment with trifluoroacetic acid gave in 99% yield the enantiopure hydroxy pyrrolidine **268**. Thus, the synthesis, based on an effective asymmetric 2 + 2 cycloaddition, produced (–)-anisomycin in approximately 8% overall yield.

2.11
Synthesis of Derivatives from Natural Anisomycin

An efficient solid-phase method for the synthesis of anisomycin derivatives has been developed based on the natural product anisomycin using Merrifield resin equipped with a dihydropyran linker (DHP HM resin) (Scheme 30) [87]. Thus, anisomycin was treated with Fmoc-OSu and triethylamine in acetonitrile to give Fmoc-anisomycin (**269**) which was subsequently loaded onto DHP HM resin to provide **270**. The Fmoc group of resin **270** was removed by treatment with 20% piperidine in DMF and the first element of diversity (R^1) was introduced at the pyrrolidine nitrogen via reaction with acylating agents, sulfonyl chlorides, isocyanates, chloroformates, and isothiocyanates to provide the corresponding amide, sulfonamide, urea, carbamate, and thiourea derivatives **271**. The acetyl group of resin **271** was then treated with sodium

DHP HM Resin   **269**   **270**

271   **272**   **273a-k**

273a R = -N   N–CH₃

273b R HN—   N—

273c R = -N   —CH₃

273d R HN—   N—   CH₃

273e

273f R = Et
273g R = Pr
273h R = i-Bu

273i R = H
273j R = o-F
273k R = p-F

Scheme 30   **a** PPTS (2 equiv.), ClCH₂CH₂Cl, 80 °C, 48 h; **b** *1.* 20% piperidine, DMF, rt, 1 h; *2.* acyl chloride (10 equiv.), or sulfonyl chloride (10 equiv.), or carbamoyl chloride (10 equiv), or chloroformate (10 equiv.), or isocyanate (10 equiv.) or isothiocyanate (10 equiv.), Et₃N (20 equiv.), DCM, rt, 18 h; **c** *1.* NaOMe (20 equiv), MeOH, THF, rt, 6.5 h; *2.* 4-nitrophenyl chloroformate (10 equiv.), Et₃N (20 equiv.), DMC, rt, 20 h; *3.* 1° or 2° amine (10 equiv.), DMC, rt, 18 h; **d** TFA, DMC, MeOH, rt, 1 h

methoxide in THF. The liberated *C*-3 hydroxyl group was treated with 4-nitrophenyl chloroformate and triethylamine in DCM to form the corresponding 4-nitrophenyl carbonate which, upon treatment with primary or secondary amines, afforded carbamates **272** with a second element of diversity (NR²R³). The resulting anisomycin derivatives were cleaved from the resin by the treatment with TFA/DCM/MeOH (2 : 2 : 1) to produce **273a–k.**

References

1. El Ashry ESH, El Nemr A (2005) Synthesis of Naturally Occurring Nitrogen Heterocycles from Carbohydrates. Blackwell Publishers, Oxford, UK, [ISBN 1-4051-2934-4], p 463
2. Sobin BA, Tanner Jr FW (1954) J Am Chem Soc 76:4053
3. Buchanan JG, Maclean KA, Wightman RH, Paulsen H (1985) J Chem Soc, Perkin Trans 1, p 1463
4. Hosoya Y, Kameyama T, Naganawa H, Okami Y, Takeuchi T (1993) J Antibiotics, p 1300
5. Lynch JE, English AR, Bauck H, Deligianis H (1954) Antibiot Chemother 4:844
6. Frye WW, Mule JG, Swartzwelder C (1955) Antibiot Ann, p 820
7. Armstrong T, Santa Maria O (1955) Antibiot Ann, p 824
8. Grollman AP (1967) J Biol Chem 242:3226
9. Jiimenez A, Vazquez D (1979) In: Hahn FE (ed) Antibiotics, vol 5. Springer, Berlin Heidelberg New York, pp 1–19
10. Beereboom JJ, Butler K, Pennington FC, Solomons IA (1965) J Org Chem 30:2334
11. Butler K (1968) J Org Chem 33:2136
12. Wong CM (1968) Can J Chem 46:1101
13. Korzybski T, Kowszyk-Gindifer Z, Kurytowicz W (1978) Am Soc Microbiol 1:343
14. Santander VM, Cue AB, Diaz JGH, Balmis FJ, Miranda GG, Urbina E, Portilla J, Plata AA, Zapata HB, Munoz VA, Abreu LM (1961) Rev Invest Bol Univ Guadalajara 1:94
15. Windholz M (ed) (1976) The Merck Index, 9th ed. Merck, New Jersey, USA, p 91
16. Grollman AP (1966) AP Proc Natl Acad Sci USA 56:1867
17. Vasquez D (1974) FEBS Lett 40:S63–S84
18. Barbacid M, Vasquez D (1974) J Mol Biol 84:603
19. Van der Bosch J, Rueller S, Schlaak M (1999) DE Patent 19744676
20. Van der Bosch J, Rueller S, Schlaak M (1999) Chem Abstr 130:291581
21. Rueller S, Stahl C, Kohler G, Eickhoff B, Breder J, Schlaak M, Van der Bosch J (1999) J Clin Cancer Res 5:2714
22. Dudai Y (2000) Nature 406:686
23. Nader K, Schafe GE, Le Doux JE (2000) Nature 406:722
24. Törocsik B, Szeberényi J (2000) Eur J Neurosci 12:527
25. Schaefer JP, Wheatley PJ (1968) J Org Chem 33:166
26. Schumacher DP, Hall SS (1982) J Am Chem Soc 104:6076
27. Hall SS, Loebenberg D, Schumacher DP (1983) J Med Chem 26:469
28. Jegham S, Das BC (1988) Tetrahedron Lett 29:4419
29. Ballini R, Marcantoni E, Petrini M (1992) J Org Chem 57:1316
30. Tokuda M, Fujita H, Miyamoto T, Suginome H (1993) Tetrahedron 49:2413
31. Shi ZC, Lin GQ (1995) Tetrahedron: Asymmetr 6:2907
32. Kim G, Hong HW, Lee SH (1998) Bull Korean Chem Soc 19:37
33. Kim G, Hong HW, Lee SH (1999) Bull Korean Chem Soc 20:321
34. Delair P, Brot E, Kanazawa A, Greene AE (1999) J Org Chem 64:1383
35. Oida S, Ohki E (1969) Chem Pharm Bull 17:1405
36. Verheyden JPH, Richardson AC, Bhatt RS, Grant BD, Flitch WL, Moffat JG (1978) Pure Appl Chem 50:1363
37. Shono T, Kise N (1987) Chem Lett, p 697
38. Takahata H, Banba Y, Tajima M, Momose T (1991) J Org Chem 56:240
39. Ikota N (1995) Heterocycles 41:983

40. Han G, LaPorte MG, Mclntosh MC, Weinreb SM, Parvez M (1996) J Org Chem 61:9483
41. Huang PQ, Wang SL, Ruan YP, Gao JX (1998) Nat Prod Lett 11:101
42. Schwartdt O, Veith U, Gaspard C, Jäger V (1999) Synthesis, p 1473
43. Wang Y, Ma D (2001) Tetrahedron: Asymmetr 12:725
44. Hulme AN, Rosser EM (2002) Org Lett 4:265
45. Chang MY, Chen ST, Chang NC (2003) Heterocycles 60:1203
46. Baer HH, Zamkanei M (1988) J Org Chem 53:4786
47. Thiem J, Wessel HP (1983) Justus Liebigs Ann Chem, p 2173
48. Green JW, Pacsu E (1937) J Am Chem Soc 59:1205
49. Pascu E (1963) Meth Carbohydr Chem 2:354
50. Felner I, Schenker K (1970) Helv Chim Acta 53:754
51. Wong CM, Ruccini J, Chang I, Te Raa J, Schwenk R (1969) Can J Chem 47:2421
52. Iida H, Yamazaki N, Kibayashi C (1986) J Org Chem 51:1069
53. Yoda H, Nakajima T, Yamazaki H, Takabe K (1995) Heterocycles 41:2423
54. Nagai I, Gaudino JJ, Wilcox CS (1992) Synthesis, p 163
55. Buchanan JG, Maclean KA, Paulsen H, Wightman RH (1983) J Chem Soc, Chem Commun, p 486
56. Hughes NA, Speakman PRH (1965) Carbohydr Res 1:171
57. Veith U, Schwardt O, Jäger V (1996) Synlett, p 1181
58. Valverde S, Herradon B, Martin-Lomas M (1985) Tetrahedron Lett 26:3731
59. Appel R, Kleinslück R (1974) Chem Ber 107:5
60. Rassu G, Casiraghi G, Pinna L, Spanu P, Ulgheri F (1993) Tetrahedron 49:6627
61. Kang SH, Choi HW (1996) Chem Commun, p 1521
62. Savage I, Thomas EJ (1989) J Chem Soc, Chem Commun, p 717
63. Hutin P, Haddad M, Larchevêque M (2000) Tetrahedron: Asymmetr 11:2547
64. Takano S, Iwabuchi Y, Ogasawara K (1989) Heterocycles 29:1861
65. Baldwin JJ, Raab AW, Mensler K, Arison BH, McClure DE (1978) J Org Chem 43:4876
66. Meyers AI, Dupre B (1987) Heterocycles 25:113
67. Reddy JS, Kumar AR, Rao BV (2005) Tetrahedron: Asymmetr 16:3154
68. Wu W, Wu Y (1993) J Org Chem 58:3586
69. Shi ZC, Zeng CM, Lin GQ (1995) Heterocycles 41:277
70. Schreiber LS, Schreiber TS, Smith OB (1987) J Am Chem Soc 109:1526
71. Shi ZC, Lin GQ (1995) Tetrahedron 51:2427
72. Baker BR, Joseph JP, Williams JH (1955) J Am Chem Soc 77:1
73. Siedel W, Sturm K, Geiger R (1963) Ber 96:1436
74. Siedel W, Sturm K, Geiger R (1963) Chem Abstr 59:2951f
75. Moersch GW, Rebstock MC, Wittle EL, Tinney FJ, Nicolaides ED, Hutt MP, Mich TF, Vandenbelt JM (1979) J Med Chem 22:935
76. Ono M, Suzuki K, Akita H (1999) Tetrahedron Lett 40:8223
77. Ono M, Tanikawa S, Suzuki K, Akita H (2004) Tetrahedron 60:10187
78. Miyazawa M, Ishibashi N, Ohnuma S, Miyashita M (1997) Tetrahedron Lett 38:3419
79. Nagel U, Kinzel E, Andrade J, Prescher G (1986) Chem Ber 119:3326
80. Kim JH, Curtis-Long MJ, Seo WD, Ryu YB, Yang MS, Park KH (2005) J Org Chem 70:4082
81. Sardina FJ, Rapoport H (1996) Chem Rev 96:1825
82. Chang KT, Jang KC, Park HY, Kim YK, Park KH, Lee WS (2001) Heterocycles 55:1173
83. Schumacher DP, Hall SS (1981) J Org Chem 46:5060
84. Ryu Y, Kim G (1995) J Org Chem 60:103
85. Huang PQ, Zheng X (2003) ARKIVOC 2:7

86. Huang PQ, Chen QF, Chen CL, Zhang HK (1999) Tetrahedron: Asymmetr 10:3827
87. Shi S, Zhu S, Gerritz SW, Esposito K, Padmanabha R, Li W, Herbst JJ, Wong H, Shu YZ, Lam KS, Sofia M (2005) J Bioorg Med Chem Lett 15:4151
88. Chandrasekhar S, Ramachandar T, Venkat Reddy M (2002) Synthesis, p 1867
89. Wong CM, Buccini J, Te Raa J (1968) Can J Chem 46:3091
90. Rosser EM, Morton S, Ashton KS, Cohen P, Hulme AN (2004) Org Biomol Chem 2:142

Top Heterocycl Chem (2007) 7: 287–323
DOI 10.1007/7081_2007_076
© Springer-Verlag Berlin Heidelberg
Published online: 9 June 2007

1,3-Dipolar Cycloadditions of Sugar-Derived Nitrones and their Utilization in Synthesis

L. Fišera

Institute of Organic Chemistry, Catalysis and Petrochemistry,
Slovak University of Technology, 81237 Bratislava, Slovak Republic
lubor.fisera@stuba.sk

Abstract This review is devoted to the stereoselectivity of intermolecular (intramolecular cycloadditions are not included) 1,3-dipolar cycloadditions of sugar-derived nitrones. Stereoselective cycloaddition (transformation of isoxazolidine followed by reduction of the N–O bond to produce both an amino and a hydroxy function) allows the synthesis of tailor-made products of possible biological interest such as polyhydroxylated pyrrolidines, pyrrolizidines, indolizidines, β-aminocarbonyl compounds, and disaccharides. Attention is focused on the preparation of isoxazolidinyl nucleosides and to the catalysis of the cycloaddition by Lewis acids. This review has concentrated on the new developments achieved from 1999 to February 2007.

Keywords Sugars · Cycloadditions · Nitrones · Nucleosides

1
Introduction

1,3-Dipolar cycloaddition of nitrones is a fascinating field with a multitude
of biological implications. Not less important is the role of isoxazolidines
in preparative organic chemistry, especially for the synthesis of γ-amino-
alcohols. 1,3-Dipolar cycloaddition between alkenes and nitrones is probably
the best method for the preparation of isoxazolidines, useful precursors in
the total synthesis of complex natural products. Stereoselective cycloaddition,
transformation of isoxazolidine followed by reduction of the $N - O$ bond to
produce both an amino and a hydroxy function, allows the synthesis of tailor-
made products of possible biological interest [1–5].

2
Value of Sugar-Derived Nitrones

Cyclic glycosides are important as enzyme inhibitors and as chiral synthons,
suitable for the synthesis of many natural products. Since the 1,3-dipolar cycloaddition has a nearly singular capability of establishing large numbers
of stereogenic centers in one synthetic step, in recent years attention has
been focused on the preparation of chiral sugar-derived nitrones. The configuration of the newly generated stereogenic centers is determined by the
nitrone. Asymmetric induction in 1,3-dipolar cycloaddition has been efficiently achieved by using nitrones with chiral groups at either the nitrogen
atom [6] or the carbon atom [7].

Among nitrones, the sugar-derived nitrones represent versatile substrates
as they provide a polyhydroxylated carbon framework with multiple avenues
of chirality, as well as an access for the amino group transformation required
for the synthesis of polyhydroxylated piperidine, pyrrolidine, pyrrolizidine,
indolizidine, and quinolizidine alkaloids. This class of compounds, commonly known as iminosugars (namely nojirimycin **1** and castanospermine
2, see Fig. 1), has attracted considerable attention because of promising glycosidase inhibitory activity and, therefore, possible therapeutic applications
as immunosuppressive, antimalarial, antiviral, anticancer, and antidiabetic
agents [8–10].

This review is devoted to the stereoselectivity of intermolecular (intramolecular cycloadditions are not included) 1,3-dipolar cycloadditions of
chiral sugar-derived nitrones. Note that the labile $N - O$ bond in so-prepared
cycloadducts can be readily cleaved under mild reduction conditions, followed by an intramolecular cyclization of the liberated amino group and the

Fig. 1 Nojirimycin and castanospermine

preexisting functionalities from either the nitrone or the dipolarophile. This review concentrates on the new developments achieved from 1999 to February 2007. There are some reviews on asymmetric 1,3-dipolar cycloadditions, including sugar-derived nitrones [11–17].

3
1,3-Dipolar Cycloadditions of Sugar-Derived Nitrones

It should be mentioned that the stereoselectivity of cycloaddition of chiral sugar-derived nitrone to an alkene is difficult to predict, and would appear to be dependent on minor structural changes in either component. Three structural features can influence the stereochemical outcome of nitrone/alkene cycloadditions: E/Z nitrone isomerization about the $C = N$ bond, alkene or/and nitrone facial selectivity, and *endo/exo* preferences [6].

In many cases the stereoselectivity of cycloadditions was dependent on the steric hindrance of the nitrone. The selectivity increases as the size of C-"chiral" group and N-alkyl group attached to the nitrone increases.

3.1
C-Glycosylnitrones

3.1.1
Pyrrolizidines

With the goal of developing a simple route to the synthesis of polyhydroxylated derivatives of pyrrolizidines displaying antiviral activities, Fišera et al. have developed 1,3-dipolar cycloadditions of D-erythrose and D-threose-derived nitrones with alkenes [18–26]. The cycloaddition of nitrones **3** derived from erythrose, which had been prepared from D-glucose, with styrene proceeded with high yields and was selective (Fig. 2). The ratio of the diastereoisomers was dependent on the substituent located on the nitrogen atom of nitrones. The best diastereoselectivity was achieved by using *N*-benzyl-nitrone **3a** [18, 19]. The X-ray analysis of product configuration of the major cycloadduct **4** revealed C-3/C-4′ *erythro* and C-3/C-5 *cis* configuration. This indicates that cycloaddition arises from the more sterically accessible

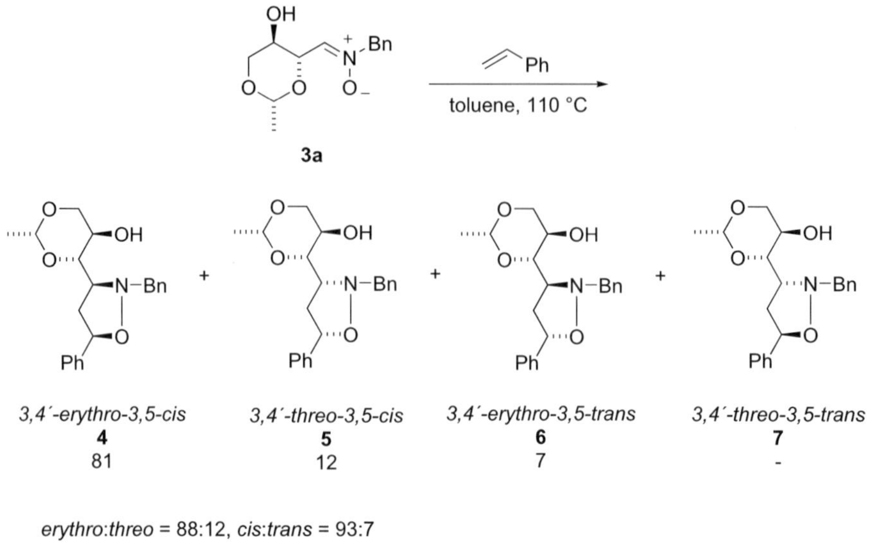

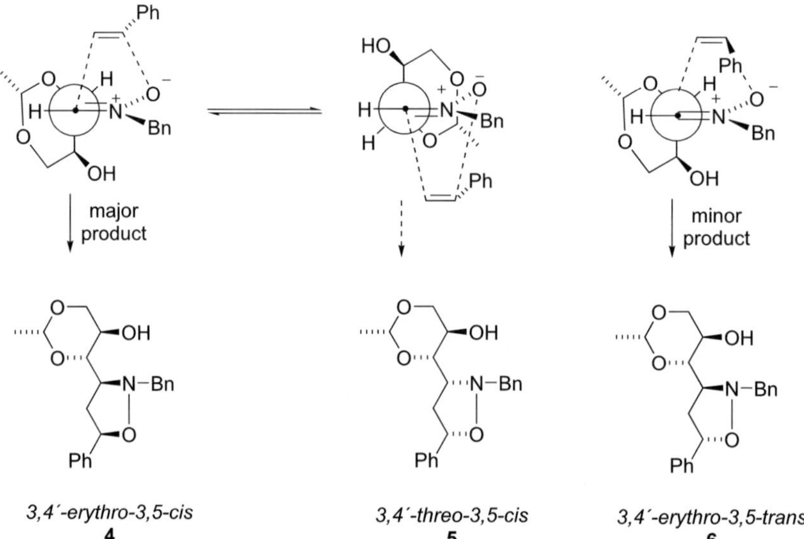

Fig. 2 1,3-Dipolar cycloaddition of D-*erythro* nitrones

face of the *Z*-nitrone **3a**, via the less-hindered *endo* transition state, with antiperiplanar relationship of the phenyl and *N*-alkyl group, and in an antiperiplanar manner with respect to the largest group of the heterocyclic acetal (Fig. 3).

Fig. 3 1,3-Dipolar cycloaddition of D-*erythro* nitrone with styrene

On the other hand, the more pronounced steric hindrance present in the approach leading to the *threo–cis* **5** and *erythro–trans* **6** diastereomers might explain the observed preference of *anti* (ratio 88 : 12) and *endo* cycloaddition (ratio 93 : 7). The cycloaddition of the D-*threo* nitrone **8a** prepared from D-galactose proceeds with the best *anti*-facial (93 : 7) and *endo*-facial (95 : 5) preference in this series (Fig. 4).

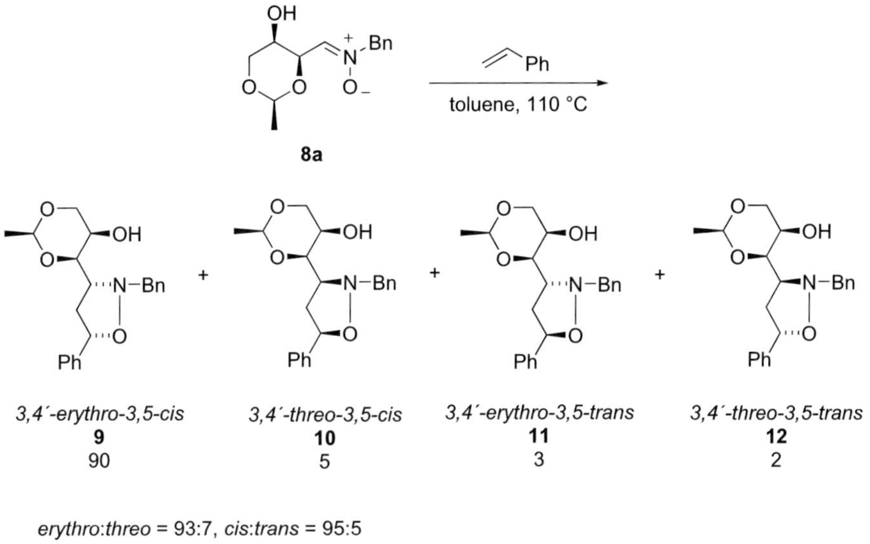

erythro:threo = 93:7, cis:trans = 95:5

Fig. 4 1,3-Dipolar cycloaddition of D-*threo* nitrone

Fišera et al. [20] also found that the cycloaddition of chiral sugar-derived nitrones **3a** and **8a** with methyl acrylate, which was very important for further synthesis, proceeded analogously, but the selectivity was decreased (Figs. 5 and 6). The MM2 modeling and AM1 calculations suggest that an increase in the size of the hydroxyl-protecting group should result in an increase of the induced stereoselectivity. The more bulky protection of the free hydroxyl group of the nitrone **8b** can efficiently hinder the approach of the alkene from behind, the front side attack is then favored and leads to the major *erythro–cis* product **18b** (Fig. 7). Indeed, the prepared β-hydroxy-protected nitrones **3b** and **8b**, which possessing the more sterically demanding *O-tert*-butyldimethylsilyl group, show a strong preference for the *erythro–cis* products **14b** and **18b**. The selectivities were increased and the cycloaddition of D-*threo* nitrone **8b** proceeded with the best *anti*-facial (99 : 1) and *endo*-facial (97 : 3) preference in this series.

In Fig. 8 is shown an efficient synthetic pathway to pyrrolizidine **24** starting from D-glucose via enantiomerically pure isoxazolidine intermediate in 14 steps, with an overall yield of 12% [21]. 1,3-Dipolar cycloaddition of sugar-

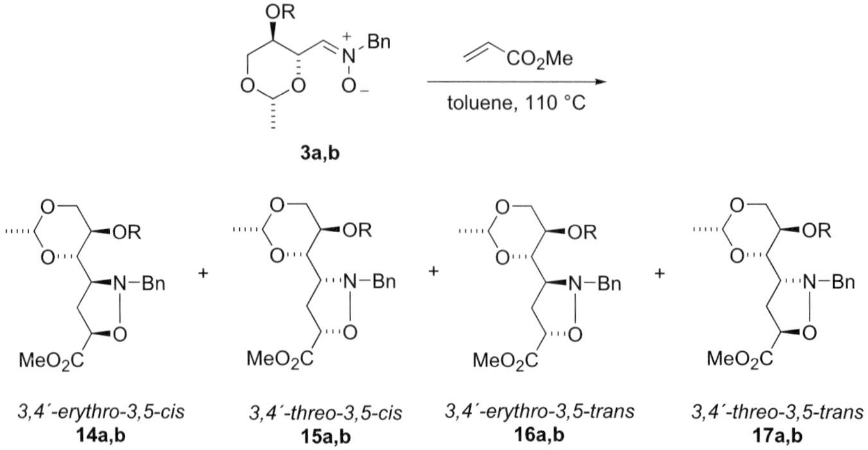

3a,b → CO$_2$Me, toluene, 110 °C

3,4′-erythro-3,5-cis **14a,b** + 3,4′-threo-3,5-cis **15a,b** + 3,4′-erythro-3,5-trans **16a,b** + 3,4′-threo-3,5-trans **17a,b**

a R = H, **14a:15a:16a:17a** = 48:18:29:5
b R = TBDMS, **14b:15b:16b:17b** = 87:5:6:2

Fig. 5 1,3-Dipolar cycloaddition of D-*erythro* nitrone with methyl acrylate

8a,b → CO$_2$Me, toluene, 110 °C

3,4′-erythro-3,5-cis **18a,b** + 3,4′-threo-3,5-cis **19a,b** + 3,4′-erythro-3,5-trans **20a,b** + 3,4′-threo-3,5-trans **21a,b**

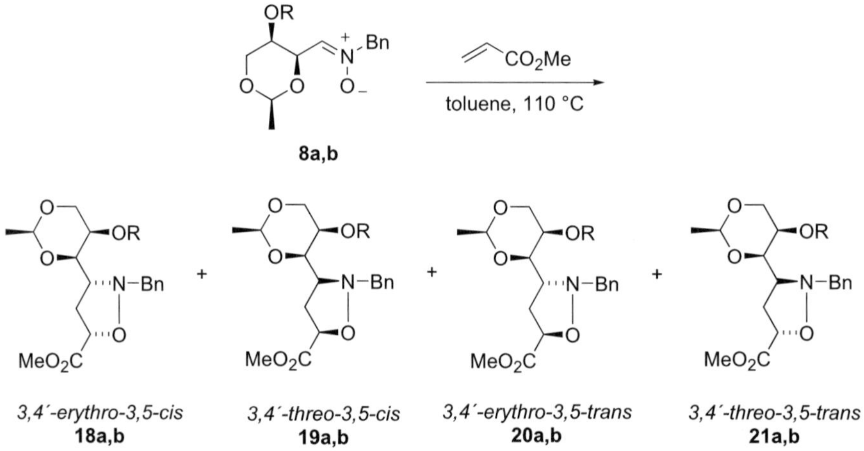

a R = H, **18a:19a:20a:21a** = 62:11:19:8
b R = TBDMS, **18b:19b:20b:21b** = 96:1:3:-

Fig. 6 1,3-Dipolar cycloaddition of D-*threo* nitrone with methyl acrylate

derived O-silylated nitrone **3b** prepared from D-glucose with methyl acrylate gave the diastereoisomerically pure isoxazolidine **14b** in 88% yield. Direct hydrogenolysis of mesylate **22** in the presence of catalyst palladium hydroxide on carbon in methanol resulted in a cascade reaction sequence involving isox-

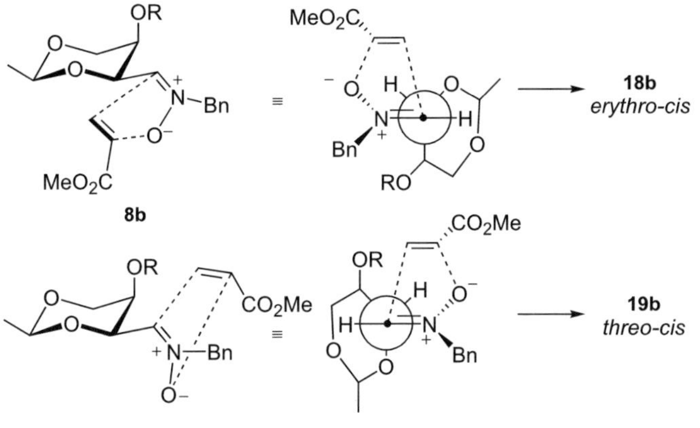

Fig. 7 1,3-Dipolar cycloaddition of D-*threo* nitrone with methyl acrylate

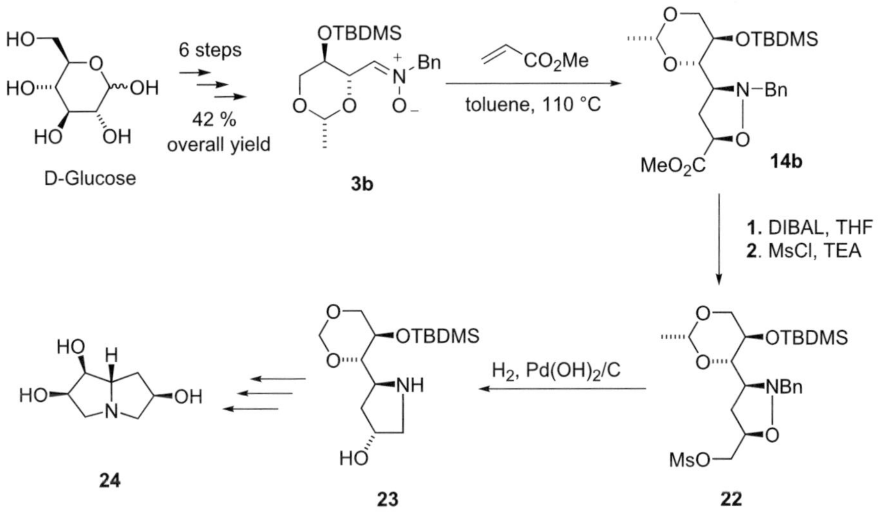

Fig. 8 Pyrrolizidine

azolidine N – O bond cleavage, N-debenzylation and spontaneous cyclization affording pyrrolidine **23**, which was subsequently converted to pyrrolizidine **24**.

3.1.2
Lewis Acid Catalyzed Cycloadditions

Fišera et al. have [22, 23] found a very unusual and rare reversal of regioselectivity of the cycloaddition of TBDPS-protected D-erythrose-derived nitrone

3b with acryloyloxazolidinone (**25**) in the presence of Lewis acid that favors the preferred formation of electronically controlled 4-substituted isoxazolidines **30–33** (Fig. 9). It should be mentioned that this is the first example of the reversal of regioselectivity caused by Lewis acids in the case of chiral nitrones. The cycloaddition with the chiral nitrone **3b** in the absence of any catalyst in CH_2Cl_2 at room temperature for 14 days is completely regioselective, with only the sterically favored 5-substituted isoxazolidines **26–29** being detected. A mixture of four diastereoisomers **26–29** was provided in a ratio of 44 : 28 : 17 : 11, with the *erythro–trans* isomer **26** being predominant. On the other hand, when the reaction was performed in the presence of

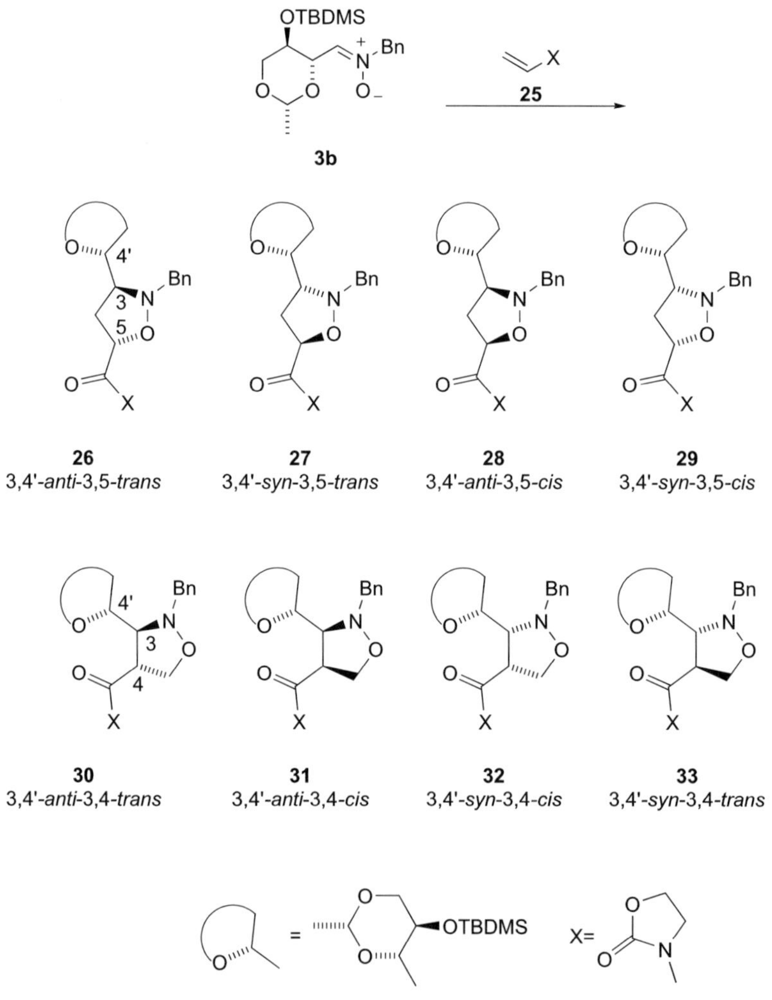

Fig. 9 Lewis acid catalyzed cycloaddition

Ti(Oi-Pr)$_2$Cl$_2$ as catalyst, the regioselectivity observed in the thermal reaction was reversed (18 : 82). Additionally, other Lewis acids (such as ZnI$_2$, ZnCl$_2$, ZnBr$_2$, and MgI$_2$-I$_2$) were found to be efficient and to reverse the regioselectivity of this reaction as compared to the uncatalyzed cycloaddition [23]. The enhanced regioselectivity in the case of Ti(OiPr)$_2$Cl$_2$ as catalyst would be due to the increase of electron deficiency at the β-carbon of the enone dipolarophile complex.

The effect of the addition of Lewis acid upon the stereoselectivity of cycloaddition of chiral nitrones **3a** and **34** to electron-rich alkene, with ethyl vinyl ether and the further transformation of so-prepared isoxazolidine **36b** to new β-amino acid ester **40**, has been also investigated by the same team (Fig. 10). The 1,3-dipolar cycloaddition of nitrone **3a** with ethyl vinyl ether under AlMe$_3$ and Et$_2$AlCl catalysis proceeded diastereoselectively and finished at – 8 °C over 20 h, providing only two diastereoisomers **35a** and **36a** in a ratio of 90 : 10 with *erythro–cis* **35a** predominant, although four diastereoisomers are possible. Indeed, cycloaddition in the absence of any Lewis acids proceeded very slowly with excess of ethyl vinyl ether at room temperature over 14 days to give a mixture of all four diastereoisomers **35–38** with a considerable decrease of the stereoselectivity in a ratio of 59 : 12 : 14 : 15, although *erythro–cis* **35a** was still the major adduct. The 1,3-

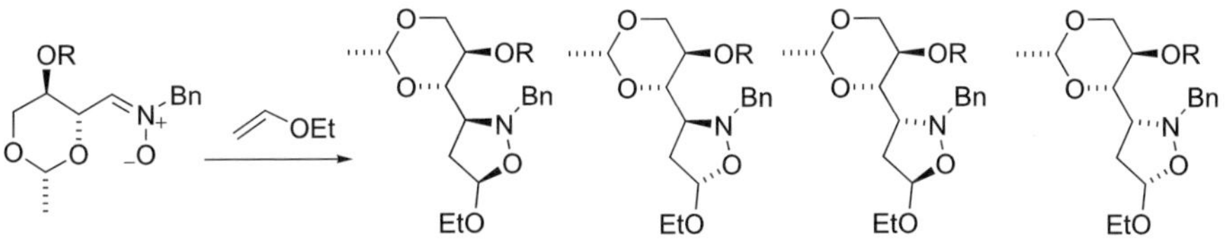

3a R = H, rt, 30 d, 59:12:14:15
3a R = H, -8 °C, Et$_2$AlCl, 24 h, 90:-:10:-
34 R = TBDPS, -15 °C, Et$_2$AlCl, 24 h, 9:91:-:-

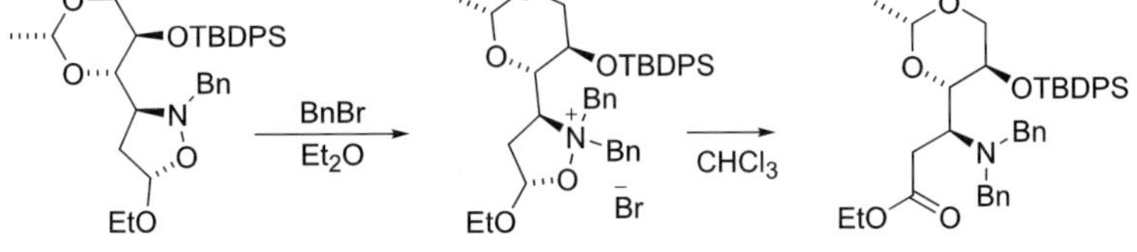

Fig. 10 Lewis acid catalyzed cycloaddition

dipolar cycloaddition of TBDPS-protected D-*erythro*-derived nitrone **34** gave similar results, except for reversal of the diastereoselectivity, probably due to the steric effects, favoring isomer *erythro–trans* **36b**. Moreover, in this case, cycloaddition in the absence of Lewis acid proceeded very slowly and gave only traces of the cycloadducts. The enhanced reactivity in the case of AlMe$_3$ and Et$_2$AlCl as catalysts could be due to the increase of electron deficiency of the dipole–Lewis acid complex.

Finally, the Murahashi oxidative ring-opening reaction of isoxazolidines for the preparation of β-amino acid esters has been used. The chiral isoxazolidine **36b** gave, after treatment with benzyl bromide, the chiral β-amino acid ester **40** in 53% yield. The extension of this methodology to other dipolarophiles such as Baylis–Hillman adducts was reported in 2004 [25, 26].

3.1.3
Pyrrolidines

The chemistry of imino disaccharides has become of great interest in recent years because they can be powerful glycosidase inhibitors, thus having an enormous potential as antibacterial or antidiabetic agents [27]. The glycosidic linkage can also be replaced by different sequences of atoms and the imino sugar can be either a piperidine or a pyrrolidine ring. An elegant strategy for the synthesis of imino-*C*-disaccharides, in which a pyranose ring is directly linked to a hydroxylated pyrrolidine based on cycloaddition chemistry, was described by Brandi and Goti et al. [28]. The same methodology was successfully applied by Merino et al. for the preparation of glycosyl pyrrolidines via 1,3-dipolar cycloaddition between *N*-benzyl-*C*-glycosyl nitrones and methyl acrylate [29]. Of the three chiral nitrones used, the furanosyl nitrones **41** and **42** showed little *endo/exo* and diastereofacial selectivities, whereas the pyranosyl nitrone **43** was much more selective and only one isomer was obtained (Fig. 11). In all cases, the chemical yield of the reaction was quantitative. The obtained isoxazolidines **44–47** were transformed into the corresponding *N*-benzyl-3-hydroxy-2-pyrrolidinones **48–51** by treatment with Zn in acetic acid. The pyrrolidinone **48** obtained from D-galactose-derived nitrone **43** was further converted into protected galactosyl pyrrolidine **52**.

The another approach for the synthesis of a protected aza-*C*-disaccharide derivative with a pyrrolidine ring as the aza component, based on a cycloaddition of an open chain nitrone and an alkene (both obtained from sugar precursors), has been developed by Argyropoulos et al. [30]. The key step is a stereoselective cycloaddition of a chiral nitrone **53** prepared from L-erythrose and a sugar alkene **54** derived from D-galactose (Fig. 12). Interestingly the cycloaddition was enhanced by the addition of a catalytic amount of triethylamine and was complete after heating for 2 days in refluxing benzene. A remarkable stereoselectivity was observed and only stereoisomer **55** was formed in very good yield. An intramolecular N-alkylation, followed by

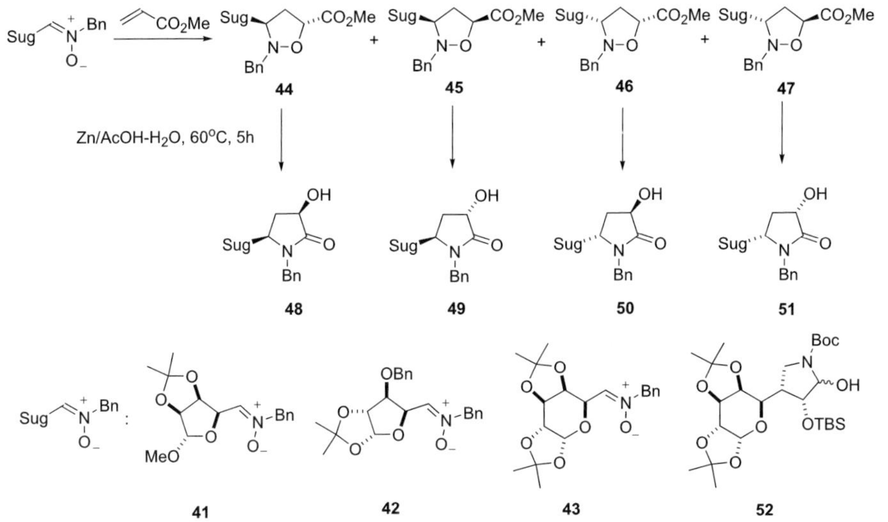

Fig. 11 Pyrrolidinone

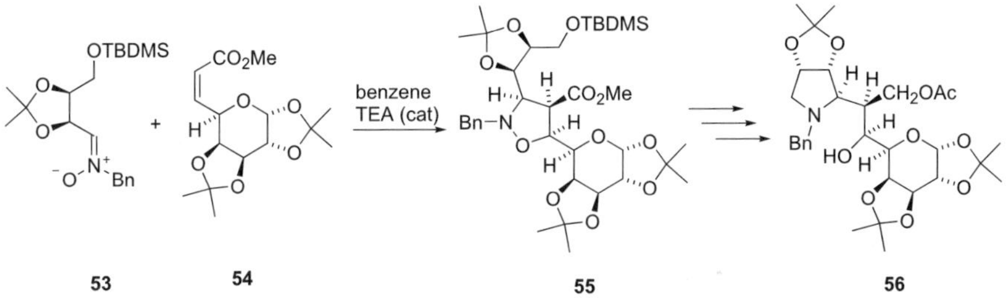

Fig. 12 Aza-*C*-disaccharide

a reductive cleavage of the isoxazolidine N – O bond, in one pot, gave the final product **56**, in a 65% overall yield.

In 2005 Goméz-Guillén et al. found a short and efficient route to enantiomerically pure pentahydroxyperhydroazulene **60**, ring homologs of castanospermine **2**, starting from the sole isoxazolidine derivative **44** obtained from the D-galactose-derived nitrone **43** and methyl acrylate [31, 32]. Cycloadduct **44** was subjected to isoxazolidine-ring cleavage by treatment with hexacarbonylmolybdenum to afford nonopyranos-uronolactam **57** in 62% yield (Fig. 13). The **57** was reduced with lithium aluminum hydride in ether, followed by protection with benzyloxycarbonyl chloride to give pyrrolidine **58** in 96% yield, which was then O-deprotected by the action of 80% TFA at room temperature, giving 95% of compound **59**. Finally, catalytic hydrogenation (Pd/C) in the presence of acetic acid furnished pure pentahydroxyperhydroazulene **60** in high yield (99%).

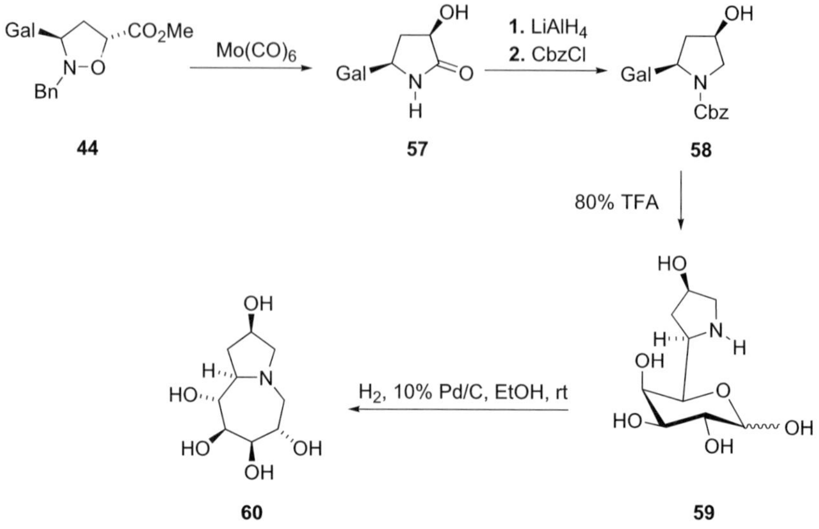

Fig. 13 Pentahydroxyperhydroazulene

3.1.4
Aminoaldehydes

Higher-carbon (C_7–C_9) amino deoxy sugars, including glycosoamino acids, are substances of high biological interest [33]. Goméz-Guillén et al. [34] have described the stereoselective preparation of protected C_7 and C_8 aminodialdoses, which could be considered intermediate precursors of β-amino glycuronic acids, from readily available monosaccharide *N*-benzyl nitrones **42**, **43** and **61**, by regio- and diastereoselective 1,3-dipolar cycloaddtion with vinyl trimethylsilane, followed by acetyl chloride-mediated cleavage of the 5-(trimethylsilyl)isoxazolidine formed (Fig. 14). The cycloaddition took place in moderate to good yields (67–74%) and showed total *endo* preference for the reaction of the D-*galacto*-configured nitrone **43** and high *endo* preference for the D-*ribo* analog **61**, but *exo* preference for the D-*xylo* configured substrate **42**. Attack on the *re* face of the nitrone was predominant in all cases. Opening of the isoxazolidine ring of compounds **62a–64a** was achieved by treatment with acetyl chloride at 0 °C to afford the respective 3-aminoaldehydes **65–67** in moderate yields.

3.1.5
Indolizidines

Dhavale and et al. [35] have synthesized polyhydroxylated indolizidine alkaloids, namely 2-hydroxy-1-deoxycastanospermine **71a,b** and 2-hydroxy-1-deoxy-8a-*epi*-castanospermine **71c,d** (Fig. 15). The key step involves the

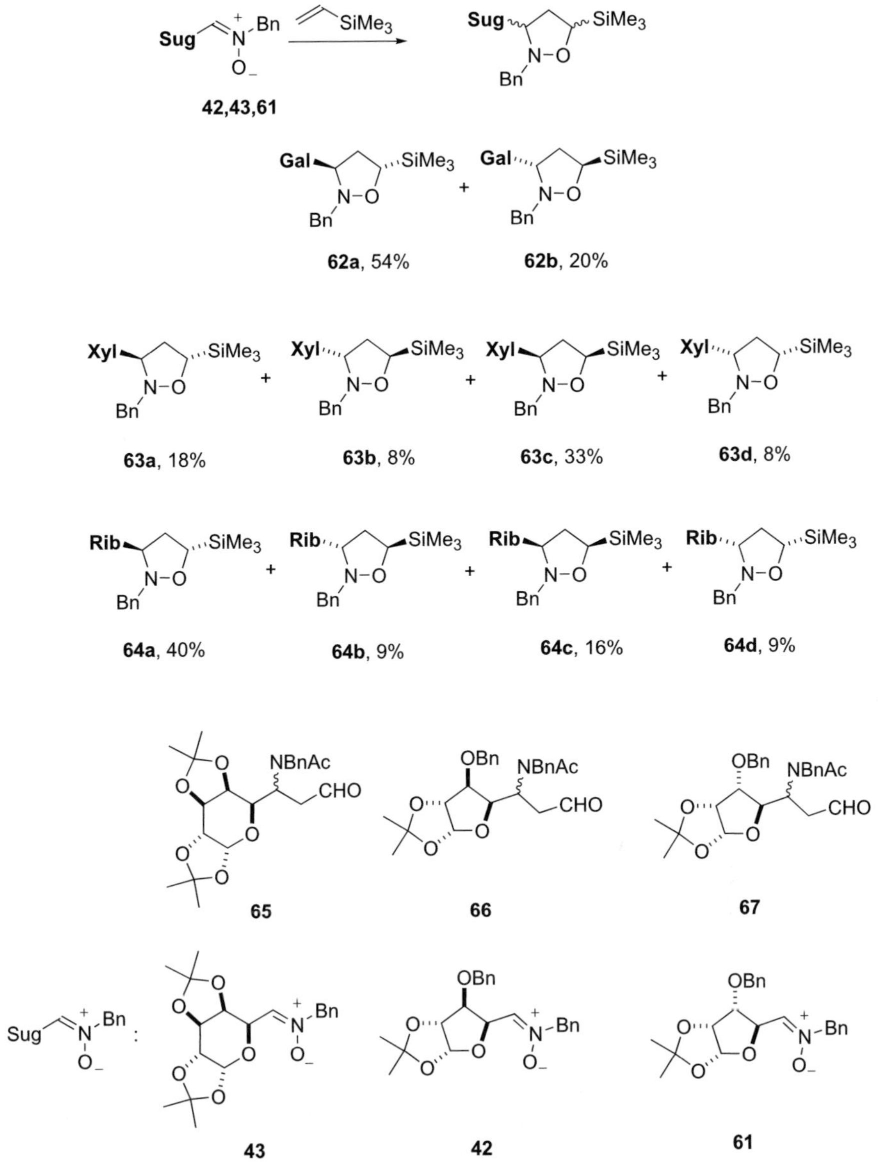

Fig. 14 3-Aminoaldehydes

1,3-dipolar cycloaddition of allyl alcohol to D-glucose-derived nitrone **68**, followed by tosylation, that afforded diastereomeric sugar-substituted isoxazolidines **69a,b** with the desired regioselectivity. The cycloaddition of nitrone **68** with allyl alcohol furnished an inseparable mixture of isoxazolidines in 95% yield, which on further treatment with *p*-toluenesulfonyl chloride in

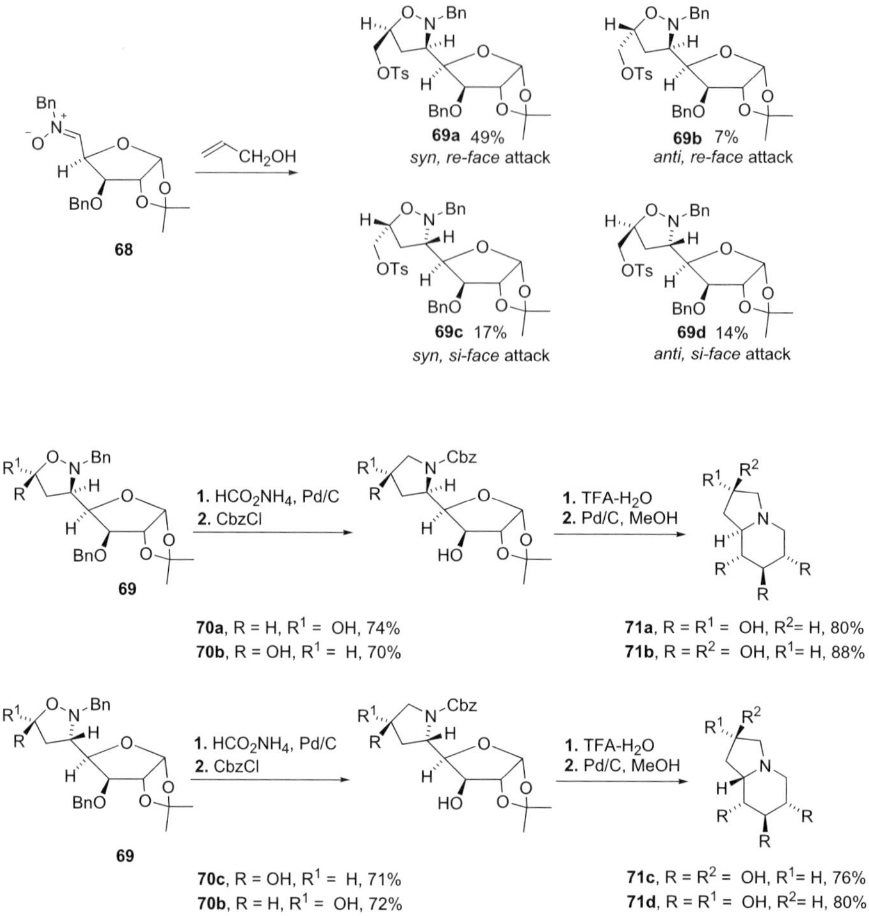

68 + CH₂OH →

69a 49%
syn, re-face attack

69b 7%
anti, re-face attack

69c 17%
syn, si-face attack

69d 14%
anti, si-face attack

69 →
1. HCO₂NH₄, Pd/C
2. CbzCl
→
1. TFA-H₂O
2. Pd/C, MeOH
→

70a, R = H, R¹ = OH, 74%
70b, R = OH, R¹ = H, 70%

71a, R = R¹ = OH, R² = H, 80%
71b, R = R² = OH, R¹ = H, 88%

69 →
1. HCO₂NH₄, Pd/C
2. CbzCl
→
1. TFA-H₂O
2. Pd/C, MeOH
→

70c, R = OH, R¹ = H, 71%
70b, R = H, R¹ = OH, 72%

71c, R = R² = OH, R¹ = H, 76%
71d, R = R¹ = OH, R² = H, 80%

Fig. 15 Polyhydroxylated indolizidines

pyridine afforded the separable tosyloxylated isoxazolidines in 87% yield (*syn : anti* 76 : 24). In the one-pot conversion of **69a,b** to pyrrolidines **70a,b** by hydrogenolysis, removal of 1,2-acetonide functionality afforded corresponding target molecules **71a,b**. Treatment of **69a** with ammonium formate and 10% Pd/C, followed by selective amine protection with benzyl chloroformate, afforded *N*-Cbz-protected diol **70a** in 70% yield. This one-pot three-step hydrogenation reaction resulted in N – O bond cleavage, intramolecular aminocyclization to form the pyrrolidine ring skeleton, and the removal of *N*- and *O*-benzyl groups. Subsequently, deprotection of 1,2-acetonide functionality in **70a** with TFA/water, followed by hydrogenation, afforded 2-(*S*)-hydroxy-1-deoxycastanospermine **71a**. The same sequence of reactions with **69b–d**, gave the corresponding 2-hydroxy-1-deoxycastanospermine analogs **71b–d**.

3.1.6
Amino-*C*-disaccharides

The stereoselective 1,3-dipolar cycloaddition of the glucose *exo*-methylene derivative **72** with a sugar nitrone **73** in refluxing toluene provided the corresponding cycloadducts, ketosyl spiroisoxazolidine disaccharides **74** and **75**, in the yields of 17% and 52%, respectively (Fig. 16). It should be noted that the cycloaddition underwent diastereoselectively and afforded only two anomeric isomers **74** and **75** possessing *R*-configuration on C-6 [36]. Followed by reductive cleavage of the N – O bond of the isoxazolidine ring with catalytic hydrogenation using $Pd(OH)_2/C$ as catalyst, the cycloadduct **74** was converted into a novel amino-*C*-disaccharide **76** possessing a ketose form.

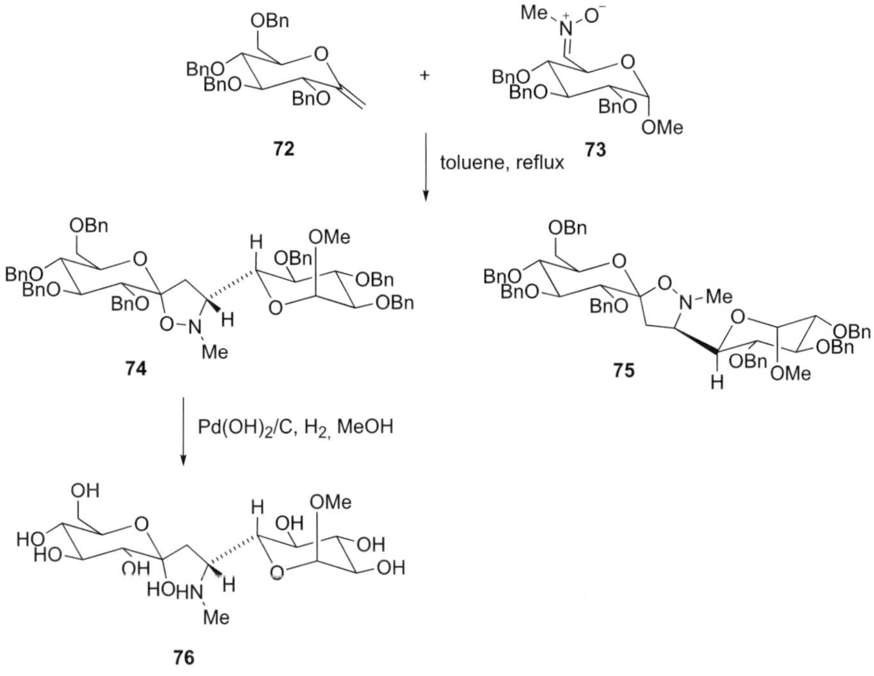

Fig. 16 Aza-*C*-disaccharide

3.1.7
Nitroisoxazolidines

Cycloaddition of above-mentioned sugar nitrones **42**, **43**, and **61** with nitroalkenes **77**, including sugar nitroolefins, led with complete regioselectivity to 4,5-*trans*-4-nitroisoxazolidines **78–80** in 51–78% global yields (Fig. 17). The *endo/exo* stereoselectivity depends on the type of sugar derivative

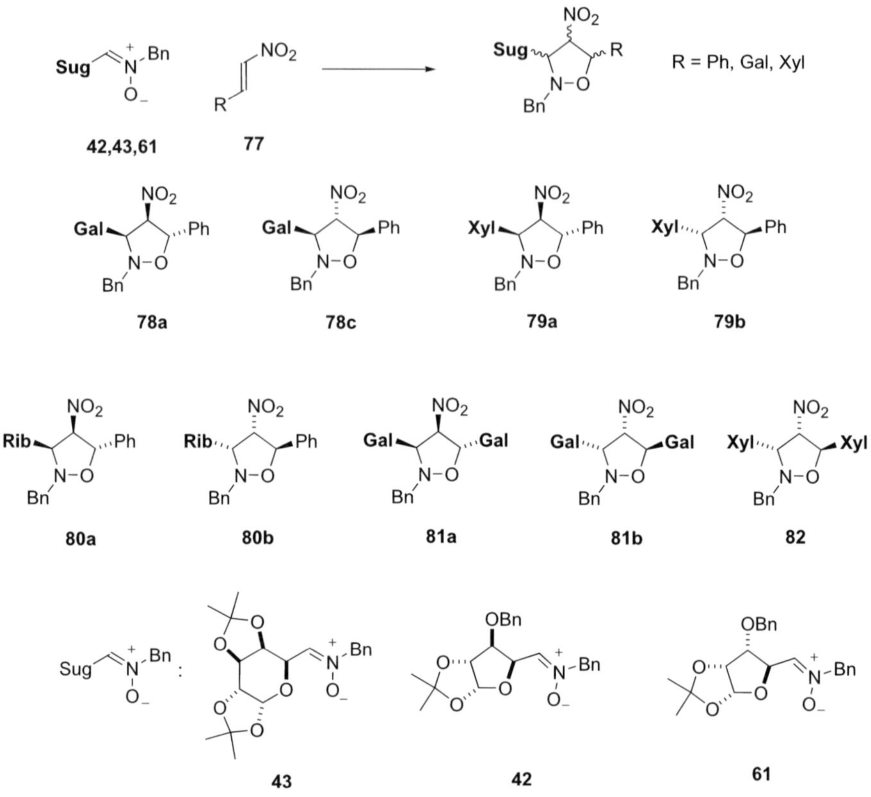

Fig. 17 Nitroisoxazolidines

used [37]. Only two diastereoisomers were formed when the glycosyl substituent is at position 3 (**78a**:**78c** = 65 : 35, **79a**:**79b** = 35 : 65, **80a**:**80b** = 39 : 61). As expected, the best diastereofacial selectivity was observed when both partners were sugar derivatives. Nitrone **43** and nitroolefin **77**, both derived from D-galactose, afforded a mixture of cycloadducts **81a** and **81b** in the ratio of 30 : 70, whereas the D-xylose derivatives **42** and **77** gave rise to one predominant diastereoisomer **82b** (> 90 : 10).

3.1.8
Chlorines

In 2002, Cavaleiro et al. [38] developed the stereoselective synthesis of glycoconjugated isoxazolidine-fused chlorines (potential photosensitizers in photodynamic cancer therapy) by 1,3-dipolar cycloadditions of *meso*-tetrakis(pentafluorophenyl)porphyrin (**83**) with glycosyl nitrones **41–43** and **84**. In all cases of sugar moieties, the configuration of the major products **85** indicated an *endo* addition, as shown in Fig. 18.

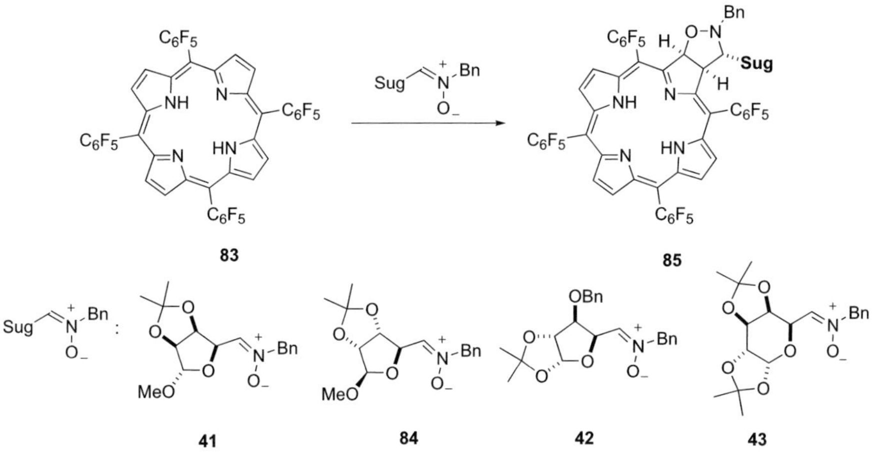

Fig. 18 Isoxazolidine-fused chlorines

3.1.9
Isoxazolidinyl Nucleosides

Many nucleoside analogs have been synthesized with the modification of base, sugar, and phosphane region. In particular, nucleoside analogs in which the furanose ring has been replaced by different carbon or heterocyclic systems, have attracted special interest by virtue of their biological action as antiviral and/or anticancer agents [39]. Among them, uracil, thymine, cytosine, and adenine nucleosides possessing an isoxazolidinyl moiety (carbocyclic-2′-oxo-3′-azanucleosides) are emerging as an interesting class of dideoxynucleoside analogs with potential pharmacological activity [39]. For the synthesis of modified isoxazolidinyl nucleosides, two strategies can be used: a one-step approach based on 1,3-dipolar cycloaddition of nitrones with vinyl nucleobases, and a two-step methodology based on the Vorbrüggen nucleosidation [40] of the 5-acetoxyisoxazolidines. With the aim of preparing some novel azanucleosides by the transformation of modified isoxazolidinyl nucleosides, Fišera et al. have prepared the appropriate sugar-derived nitrones possessing structures suitable for the building of pyrrolidine [41].

Cycloadditions of chiral nitrones **86** and **87**, easily prepared from D-xylose, to vinylated nucleobases **88** derived from uracil and adenine proceeded regioselectively and led to the isoxazolidines **89–91** as a mixture of four diastereoisomers in all cases (Fig. 19). In the case of cycloadditions of nitrones **86** and **87** with 9-vinyladenine **88b** an inseparable mixture of the adenosine diastereoisomers was obtained. To improve the separation, the free amino group of the adenine moiety was protected. The cycloaddition of nitrone **86** with N,N-dimesylated 9-vinyladenine **88b** proceeded with better selectivity in favor of the *anti–cis* isomer **91a**.

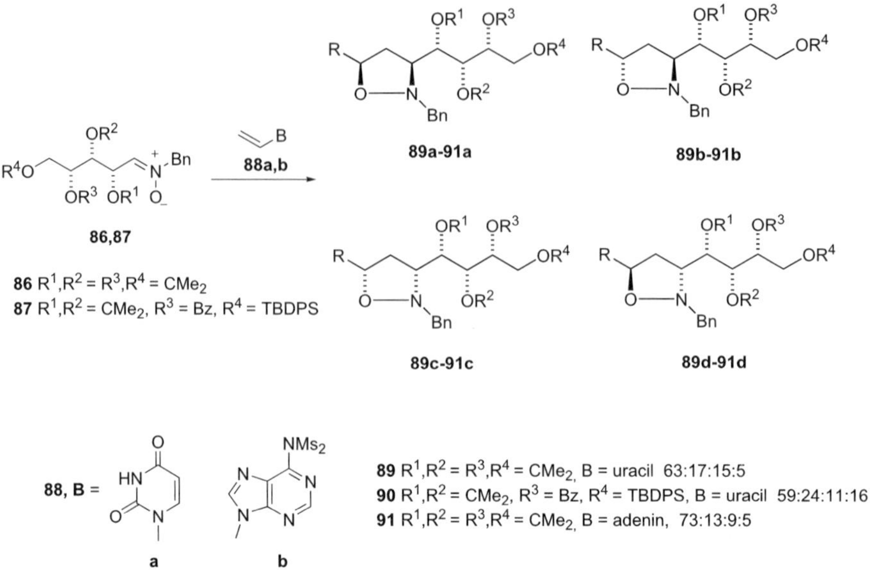

Fig. 19 1,3-Dipolar cycloaddition of D-*xylo* nitrone with vinyl nucleobases

X-ray analysis of the major diastereoisomer **89a** prepared from uracil reveals a C-3/C-1' *anti* and C-3/C-5 *cis* configuration and therefore indicates that cycloaddition arises from the more sterically accessible *si* face of the Z-nitrone **86**, via an *exo* transition state for the *anti–cis* diastereoisomer **89a** and via an *endo* transition state for the *anti–trans* **89b** isomer. As an extension of the synthesis of novel 4'-aza-2',3'-dideoxyfuranosyl nucleosides having potential antiviral activity, the 1,3-dipolar cycloaddition of readily available chiral sugar-derived D-lyxosyl nitrone **41** with N,N-dibenzyl-9-vinyl adenine has been also investigated by the same group [42]. Nitrone **41** reacted with N,N-dibenzyl-9-vinyl adenine to give mixture of all possible 5-substituted isoxazolidines **92a–d** in the ratio of 39 : 27 : 19 : 15 in 98% yield, the *anti–cis* isomer **92a** being the major product (Fig. 20). Two of these with a C-3/C-5 *trans* configuration are α-anomers. Two others possessing a C-3/C-5 *cis* configuration are β-anomers. Whereas the regioselectivity of the reaction was very high, the corresponding 4-substituted regioisomers were not detected, both the *cis/trans* selectivity (*cis* : *trans* ratio 58 : 42) and the diastereofacial selectivity (*Re* : *Si* ratio 66 : 34) were rather low.

With the goal of comparing the alternative synthesis of isoxazolidinyl nucleosides, a two-step methodology based on the Vorbrüggen nucleosidation of the 5-substituted acetoxyisoxazolidines has also been studied [42] (also, Hýrošová et al. unpublished results). The 1,3-dipolar cycloaddition of nitrones **86** and **87** with vinyl acetate leads to the 5-acetoxysubstituted isoxazolidines as a mixture of diastereoisomers **93a–d** and **94a–c**, respectively (Fig. 21). In contrast to the cycloadditions with vinylated nucleobases,

NBn2
toluene, 110 °C
OCH3
41
39 : 27 : 19 : 15
anti/syn 66 : 34
cis/trans 58 : 42
Bn2N
Bn
92a
92b
92c
92d

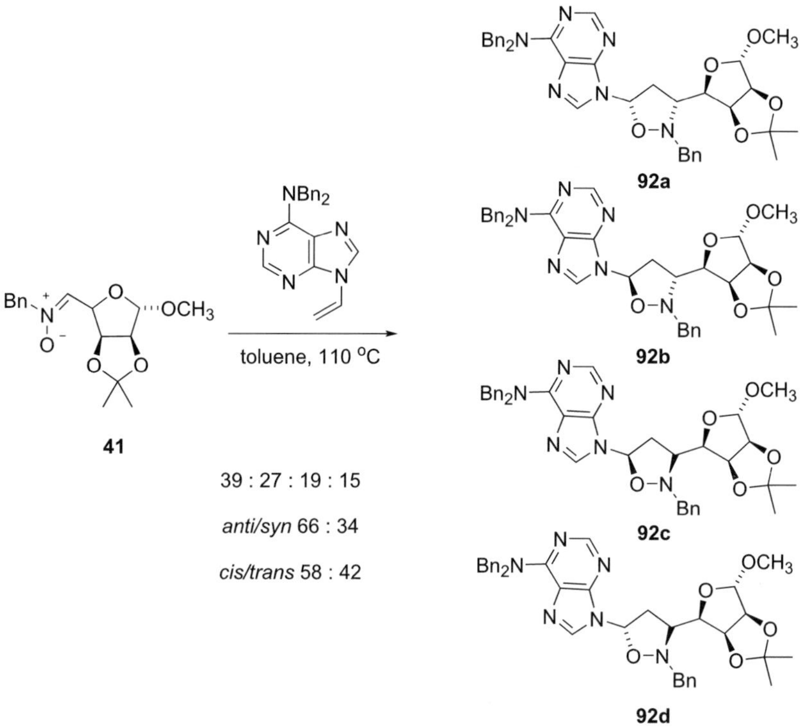

Fig. 20 Isoxazolidinyl nucleosides

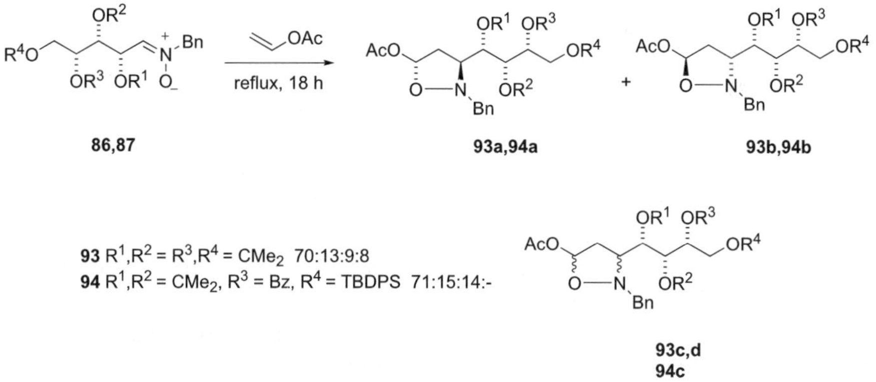

OAc
reflux, 18 h
86,87
93a,94a
93b,94b
93 R1,R2 = R3,R4 = CMe2 70:13:9:8
94 R1,R2 = CMe2, R3 = Bz, R4 = TBDPS 71:15:14:-
93c,d
94c

Fig. 21 1,3-Dipolar cycloaddition of D-*xylo* nitrone with vinyl acetate

the *anti–trans* isoxazolidine **93a** and **94a**, were the major products, respectively. The attack of the dipolarophile on the *Z*-configuration of the nitrones **86** and **87** proceeded preferentially through an *endo* transition state. Cycloaddition of nitrone **41** with vinyl acetate afforded a mixture of four diastereoisomers **95a–d** in the ratio 47 : 21 : 18 : 14 in 86% yield. Compared to the cycloaddition with vinyl adenine the reaction proceeded with better selectivity in favor of the major isomer **95a** possessing *anti–cis* configuration [42] (Fig. 22).

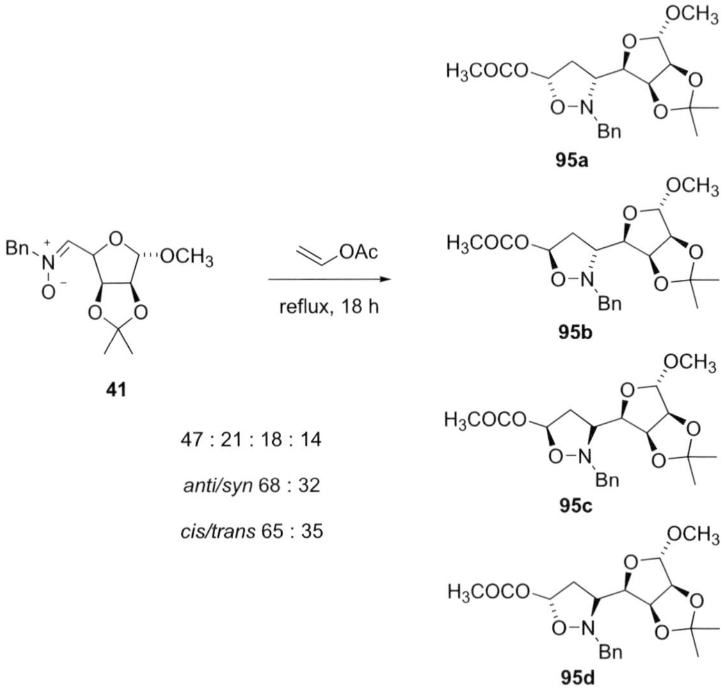

Fig. 22 1,3-Dipolar cycloaddition of D-*lyxo* nitrone with vinyl acetate

The condensation with silylated uracil, thymine, and *N*-acetylcytosine proceeded with good yields from moderate to excellent stereoselectivity under the formation of the expected β- and α-isoxazolidinyl nucleosides. The research groups of Merino, Chiacchio, and Romeo [39] have reported that the anomeric distribution obtained by the Vorbrüggen nucleosidation with chiral 5-acetoxyisoxazolidines depends on the attacking nucleobase. The attack on the intermediate oxonium from either the α or β side is possible, and hence the product distribution is sensitive to structural changes of the reactants. On the other hand, in the case of 3-*C*-glycosyl-5-acetoxyisoxazolidines **93–95** the stereoselectivity of the addition of silylated nucleobase is more dependent on the structure of the substituent at C-3 originated from the starting chiral nitrone (Hýrošová et al. unpublished results).

For the *anti–trans* cycloadduct **93a**, by the Vorbrüggen nucleosidations with uracil, thymine, and *N*-acetylcytosine, the β-anomers predominate (Hýrošová et al. unpublished results), but a significant amount of

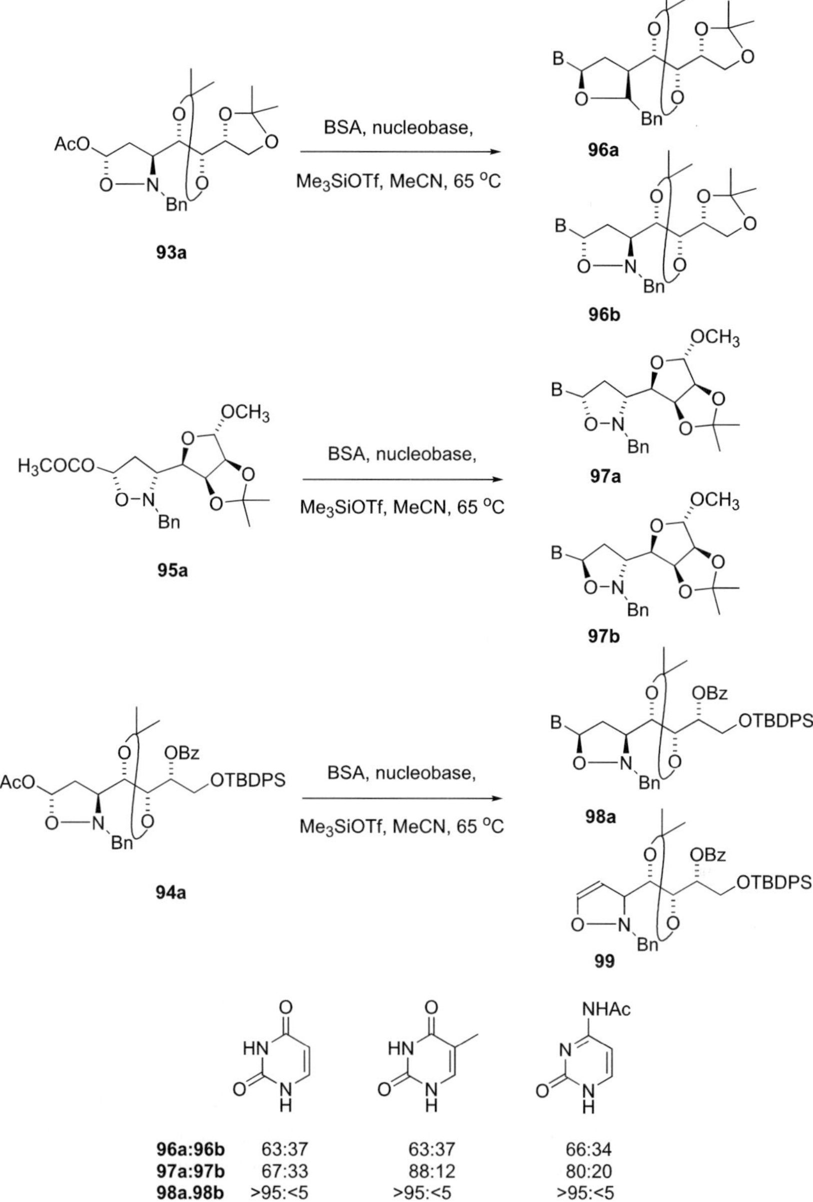

	96a:96b	63:37	63:37	66:34
	97a:97b	67:33	88:12	80:20
	98a.98b	>95:<5	>95:<5	>95:<5

Fig. 23 Isoxazolidinyl nucleosides

α-anomers has been obtained (Fig. 23). Similarly, for the *anti–cis* cycloadduct **95a,** by the Vorbrüggen nucleosidations with thymine and *N*-acetylcytosine, the β-anomers clearly predominate, while in the case of uracil a significant amount of α-anomer has been achieved [42]. These results are fully in accord with the data obtained for the related Vorbrüggen nucleosidations. On the other hand, fully surprising and unexpected, is the fact that in the case of the *anti–trans* 5-acetoxysubstituted cycloadduct **94a**, the Vorbrüggen nucleosidations proceeded with excellent stereoselectivity and the corresponding β-anomers are now the exclusive nucleosides for the all of the used nucleobases. Moreover, in this case surprisingly the product of elimination, the corresponding isoxazoline derivative **98**, was also isolated (Hýrošová et al. unpublished results).

3.2
N-Glycosylnitrones

N-Glycosylnitrones have received considerable attention as chiral equivalents in enantioselective synthesis because the glycosyl moiety acts as an auxiliary, which can be easily removed by acid treatment at an advanced stage of a synthetic sequence. Concerning cycloadditions, it has mainly been *N*-glycosylnitrones derived from formaldehyde (C-unsubstituted) [43–45], from glyoxylates (*C*-alkoxycarbonyl) [46–51], or from aromatic aldehydes (*C*-aryl) [52, 53] that have been used, and these compounds usually showed very interesting stereoselectivities.

3.2.1
Enantioselective Synthesis of Isoxazolidines

For the enantioselective synthesis of isoxazolidines, Goti et al. [54, 55] have utilized the highly stereoselective cycloaddition of *C*-phenyl-*N*-glycosylnitrones **100** and **103** to dimethyl maleate to afford trisubstituted isoxazolidines, with simultaneous formation of three new stereogenic centers (Fig. 24). The reaction of nitrone **100** with dimethyl maleate gave, in high yield and selectivity, the isoxazolidine **101**, which was by far the major isomer (> 94% diastereoselectivity). On the other hand, the cycloaddition of D-mannosylnitrone **103** afforded, in good yield, a mixture of the two diastereomeric isoxazolidines **104a** and **104b** (in a 92 : 8 ratio) with reversed stereoselectivity. The opposite chiral induction achieved by use of *N*-D-erythrosyl and *N*-D-mannosyl auxiliaries has been proved by removal of the glycosyl moiety. The cycloadduct **104a** was converted to enantiomeric pure isoxazolidine **105**, which is an oxa-analog of proline diester derivatives. Goti et al. proposed that the different diastereofacial selectivity of the aforementioned nitrones **100** and **103** is determined mainly by steric factors, since the early nature of transition state, with negligible nitrogen pyramidalization, appears to make stereoelectronic factors less important. Moreover, the sense of chiral induction in the final isoxazolidine is easily predictable on the basis of the configuration at C-2 and C-3 atoms of the sugar, locked in its furano-

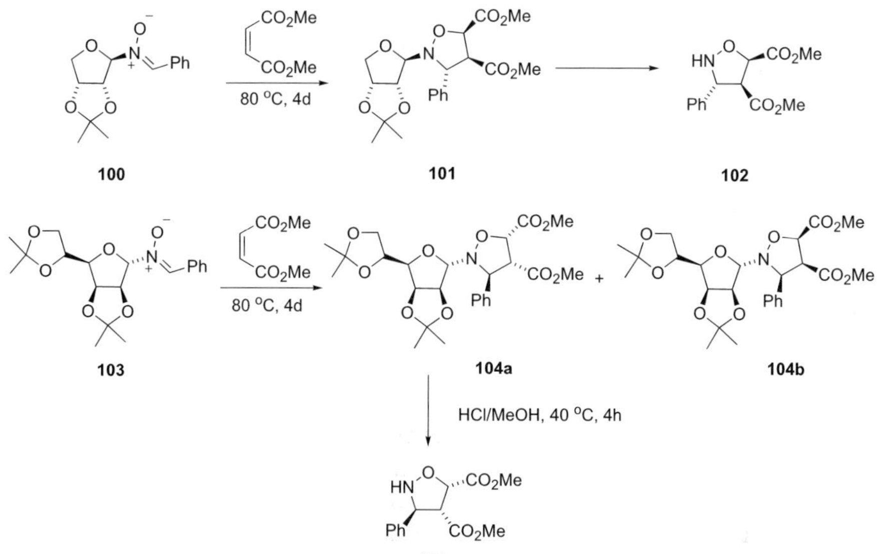

Fig. 24 Enantioselective synthesis of isoxazolidines

sidic form, as proved by the complementary behavior of N-D-erythrosyl and N-D-mannosyl nitrones.

3.2.2
Pyroglutamates

Merino and Chiacchio [56] have found an enantioselective synthesis of 4-hydroxy-D-pyroglutamic acid derivatives by an asymmetric 1,3-dipolar cycloaddition of a chiral nitrone derived from glyoxylic acid and protected D-ribosyl hydroxylamine (Fig. 25). Nitrone **106** reacted with methyl acrylate in a sealed tube for 18 h under the formation of two adducts **107a** and **108a** in a 67 : 33 ratio, and 80% chemical yield. When the reaction was carried out with the Oppolzer's sultam-derived acrylamide, the **107b** : **108b** ratio increased to 95 : 5. The excellent stereoselectivity was explained by the preferential diastereofacial *Re* selectivity induced by the sugar and sultam moiety. Isoxazolidines **107a,b** and **108a,b** were converted to the target 4-hydroxy-D-pyroglutamic acid derivatives in a one-pot procedure consisting of elimination of the sugar moiety by acidic hydrolysis and subsequent N – O cleavage by hydrogenolysis to give unprotected ethyl 4-hydroxy-D-pyroglutamates **109** and **110**.

In 2004 Tamura et al. [57] synthesized $(3'R,5'S)$-3'-hydroxycotinine [(+)-**114**], the main metabolite of nicotine, which is found in urine of smokers, by using cycloaddition of N-(L-gulosyl)-C-(3-pyridyl)nitrone **111** with $(2S)$-N-(acryloyl)bornane-10,2 sultam (**112**) (Fig. 26). Reaction of nitrone **111** with

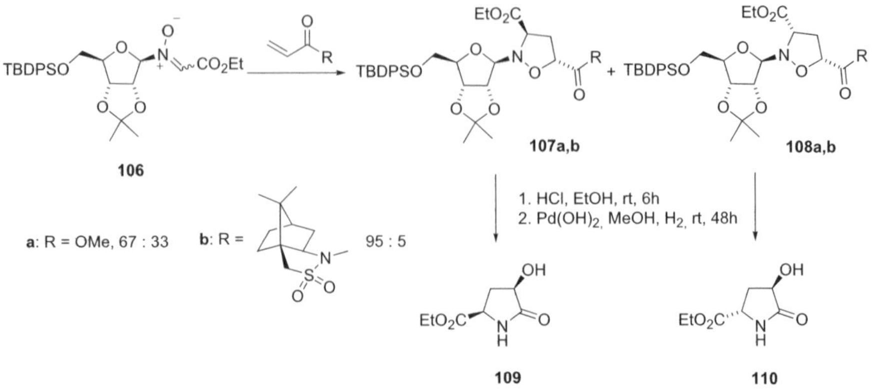

Fig. 25 Pyroglutamates

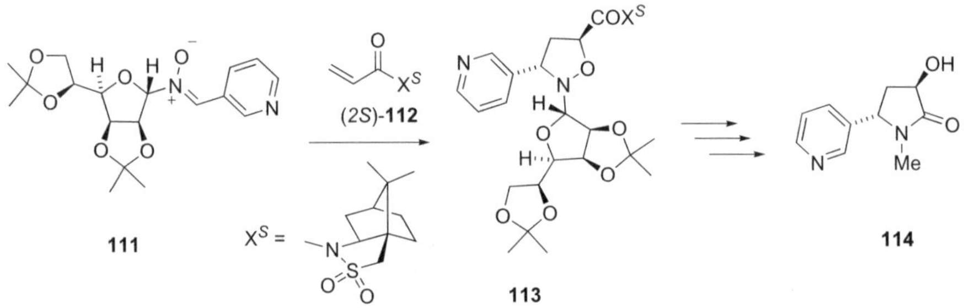

a: R = OMe, 67 : 33   b: R = [camphorsultam]   95 : 5

1. HCl, EtOH, rt, 6h
2. Pd(OH)$_2$, MeOH, H$_2$, rt, 48h

Fig. 26 1,3-Dipolar cycloaddition with double asymmetric induction

(2S)-**112** in refluxing CH$_2$Cl$_2$ afforded cycloadduct **113** as the major product along with small amounts of isomers with high selectivity (90 : 10). On the other hand, cycloaddition of nitrone **111** to (2R)-**112** furnished a complex mixture of cycloadducts. These results clearly showed a combination of nitrone **111** and (2R)-**112** to be a mismatched pair and that of nitrone **111** and (2S)-**112** to be a matched pair. The cycloadduct **113** underwent hydrolytic removal of the L-gulosyl group by treatment with hydrochloric acid to give N-free isoxazolidine, which was elaborated to (+)-**114**.

3.2.3
Isoxazolidinyl Nucleosides

The research groups of Chiacchio, Romeo and Merino have extended the 1,3-dipolar cycloaddition methodology for the enantioselective synthesis of N,O-nucleosides by using of N-glycosylnitrones, which can serve as versatile building blocks in the preparation of the target modified nucleosides [58]. The chiral auxiliary can be easily introduced before the cycloaddition process and removed subsequently to give *N,O*-nucleoside unsubstituted an the nitrogen atom. From D-ribosyl hydroxylamine, in situ-prepared chiral nitrone **106** reacted with vinyl acetate at 60 °C for 14 h under the formation of two homochiral isoxazolidines **115a** and **115b**, epimeric at C-5 in a relative ratio of *trans*:*cis* 50 : 50 (Fig. 27). In contrast to the poor *cis/trans* diastereoselectivity, the diastereofacial selectivity of the cycloaddition was high. Subsequent coupling of **115a** and **115b** with silylated thymine afforded the *N,O*-nucleosides **116a** (β) and **116b** (α) in a ratio of 28 : 72 in enantiomerically pure form. Finally, the selective cleavage of the sugar moiety in **116b** furnishes the N-unprotected nucleoside **117** in 60% yield.

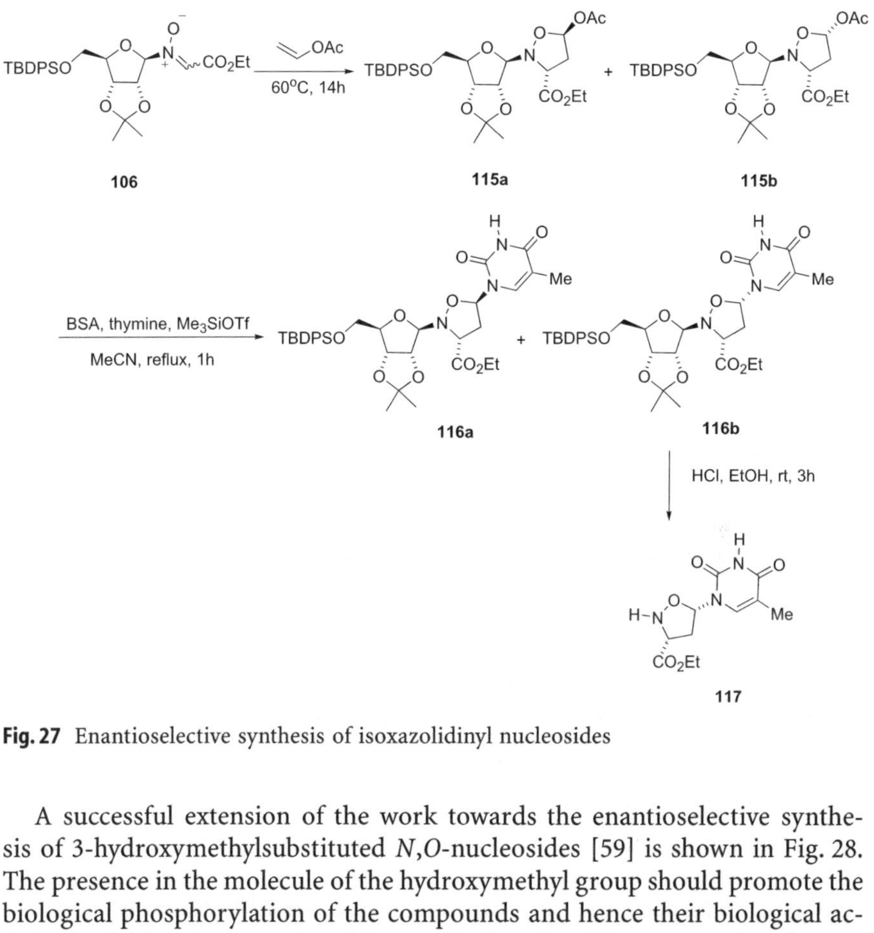

Fig. 27 Enantioselective synthesis of isoxazolidinyl nucleosides

A successful extension of the work towards the enantioselective synthesis of 3-hydroxymethylsubstituted *N,O*-nucleosides [59] is shown in Fig. 28. The presence in the molecule of the hydroxymethyl group should promote the biological phosphorylation of the compounds and hence their biological activity [60]. The aforementioned homochiral nucleosides **115a** and **115b** were converted by the Vorbrüggen nucleosidation with moderate selectivity to the expected α and β nucleosides **116b,c** and **117b,c** in a 68 : 32 ratio. The reduction of the ester function with NaBH$_4$ in dioxane/H$_2$O followed by removal

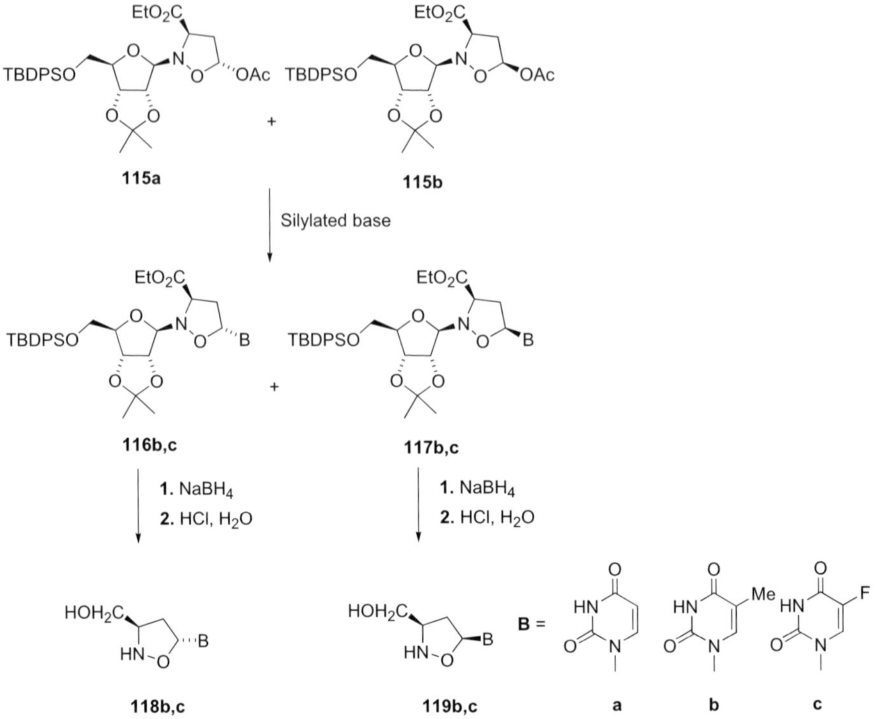

Fig. 28 Enantioselective synthesis of isoxazolidinyl nucleosides

of the sugar moiety afforded the target nucleosides α-**118b,c** and β-**119b,c** in enantiomerically pure form, with a satisfactory global yield starting from nitrone **106**. However, all the attempts to obtained uracil derivatives **116a** and **117a** failed. On the other hand, the cycloaddition of nitrone **106** with vinylated nucleobase **88a** was successful and α-**116a** and β-**117a** were formed in a 42 : 58 ratio (global yield 48%); in fact, β-nucleosides **117a** are the major products.

Next the research groups of Chiacchio and Romeo [45, 61] reported closely related results for the preparation of enantiomers of 4′-aza-2′,3′-dideoxynucleosides **124** and **125** possessing uracil, fluorouracil, thymine, cytosine, fluorocytosine, and adenine moieties by two different synthetic approaches on the basis of in situ-prepared D-ribosylnitrone **120** (Fig. 29). Two homochiral isoxazolidines **121** and **122**, epimeric at C-5′, were prepared by a one-step process through the cycloaddition of nitrone **120** with vinyl bases **88** in chloroform at 60 °C in a relative ratio 60 : 40, or by a two-step sequence based on the cycloaddition of nitrone **120** with vinyl acetate, followed by the Vorbrüggen nucleosidation with silylated nucleobases. The C-5 epimeric acetoxy-substituted isoxazolidines **123a** and **123b** were formed in a relative

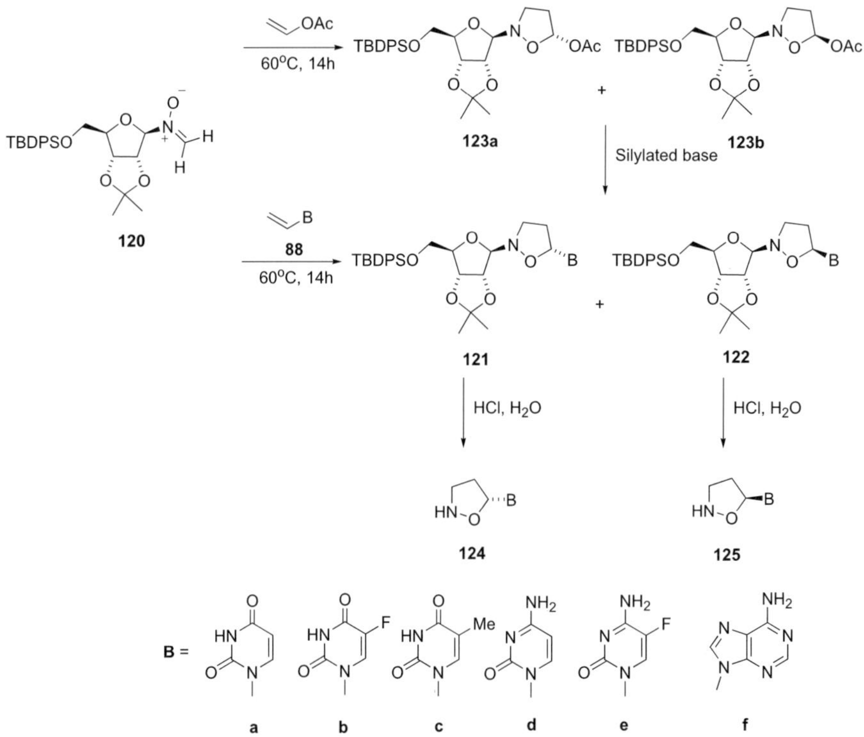

Fig. 29 Enantioselective synthesis of isoxazolidinyl nucleosides

ratio of 60 : 40 (90% yield) and the subsequent coupling with silylated nucleobases, in the presence of TMSOTf at 0 °C, occurred with 80% yield and led to nucleosides **121** and **122** in a ratio of 58 : 42. The synthesized isoxazolidinyl nucleosides **124** and **125** have shown cytotoxicity and apoptotic activity.

3.2.4
Thiazoles

In 2005, an efficient synthetic route to isoxazolidinyl analogs of tiazofurin was developed by Merino et al. [62]. Tiazofurin is a thiazole-containing *C*-nucleoside that has demonstrated significant activity in vitro against a number of model tumor systems [63]. The key step consisted of using the *N*-ribosyl nitrone **127** as starting material, in which the carbohydrate unit acted both as a chiral auxiliary and as protecting group. The cycloaddition of in situ-prepared nitrone **127** with acrylonitrile at 70 °C for 12 h in a sealed tube afforded four adducts **128a–d** in a isomeric ratio of 30 : 7 : 3 : 60 in a combined yield of 86%. The major product, 2-cyano isoxazolidine **128d**, was

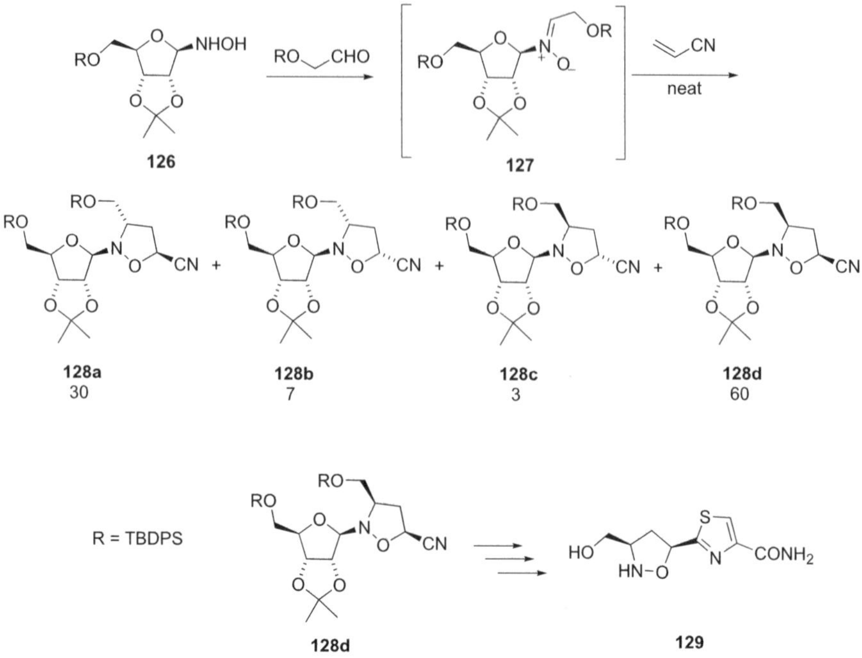

Fig. 30 Thiazole

converted into the enantiomeric target compound **129** by constructing the thiazole ring via condensation with L-cysteine.

3.2.5
Spiroisoxazolidines

In contrast to the broad synthetic applications described for chiral nitrones derived from aldehydes, use of chiral ketonitrones has been rare. Such a trend is especially pronounced in the carbohydrate field where intermolecular cycloadditions are very rare. Alonso et al. [64] have studied the cycloaddition of some sugar-derived ketonitrones. The reaction of nitrone **130** with ethyl vinyl ether in refluxing benzene for 8 h took place with complete regioselectivity, a result expected both from electronic and steric effects, to give cycloadducts **131** and **132** in an acceptable 73% combined yield (Fig. 31). Formation of the nitrogenated quaternary center proceeded, however, with modest stereoselectivity (**131** : **132**, 85 : 15). On the other hand, the rigid ketonitrone **133** prepared from cheap commercial D-glucurono-6,3-lactone reacted with allyl alcohol with complete regio- and stereoselectivity giving cycloadduct **134** in 81% yield.

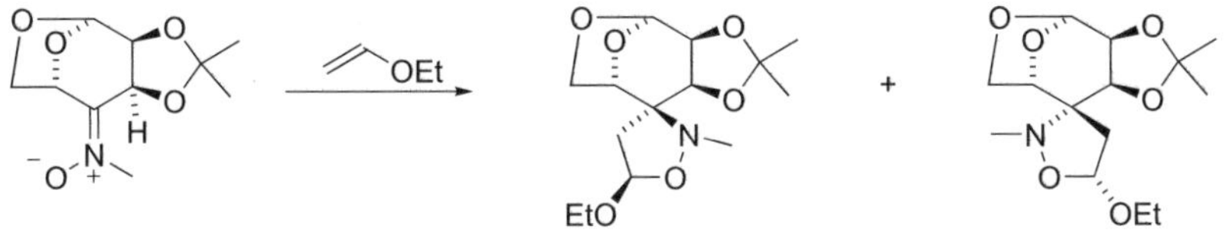

130   **131**, 62%   **132**, 11%

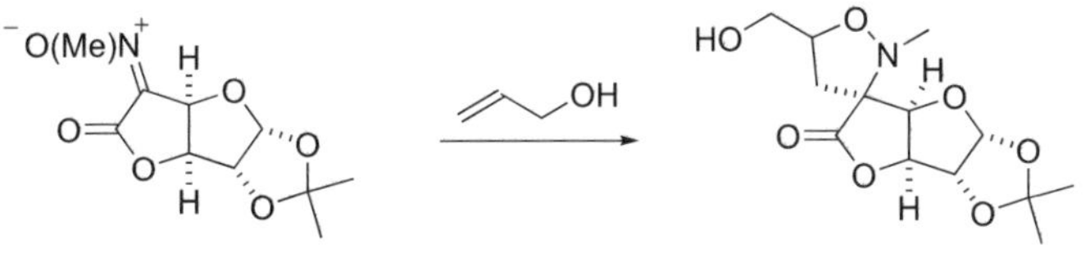

133   **134**

Fig. 31 Spiroisoxazolidines

3.3
Cyclic Nitrones

Chiral cyclic nitrones are recognized attractive synthetic intermediates, especially for optically active alkaloids and amino sugars. Special notice should be given to enantiomerically pure polyhydroxylated five-membered cyclic nitrones and to their potential in the preparation of carbohydrate-derived heterocycles. These compounds, accessible from sugars, have shown remarkable reactivity as 1,3-dipoles in cycloadditions toward alkenes. This type of reaction has been used in the synthesis of polyhydroxylated pyrrolizidines, which constitute an important class of glycoprotein-processing glycosidases and consequently display a range of important biological activities and have potential as chemotherapeutic agents. In this way the stereochemistry of the starting nitrone is directly transferred to the final products. Tamura et al. [65] have described a stereoselective cycloaddition of nitrone **135** derived from D-erythrose; the reaction with styrene in refluxing toluene afforded cycloadducts **136a** and **136b** in a ratio of 95 : 5 in 89% yield (Fig. 32).

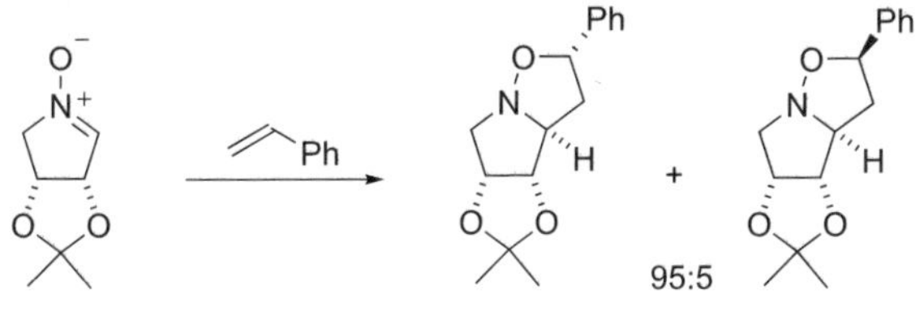

135   **136a**   95:5   **136b**

Fig. 32 1,3-Dipolar cycloaddition of cyclic nitrone with styrene

3.3.1
Pyrrolidines

In 2007 Argyropoulos et al. [66] studied the cycloaddition of the enantiomerically pure cyclic nitrones **137** and **138** (*ent*-**137**) synthesized starting from D-ribose with mono- and disubstituted alkenes. The reactions with methyl acrylate and acryloyloxazolidinone were performed with both the enantiomeric nitrones **137** and **138**, affording the diastereomeric isoxazolidines **139a,b**, **140a,b**, and **141** (*ent*-**139**) and **142** (*ent*-**140**), respectively, in a ratio of 75 : 25 for the reactions with methyl acrylate and 67 : 33 for the reactions with acryloyloxazolidinone and 80–90% total yield (Fig. 33). The cycloaddition of

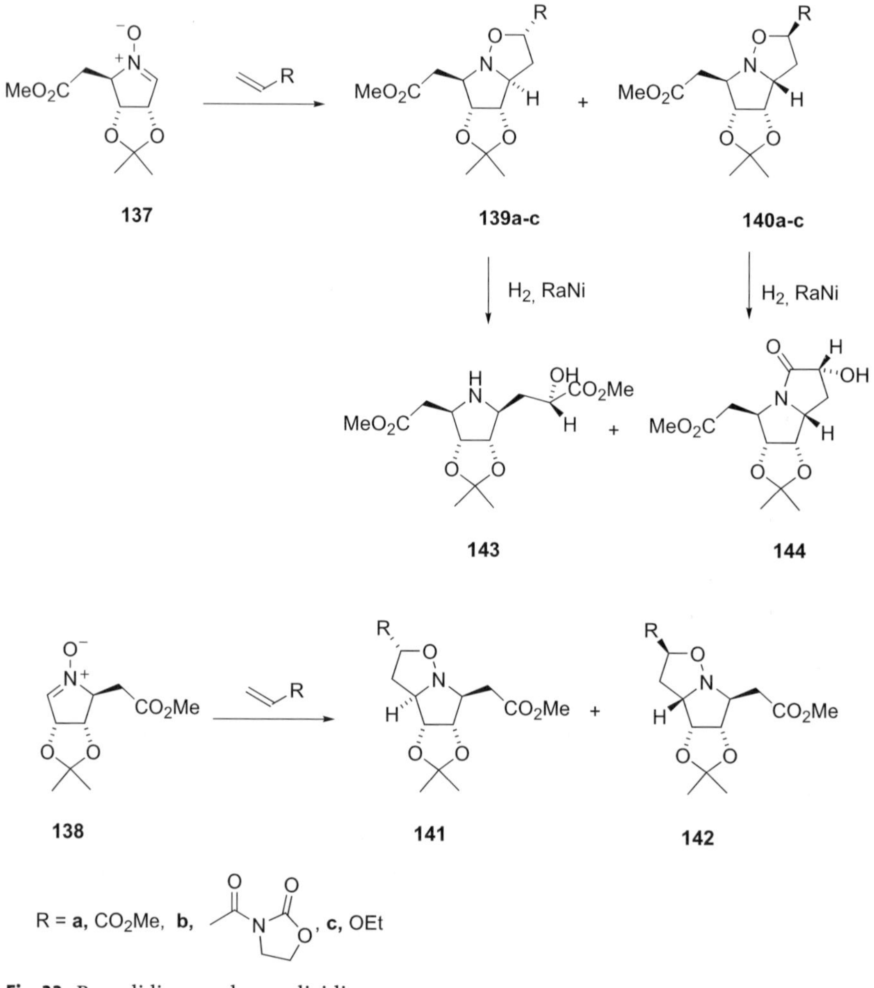

Fig. 33 Pyrrolidines and pyrrolizidines

nitrone **137** with ethyl vinyl ether also gave two diastereomeric cycloadducts **139c** and **140c** in a ratio 67 : 33 and 94% total yield. In all these reactions the preferred stereochemical outcome of the cycloaddition step comes from a 5-*exo–anti* transition state. Reductive cleavage of the obtained isoxazolidines **139** and **140** depends on the kind and geometry of the preexisting alkene substituent and can lead to pairs of enatiomerically pure pyrrolidine **143** or pyrrolizidine **144** derivatives.

Defoin et al. has developed the synthesis of hydroxy-ethane derivative **148**, a nanomolar α-L-fucosidase inhibitor, starting from chiral nitrone **145**, an analog of 5-deoxy-L-lyxose that was derived from D-ribose [67]. Nitrone **145** reacted easily in C_2Cl_4 at 50 °C with phenyl vinyl ether to give cycloadducts **146a** and **146b** quantitatively in a 85 : 15 ratio (Fig. 34). The major isomer **146a** was converted via pyrrolidine **147** to the C-α-L-fucoside **148** in four steps in an overall yield of 59%.

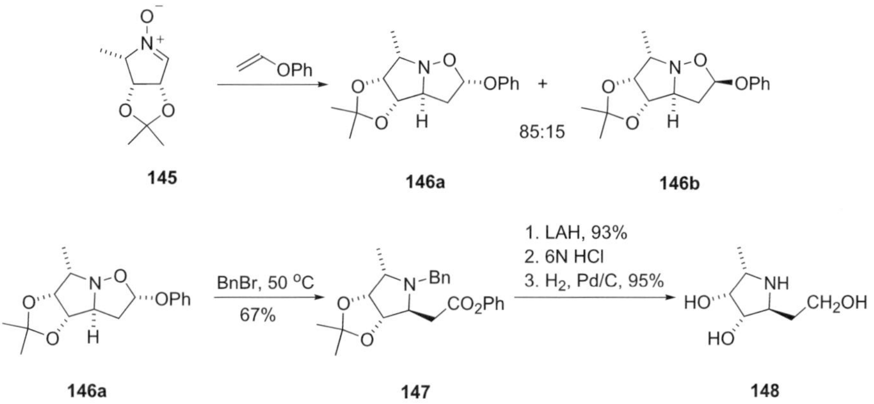

Fig. 34 Pyrrolidines

3.3.2
Pyrrolizidines

For the syntheses of natural pyrrolizidine alkaloids such as alexine, australine, causarine, and their analogs that have a hydroxymethyl group at C-3 the corresponding starting nitrones have been prepared and their 1,3-dipolar cycloadditions studied [68]. In 2003 Goti et al. reported a straightforward and completely stereoselective entry to polyhydroxypyrrolizidines, as demonstrated by the total syntheses of 7-deoxycausarine (**149**) and hyacinthacine A_2 (**150**) in only seven steps and 24% overall yield, and eight steps and 13% overall yield, respectively (Fig. 35). The key step was the highly stereoselective 1,3-dipolar cycloaddition of trisubstituted chiral nitrone **151**, which has the correct and absolute stereochemistry at C-1, C-2, and C-3, as required for

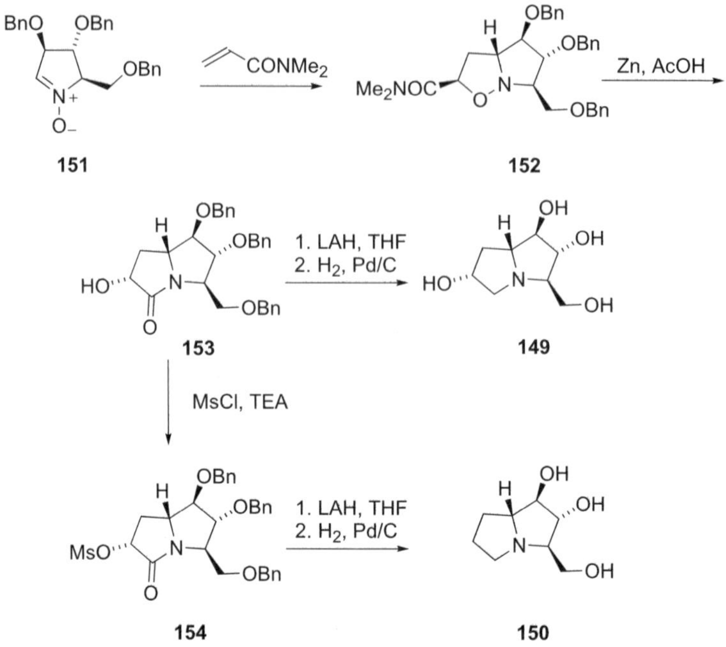

Fig. 35 7-Deoxycausarine and hyacinthacine A$_2$

the above-mentioned pyrrolizidine alkaloids. The nitrone **151** that has been prepared starting from L-xylose or D-arabinose, reacted with *N,N*-dimethyl acrylamide in CH$_2$Cl$_2$ at room temperature to afford exclusively the *anti–exo* adduct **152**, possessing the stereochemistry required by casuarine at C-6. The pyrroloisoxazolidine **152** was converted into the desired pyrrolizidine **149** and **150** by simple three- and four-step sequences, respectively.

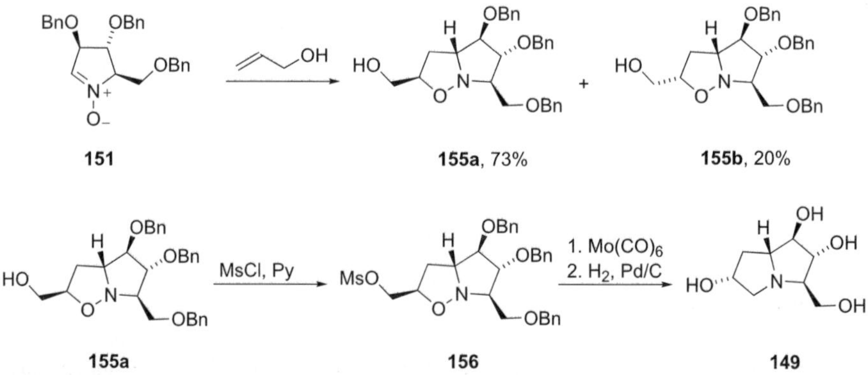

Fig. 36 7-Deoxycausarine

In the same year, Carmona et al. [69], independently from Goti, utilized the nitrone **151** for the synthesis of 7-deoxycasuarine (**149**). Allyl alcohol was used instead of acryl amide. Heating a mixture of nitrone **151** and allyl alcohol in toluene under reflux led to the formation of cycloadducts **155a** and **155b** (78 : 22) in 93% yield. The major adduct **155a** was converted to 7-deoxycasuarine (**149**) according the reaction sequence given in Fig. 36.

3.3.3
Indolizidines

Vasella and Peer have developed a regioselective synthesis of the protected six-membered cyclic L-fucose-derived nitrone **157** in view of the synthesis of α-L-fucosidase inhibitors [70]. The nitrone **157** has been transformed into the indolizidines **160** and **161** via a cycloaddition to an acrylate (Fig. 37). The reaction proceeded regioselectively and led to a 60 : 40 mixture of the diastereomeric isoxazolidines **158** (55%) and **159** (34%) by exclusively axial C – C bond formation. Reductive cleavage of the N – O bond of the isoxazolidines **158** and **159**, followed by reduction with BH_3 in THF with subsequent hydrogenolytic debenzylation in MeOH/HCl, afforded the indolizidines **160** and **161** as their hydrochlorides.

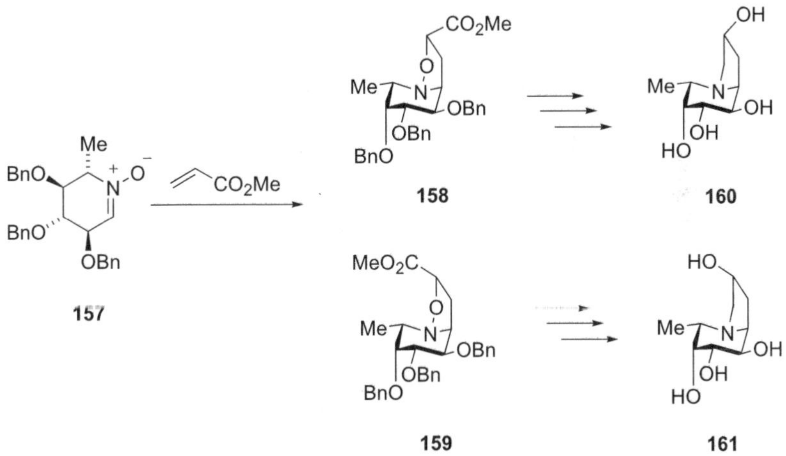

Fig. 37 Indolizidines

3.3.4
Nucleosides

The key step for the synthesis of uridine diphosphono-β-Gal*f* mimics **162**, based on an iminosugar skeleton linked to UMP by a 2-hydroxypropyl tether, is a highly regio- and stereoselective cycloaddition of an original uridin-5′-yl allylphosphonate **163** with the *galacto*-configured cyclic nitrone **164** prepared from tetra-*O*-benzyl-D-glucofuranose, followed by the reductive elaboration of the cycloaddition product [71]. The cycloaddition of nitrone **164** with **163** at 90 °C in tetrachloroetylene for 16 h afforded the isoxazolidine **165** in an isolated yield of 80% (Fig. 38). A three-step sequence (Zn/AcOH then BCl_3 followed with an excess $BrSiMe_3$) gave the original iminogalactose β-linked UDP-Gal*f* mimic **162** in excellent yield.

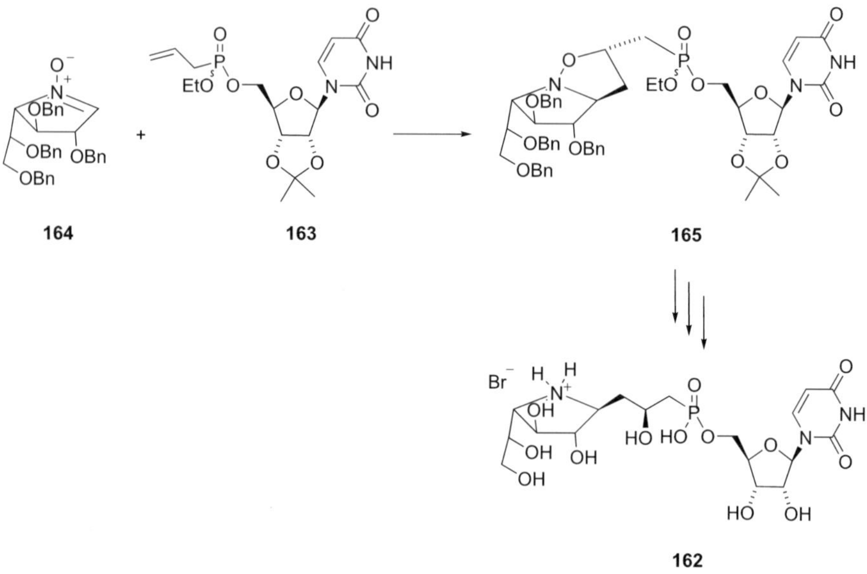

Fig. 38 Nucleosides

3.3.5
Disaccharides

Stereoselective cycloaddition between functionalized six-membered cyclic nitrone **166** derived from D-lyxose and sugar alkenes was employed for the synthesis of aza-*C*-disaccharides [72]. Reaction of nitrone **166** and alkene **167** prepared from methyl α-D-mannopyranoside in toluene at reflux led to the isolation of a crystalline cycloadduct **168** in 84% yield. The isoxazolidine **168** was acetylated and cleaved using $Mo(CO)_6$. This was followed by deoxygenation and routine deprotection to give the aza-*C*-disaccharide **169**, isolated as its hydrochloride (44% overall from **168**).

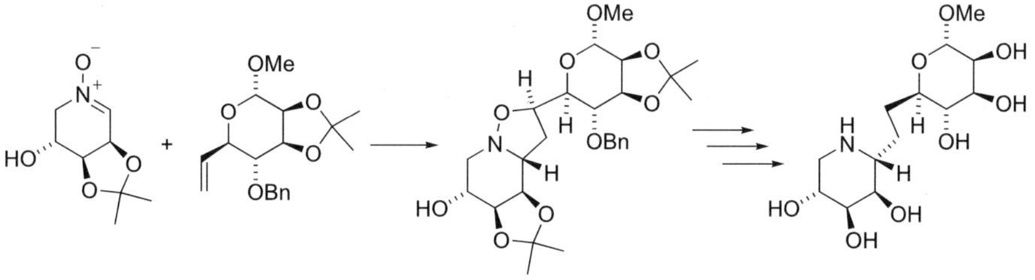

Fig. 39 Aza-*C*-disaccharides

Acknowledgements The author is grateful to the Slovak Grant Agency (No. 13549/06) and APVV (No. 20/000305).

References

1. Tufariello JJ (1984) In: Padwa A (ed) 1,3-Dipolar cycloaddition chemistry. Wiley-Interscience, New York, p 83
2. Torssell KBG (1988) Nitrile oxides, nitrones, and nitronates in organic synthesis; novel strategies in synthesis. VCH, New York
3. Grünanger P, Vita-Finzi P (1991) Isoxazoles, part 1. In: Taylor EC, Weissberger E (eds) The chemistry of heterocyclic compounds. Wiley, New York
4. Chiacchio U, Genovese P, Iannazzo D, Librando V, Merino P, Rescifina A, Romeo R, Procopio A, Romeo G (2004) Tetrahedron 60:441
5. Dondoni A, Franco S, Junquera F, Merchán F, Merino P, Tejero T (1994) Synth Commun 24:2537
6. Fišera L, Al-Timari UAR, Ertl P (1992) In: Gulliano E (ed) Cycloadditions in carbohydrate chemistry, ACS Monograph. Am Chem Soc, Washington, p 158
7. Merino P (2004) In: Padwa A (ed) Sciences of synthesis, vol 1. Thieme, Stuttgart, p 511
8. Tyler PC, Winchester BG (1999) In Stütz AE (ed) Iminosugars as glycosidase inhibitors. Nojirimycin and beyond. Wiley-VCH, Weinheim, p 125
9. El Ashry ESH, Rashed N, Shobier AHS (2000) Pharmazie 55:331
10. Asano N (2003) Glycobiology 13:93R
11. Frederickson M (1997) Tetrahedron 53:403
12. Gothelf KV, Jørgensen KA (1988) Chem Rev 98:86
13. Osborn HMI, Gemmel N, Harwood LM (2002) J Chem Soc Perkin Trans I 22:2419
14. Koumbis AE, Gallos JK (2003) Current Org Chem 7:585
15. Fišera L, Ondruš V, Kubán J, Micúch P, Blanáriková I, Jäger V (2000) J Heterocyclic Chem 37:551
16. Jones RCF, Martin JN (2002) Synthetic applications of 1,3-dipolar cycloaddition chemistry toward heterocycles and natural products. In: Padwa A, Pearson WH (eds) The chemistry of heterocyclic compounds, vol 59. Wiley, New York, p 1
17. Pellissier H (2007) Tetrahedron 63:3235
18. Kubán J, Blanáriková I, Fengler-Veith M, Jäger V, Fišera L (1998) Chem Papers 52: 780

19. Kubán J, Blanáriková I, Fišera L, Jarošková L, Fengler-Veith M, Jäger V, Kozíšek J, Humpa O, Prónayová N, Langer V (1999) Tetrahedron 55:9501

20. Kubán J, Kolarovic A, Fišera L, Jäger V, Humpa O, Prónayová N, Ertl P (2001) Synlett, p 1862

21. Kubán J, Kolarovic A, Fišera L, Jäger V, Humpa O, Prónayová N (2001) Synlett, p 1866

22. Dugovic B, Fišera L, Hametner C (2004) Synlett, p 1569

23. Dugovic B, Fišera L, Cyranski MK, Hametner C, Prónayová N, Obranec M (2005) Helv Chim Acta 88:1432

24. Dugovic B, Wiesenganger T, Fišera L, Hametner C, Prónayová N (2005) Heterocycles 65:591

25. Dugovic B, Fišera L, Hametner C, Prónayová N (2003) ARKIVOC xiv:162

26. Dugovic B, Fišera L, Cyranski MK, Hametner C, Prónayová N (2004) Monatsh Chem 135:685

27. Liu PS (1987) J Org Chem 52:4717

28. Cardona F, Salanski P, Chmielewski M, Valenza S, Goti A, Brandi A (1998) Synlett, p 1444

29. Merino P, Franco S, Merchan FL, Romero P, Tejero T, Uriel S (2003) Tetrahedron Asymmetry 14:3731

30. Argyropoulos NG, Sarli VC (2004) Tetrahedron Lett 45:4237

31. Torres-Sánchez I, Borrachero P, Cabrera-Escribano F, Gómez-Guillén M, Angulo-Álvarez M, Diánez J, Estrada D, López-Castro A, Pérez-Garrido S (2005) Tetrahedron Asymmetry 16:3897

32. Borrachero P, Cabrera-Escribano F, Gómez-Guillén M, Torres M (2004) Tetrahedron Lett 45:4835

33. Schweizer F (2002) Angew Chem Int Ed 41:230

34. Borrachero P, Cabrera-Escribano F, Diánez J, Estrada D, Gómez-Guillén M, López-Castro A, Pérez-Garrido S, Torres I (2002) Tetrahedron Asymmetry 13:2025

35. Karanjule NS, Markad SD, Sharma T, Sabharwal SG, Puranik VG, Dhavale DD (2005) J Org Chem 70:1356

36. Li X, Takahashi H, Ohtake H, Ikegami S (2004) Tetrahedron Lett 45:4123

37. Borrachero P, Cabrera F, Diánez J, Estrada D, Gómez-Guillén M, López-Castro A, Moreno J, de Paz JL, Pérez-Garrido S (1999) Tetrahedron Asymmetry 10:77

38. Silva AMG, Tomé AC, Neves MGPM, Silva AMS, Cavaleiro JAS, Perrone D, Dondoni A (2002) Tetrahedron Lett 43:603

39. Merino P (2002) Curr Med Chem Anti-Infective Agents 1:389

40. Vorbrüggen H, Krolikiewicz K. Bennua B (1981) Chem Ber 114:1234

41. Fischer R, Drucková A, Fišera L, Rybár A, Hametner C, Cyrañski MK (2002) Synlett, p 1113

42. Hýrošová E, Fišera L, Jame RMA, Prónayová N, Medvecký M, Koóš M (2007) Chem Heterocyclic Compounds, p 80

43. Vasella A (1977) Helv Chim Acta 60:426

44. Mzengeza S, Whitney RA (1988) J Org Chem 53:4074

45. Chiacchio U, Rescifina A, Corsaro A, Pistara V, Romeo G, Romeo R (2000) Tetrahedron Asymmetry 11:2045

46. Vasella A, Voeffray R (1982) Helv Chim Acta 65:1134

47. Vasella A, Voeffray R, Pless J, Huguenin R (1983) Helv Chim Acta 66:1241

48. Kasahara K, Iida H, Kibayashi C (1989) J Org Chem 54:2225

49. Machetti F, Cordero FM, De Sarlo F, Guarna A, Brandi A (1996) Tetrahedron Lett 37:4205

50. Tamura O, Mita N, Kusaka N, Suzuki H, Sakamoto M (1997) Tetrahedron Lett 38:429

51. Chiacchio U, Corsaro A, Gumina G, Rescifina A, Iannazzo D, Piperno A, Romeo G, Romeo R (1999) J Org Chem 64:9321
52. Borthakur DR, Prajapati D, Sandhu JS (1988) Ind J Chem 27B:724
53. Fišera L, Al-Timari UAR, Ertl P, Prónayová N (1993) Monatsh Chem 124:1019
54. Cicchi S, Marradi M, Corsi M, Faggi C, Goti A (2003) Eur J Org Chem 4152
55. Cicchi S, Corsi M, Marradi M, Goti A (2002) Tetrahedron Lett 43:2741
56. Merino P, Revuelta J, Tejero T, Chiacchio U, Rescifina A, Piperno A, Romeo G (2002) Tetrahedron Asymmetry 13:167
57. Tamura O, Kanoh A, Yamashita M, Ishibashi H (2004) Tetrahedron 60:9997
58. Chiacchio U, Corsaro A, Gumina G, Rescifina A, Iannazzo D, Piperno A, Romeo G, Romeo R (1999) J Org Chem 64:9321
59. Chiacchio U, Corsaro A, Iannazzo D, Piperno A, Pistara V, Rescifina A, Romeo R, Sindona G, Romeo G (2003) Tetrahedron Asymmetry 14:2717
60. Challand R, Young R (1998) Antiviral chemotherapy. Oxford University Press, Oxford
61. Chiacchio U, Corsaro A, Iannazzo D, Piperno A, Pistara V, Rescifina A, Romeo R, Valveri V, Mastino A, Romeo G (2003) J Med Chem 46:3696
62. Merino P, Tejero T, Unzurrunzaga FJ, Franco S, Chiacchio U, Saita MG, Iannazzo D, Piperno A, Romeo G (2005) Tetrahedron Asymmetry 16:3865
63. Yalowitz JA, Pankiewicz K, Patterson SE, Jayaram HN (2002) Cancer Lett 181:31
64. Torrente S, Noya B, Branchadell V, Alonso R (2003) J Org Chem 68:4772
65. Tamura O, Toyao A, Ishibashi H (2002) Synlett, p 1344
66. Argyropoulos NG, Panagiotidis T, Coutouli-Argyropoulos E, Raptopoulou C (2007) Tetrahedron 63:321
67. Chevrier C, Le Nouen D, Neuburger M, Defoin A, Tarnus C (2004) Tetrahedron Lett 45:5363
68. Cardona F, Faggi E, Liguori F, Cacciarini M, Goti A (2003) Tetrahedron Lett 44:2315
69. Carmona T, Whigtman RH, Robina I, Vogel P (2003) Helv Chim Acta 86:3066
70. Peer A, Vasella A (1999) Helv Chim Acta 82:1044
71. Liauturd V, Christina AE, Desvergnes V, Martin OR (2006) 71:7337
72. Duff FJ, Vivien V, Wightman RH (2000) Chem Commun, p 2127

Top Heterocycl Chem (2007) 7: 325–346
DOI 10.1007/7081_2007_078
© Springer-Verlag Berlin Heidelberg
Published online: 20 June 2007

Anhydro Sugars:
Useful Tools for Chiral Syntheses of Heterocycles

Khalid Mohammed Khan[1] · Shahnaz Perveen[2] · Wolfgang Voelter[3] (✉)

[1]H.E.J. Research Institute of Chemistry,
 International Center for Chemical and Biological Sciences, University of Karachi,
 75270 Karachi, Pakistan

[2]PCSIR Laboratories Complex Karachi, Shahrah-e-Dr. Salim-uz-Zaman Siddiqui,
 Karachi 75280, Pakistan

[3]Interfakultäres Institut für Biochemie, Hoppe-Seyler-Str. 4, 72076 Tübingen, Germany
 wolfgang.voelter@uni-tuebingen.de

Abstract Anhydro sugars have proved to be versatile chiral synthons for the synthesis of chiral heterocycles including pyradazines, oxazines, triazines, chiral morpholines, tetrahydroquinoxalines, benzodioxins, oxazolidinones, episulfides, thiazolines, and cyclic trithiocarbonates. Selected examples are described in this chapter.

Keywords Anhydro sugars · Chiral heterocycles · Synthesis

1
Introduction

During the last few decades, a series of communications have appeared from our laboratories reporting frequent use of anhydro sugars as versatile chiral synthons [1–28]. In the present article, we would like to focus on the importance of anhydro sugars in the syntheses of chiral heterocycles. In the last decade, we significantly used anhydro sugars in our laboratories for the synthesis of useful classes of chiral heterocycles including pyradazines, oxazines, thiazolines, triazines, morpholines, tetrahydroquinoxalines, fused oxygen heterocycles, benzodioxins, oxazolidinones, episulfides, and cyclic trithiocarbonates.

The trifluoromethanesulfonates, commonly referred to as triflates, have excellent leaving group properties and serve in numerous transformations [29] and in carbohydrate chemistry [30], and we have carried out a series of reactions involving displacement of the triflate group from C-4 of 2,3-anhydro sugars by a wide variety of nucleophiles. The displacement reactions at this position are not only facilitated by the neighboring oxirane ring [31], but also by the favorable geometry [2]. The sugar triflates are prepared from the corresponding anhydro sugars by treating them with 1.1 equivalents of triflic anhydride in pyridine at low temperature.

2
Fused Oxygen Heterocycles

2.1
Tetrahydrofurylidenes

The occurrence of highly functionalized furanoid systems either as a key entity or key substructure of more complex molecules in various biologically active natural products, such as the entire class of polyether antibiotics, prostaglandins, prostacyclins, and marine *Laurencia* metabolites, initiated new developments in their syntheses.

Ambient nucleophiles (β-diketones or β-ketoesters) are known to attack always by the more basic γ-carbon atom. A difficulty, frequently encountered in intramolecular alkylation of β-dicarbonyl compounds, is the concurrent formation of both C- and O-alkylated products. It is, however, normally possible to direct the alkylation toward carbon or oxygen by proper selection of (1) the solvent, (2) the enolate counter ion, and (3) the leaving group.

It is well known that five-membered ring formation, involving intramolecular oxygen alkylation of delocalized ketone and ester enolates under kinetic conditions, is seen with almost total exclusion of the thermodynamically preferred carbon alkylation reaction. Baldwin has provided a satisfactory explanation [32] for the exceptional preference for intramolecular oxygen alkylation, which features bond formation via the use of a geometrically more accessible in-plane lone-pair electron on the oxygen atom. For example, exposure of **1** and **2** to the dianion of *tert*-butyl acetoacetate at $-78\,^\circ$C in THF delivered the (*Z*)-tetrahydrofurylidenes **3** and **5** in 78 and 76% yields, respectively (Scheme 1). The alternate (*E*)-isomers **4** and **6** were obtained after geometric isomerization of the (*Z*)-isomers **3** and **5** upon treatment with 1% TFA/CH$_2$Cl$_2$ at 0 $^\circ$C to room temperature for 6 h [20, 28].

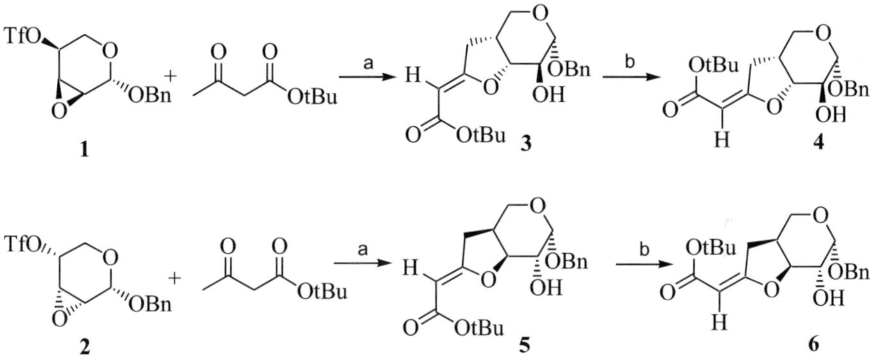

Scheme 1 *a* NaH, *n*-BuLi, THF, $-78\,^\circ$C; *b* 1% TFA, CH$_2$Cl$_2$ [28]

For further application of this methodology to the synthesis of a specific target, an alkyl group at C-6 needs to be introduced, since the entire class of polyether antibiotics contains at least one methyl moiety on a tetrahydrofuran ring. Alkylation of the enolate of the α,β-unsaturated ester was expected to yield an α-alkyl-β,γ-unsaturated ester. Indeed, the dianion of methyl propionyl acetate (**7**) reacted with benzyl 2,3-anhydro-β-L-ribopyranoside (**1**) in THF at $-78\,^\circ$C to afford a 1 : 1 mixture of the diastereoisomers **8** and **9** in 90% combined yield (Scheme 2). Due to the formation of a mixture, 50% of the desired product got lost. Treatment of the mixture of **8** and **9** with 1% TFA in CH$_2$Cl$_2$ caused epimerization and afforded a 9 : 1 mixture of (6*S*)-**8** and (6*R*)-**10** (Scheme 2) [33].

Furthermore, it was found that the *tert*-butyl acetoacetate dianion almost quantitatively opens in a *trans*-diaxial manner the epoxy ring of **11** at C-3, resulting in the formation of the pyranoside **13** (Scheme 3). Conventional mesylation of the branched-chain sugar **13**, followed by base treatment, afforded quantitatively the *cis*-fused bicyclic system **14** via 5-enol *exo-exo-tet* ring closure, whereby O alkylation is preferred over C alkylation due to stereoelectronic considerations [26, 34]. Interestingly, refluxing the branched-chain sugar **13** in toluene in the presence of NaH resulted in the isolation of the *trans*-fused γ-lactone **15**. The reaction was also performed with anhydro pyranoside **12**, from which two regioisomers of the branched-chain sugars

Scheme 2 *a* NaH, *n*-BuLi, THF, –78 °C; *b* 1% TFA, CH$_2$Cl$_2$ [33]

14: R = H, R^1 = ^{t}Bu
20: R = CH$_3$, R^1 = Me

13: R = H, R^1 = ^{t}Bu
19: R = CH$_3$, R^1 = Me

Scheme 3 *a tert*-Butyl acetoacetate (R = H) or methyl propionyl acetate (R = Me), NaH, *n*-BuLi (4 eq.), THF, 0 °C to rt, 5–6 h (R = H) or 12 h (R = Me); *b* MsCl, Et$_3$N, CH$_2$Cl$_2$, –15 °C, 1 h; *c* NaH, THF, 0 °C to rt, 4–5 h; *d* NaH, toluene, reflux, 12 h [26]

16 and **17** were isolated in a ratio of 2 : 8. Subsequent mesylation of the major isomer **17**, followed by base treatment, afforded the *cis*-fused furanoid **18**. Analogously, compound **19** was converted into **20** in 92% yield (Scheme 3).

The chiral alkylidine tetrahydrofuranoids are excellent starting materials for the corresponding γ-butyrolactones, and the syntheses of these systems are easily achieved by ozonolysis of the tetrahydrofurylidene systems **21–24**, to afford different regio- and diastereoisomers of the γ-lactone frameworks **25–28** (Scheme 4), depending on the starting anhydro sugars [35].

Tricyclic structures possessing the furopyran ring were prepared by the reaction of the dianions generated from cyclic β-ketoesters of different ring sizes (five-, six-, and seven-membered rings) with triflates **1** and **2**, affording tricyclic furanoids in good yields (Scheme 5). Interestingly, in each case only one tricyclic furanoid was obtained (**29–34**). It is worth mentioning that the reactions of the five- and six-membered rings of β-ketoesters with **1** and **2** were more straightforward than the same reaction with a seven-membered ring system [35].

Reaction of **11** and **12** with five- and seven-membered cyclic β-ketoesters did not lead to oxirane ring opening, whereas reaction of six-membered

Scheme 4 *a* O$_3$, CH$_2$Cl$_2$, –78 °C, 0.5 h, and then Ph$_3$P [35]

Scheme 5 *a* NaH, THF, cyclic β-ketoester systems, *n*-BuLi, −78 to 0 °C, 3–7 h [35]

cyclic β-ketoester led successfully to a regioselective opening of the oxirane ring in 50% yield. Subsequent mesylation of the free hydroxyl group followed by treatment with sodium hydride yielded the tricyclic systems **35** and **36** in an enantioselective manner (Scheme 6).

Scheme 6 *a* NaH, THF, 2-ethoxycarbonylcyclohexanone, *n*-BuLi, rt, 1 h, reflux 4 h; *b* MsCl, Et_3N, CH_2Cl_2, 0 °C, 0.5 h; *c* NaH, THF, rt, 12 h [35]

2.2
Dihydrofuran Systems

The monoanions of 1,3-dicarbonyl compounds react smoothly with the *cis*-oriented epoxy triflate 1 to give the intermediate *I* which, after further base treatment, leads to a dihydrofuran system [20] (Scheme 7). After treatment of 1 with dianions of 1,3-dicarbonyl compounds, tetrahydrofurylidine formation is observed under kinetic conditions [20] (Scheme 7).

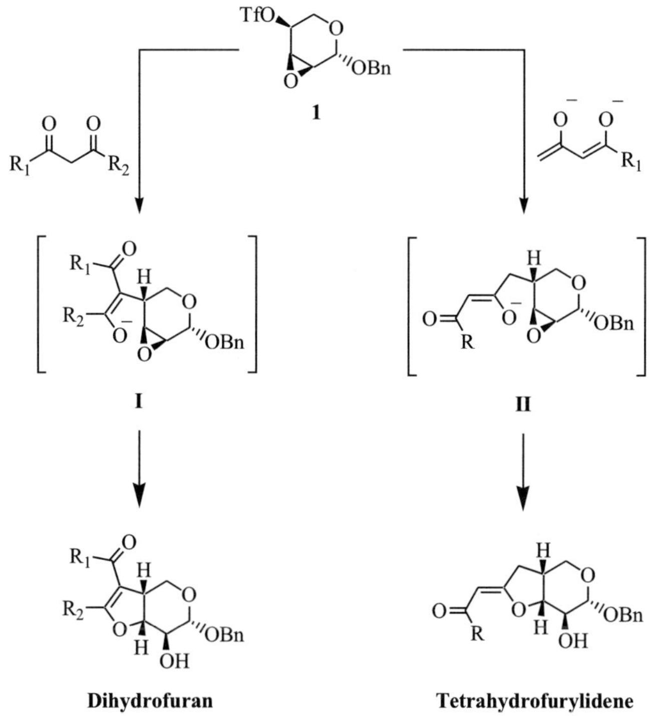

Scheme 7 [20]

To extend this powerful new dihydrofuran synthesis to more complex systems, related reactions with branched β-dicarbonyl systems were investigated [20]. In a similar fashion, the *cis*-fused furanoids **40** and **41** were prepared from hexane-2,4-dione (**37**), ethyl isobutyryl acetate (**38**), and **2**. These targets are important chiral synthons, since there are many natural products bearing ethyl and isopropyl residues on furanoid rings. The flexibility of this method was also tested with the aromatic β-diketone **39** yielding furanoid **42**, which is an important system for tetralone synthesis (Scheme 8).

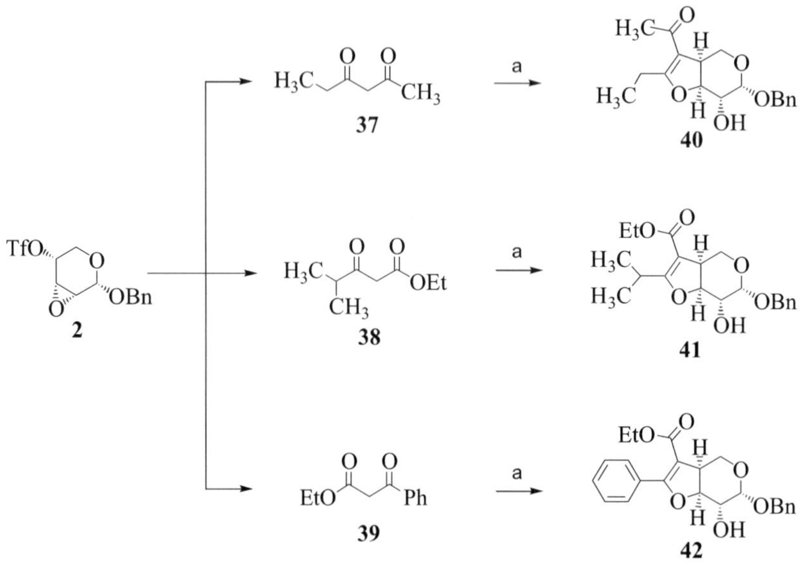

Scheme 8 *a* NaH (5 eq.), 15-crown-5, THF, β-dicarbonyl compound (5 eq.), 0 °C → rt, 2–3 h [20]

2.3
Polysubstituted Chiral Butanolides

Butanolides are ubiquitous subunits of a large number of bioactive compounds. Epoxy triflate sugars **1** and **2** open an easy access to selected polysubstituted chiral butanolides [36, 37], by a two-step sequence. The first step is displacement of the triflyl group of **1** and **2** under extremely controlled conditions with a very reactive enolate of *tert*-butyl acetoacetate followed by cyclization under acid catalysis, which furnished **45** and **46** [36] (Scheme 9).

By reacting the sodium salt of ethyl acetoacetate with the triflates **1** and **2** followed by acidic workup, the primary displacement product cyclized to af-

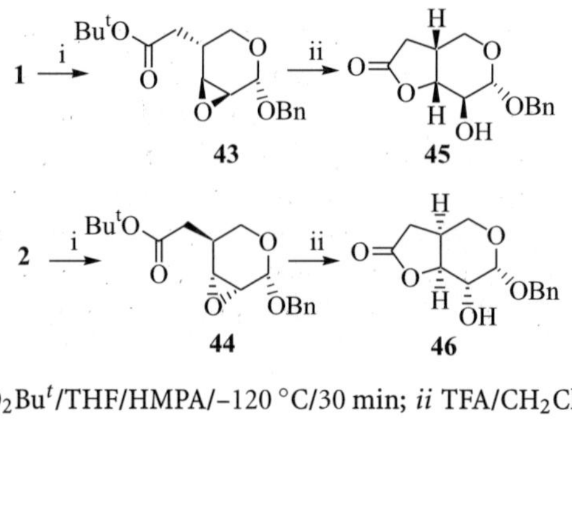

Scheme 9 *i* LiCH₂CO₂Buᵗ/THF/HMPA/–120 °C/30 min; *ii* TFA/CH₂Cl₂/rt/20 min [36]

ford the crystalline trisubstituted butanolides **47** and **48** as the only products on the off-template chiral center [36, 37] (Scheme 10).

Scheme 10 *i* NaCH(CO$_2$Et)$_2$/THF/EtOH/rt/–30 min [36]

β-Oxy-γ-butyrolactones are substructural units in many biologically interesting natural products, accessible by nucleophilic addition of the enolates of *tert*-butyl acetate on epoxy pentuloses. The lithium enolate of *tert*-butyl acetate is reacted with the epoxy ketose **50** (generated from **49**) in a mixture of THF/HMPA (3 : 1) to afford **51** and **52**. However, the *trans*-oriented isomer **52** can only be converted to the desired lactone **54** on treatment with TFA in CH$_2$Cl$_2$ at room temperature for 30 min [18] (Scheme 11).

Scheme 11 *i* LiCH$_2$CO$_2$Bu^t/THF/HMPA/–120 °C/30 min; *ii* TFA/CH$_2$Cl$_2$/rtf/20 min [36]

α-Methylene-γ-lactone fusion to the sugar template can be achieved via the Grieco and Miyashita procedure [38]. The success of this procedure depends on the proper stereochemical relationship between the α-phenylseleno substituent and the proton located in the β position to the carbonyl function. The dilithium enolate of α-phenylselenopropionic acid reacts with **1** to yield **55** and **56**. The seleno function in each individual isomer is oxidatively eliminated using H$_2$O$_2$/AcOH at 0 °C to yield **57** and **58**, respectively [18] (Scheme 12).

Scheme 12 *i* LiCCH$_3$(SePh)CO$_2$Li/THF/HMPA, –78 °C; *ii* H$_2$O$_2$/AcOH/THF, 0 °C [18]

2.4
Annulated Pyranopyrans

Substituted δ-lactones are valuable synthetic intermediates for natural product synthesis, not only because of the fact that they represent substructures of a variety of biologically active natural products, but also they can also serve as precursors of other functionalities [19, 36]. Recently, the synthesis of the chiral δ-lactone **62** was described. Cleavage of the parent sugar ring of such lactones would generate chiral δ-lactones with side chains bearing chemically differentiable functionalities for further elaboration. Furthermore, the presence of unsaturation in such lactones provides additional opportunities for further manipulations. Starting from the reaction of **59** with Li[(Me$_3$)$_3$AlCCCH$_2$OTHP] affords **60** as a single product in 92% yield. Hydro-

Scheme 13 *i* Li[(CH$_3$)$_3$AlCCCH$_2$OTHP]/toluene, 50 °C; *ii* H$_2$/Lindlar's cat./EtOH; *iii* PPTs/ MeOH, 50 °C, 3 h; *iv* MnO$_2$/acetone, 3 h [19]

genation over Lindlar's catalyst and cleavage of the tetrahydropyranyl group gave quantitatively the (Z)-allylic alcohol. Oxidation of **61** with MnO_2 afforded **62** in 93% yield. As mentioned before, such compounds play a pivotal role in strategies toward complex natural products bearing several contiguous chiral centers (Scheme 13).

2.5
Canadensolide

As shown before, 4-O-triflyl-2,3-anhydropentopyranosides are appropriate precursors for the synthesis of highly oxygenated γ-butyrolactones, serving also as starting material for the densely functionalized antibiotic canadensolide [36, 39]. Several syntheses of this target have been reported, but are relatively lengthy and low yielding [39].

Thus, hydrogenolysis of **63** affords the pyranose tautomer **64**. Wittig reaction of **64** using propyltriphenylphosphonium bromide and potassium *tert*-butoxide in dry THF at room temperature for 30 min and then at 80 °C for 20 min affords an E/Z mixture of the alkenes **65** in 58% yield. Hydrogenation of **65** followed by Swern oxidation of the reduced product affords the lactol **67** which has been previously transformed in three steps to canadensolide (**68**) (Scheme 14) [40].

Scheme 14   *i* H_2/Pd(OH)$_2$/C;   *ii* C_2H_5CHPPh$_3^+$Br$^-$/K^{t}BuO;   *iii* H_2/Pd/C;   *iv* Ac$_2$O/ DMSO [36, 37]

3
Pyradazines and Oxazines

Heterocyclic ring systems of rich stereochemical functionality, such as oxazolidines, pyrazolidines, and oxazines [41], have been found to be potent HIV reverse transcriptase inhibitors [42–45]. However, so far their syntheses suffer from the formation of different regioisomers, diastereomers, and the obligatory presence of certain substituents in the precursors [42–48]. The previously described highly functionalized long-chain sugars **13**, **16**, and **17** (Schemes 15, 16) proved to be starting materials of highly functionalized, enantiomerically pure pyridazines, e.g., treating the C-4 masked oxirane **11** with 3 equiv. of the dianion of *tert*-butyl acetoacetate delivers regioselectively compound **13** in 85% yield (Scheme 15). Mesylation, followed by treatment with hydrazine in refluxing ethanol, affords the intended pyridazine skeleton **69** in 90% yield (two steps).

Scheme 15 *a tert*-Butyl acetoacetate, NaH, *n*-BuLi (3 equiv.), THF, 0 °C to rt, 5 h; *b* MsCl, Et₃N, CH₂Cl₂, –15 °C, 2 h; *c* NH₂NH₂, pyridine, EtOH, reflux, 4 h [41]

The above described strategy was also tested with the isomeric epoxy sugar **12**. Treatment of **12** with the dianion of *tert*-butyl acetoacetate affords the regioisomers **16** and **17** in a ratio of 9 : 1 (75% combined yield). Subsequent mesylation followed by reaction with hydrazine yields the regioisomeric pyradazine derivatives **70** and **71**, respectively (Scheme 16).

Starting from anhydrotriflate **1**, a new and efficient strategy for the synthesis of oxazine derivatives was also developed. The oxime **72** is easily accessible via the condensation of *tert*-butyl acetoacetate and NH₂OH.HCl in ethanol/pyridine (Scheme 17). Conventional formation of its trianion by *n*-BuLi, followed by the addition of the ambident electrophile **1**, delivers the oxazine **73** in 70% yield (Scheme 17). Due to the difference in the reactivity between the triflyl leaving group and the oxirane ring toward ambident nucleophiles, **73** is isolated as a single product in a one-pot reaction [41].

The formation of an oxygen bridge with C-3 of the pyranoside ring rather than a carbon bridge might be attributed to stereoelectronic effects of the transition state leading to the oxazine framework, where the oxygen anion can adopt colinearity with the oxirane ring. The formation of oxazine **75** is

Scheme 16 *a tert*-Butyl acetoacetate, NaH, *n*-BuLi (3 equiv.) THF, 0 °C to rt, 5 h; *b* MsCl, Et₃N, CH₂Cl₂, –15 °C, 2 h; *c* NH₂NH₂, pyridine, EtOH, reflux, 4 h [41]

Scheme 17 *a* THF, 3.3 equiv. *n*-BuLi, 0 °C, 1 h then –78 °C, **1**, 3 h; *b* THF, 2.2 equiv. LDA, –78 °C, 1 h then **1**, 3 h [41]

also accomplished by generation of the dianion of **74** followed by addition of **1**.

4
Triazines

Substituted 1,4,5,6-tetrahydro-*as*-triazines [49] have insecticidal, herbicidal, and fungicidal activities [50, 51]. Therefore, convenient synthetic approaches are still needed. Benzyl 4-amino-2,3-anhydro-4-deoxy-α-D-lyxopyranoside (**78**) and benzyl 4-amino-2,3-anhydro-4-deoxy-β-L-lyxopyranoside (**79**) with an amino group adjacent to an oxirane ring display electrophilic and nucleophilic reactivity, and their primary amino groups add rapidly to the nitrile

Ar	a	b	c	d
	C_6H_5	p-OMeC$_6$H$_4$	p-FC$_6$H$_4$	m-ClC$_6$H$_4$

Scheme 18 *i* Et$_3$N, MeOH, 0 °C; *ii* **76a–d**, Et$_3$N, MeOH, 0 °C and then rt, 4–6 h [49]

imines **77a–d** (generated in situ from the reaction of triethylamine with the hydrazonyl chlorides **76a–d** [52]) to yield the intermediate (Z)-amidrazone adducts **80a–d** [53] as the kinetically controlled products. The latter transient acyclic adducts undergo cyclization by opening the epoxide ring to yield pyrano[4,3-*e*]-1,2,4-triazines **81a–d** in good to excellent yields in a one-step reaction. In a similar fashion, **79** reacts with the generated nitrile imines **77a–d** to produce the isomeric 1,2,4-triazines **83a–d** (Scheme 18).

5
Pyrano Morpholines

A number of morpholine-based drugs [54], such as amorolfine [55], pravadoline [56], and timolol [57], are widely used for medical treatments. In addition, there are several molecules demonstrating the presence of a morpholine ring that may increase and/or modify the desired biological activities, e.g., the antitumor activities of 4- and 5-substituted derivatives of isoquinoline-L-carboxaldehyde thiosemicarbazone [58] or morpholinyl anthracylines [59], the analgesic activities of some aminobenzylcyclanols [60], or the cannabinoid agonist activities of aminoalkylindole derivatives [61]. In the case of chiral morpholine, gastrokinetic [62], dopamine agonist [63],

α_1-adrenergic [64], antioxidant [65], and antifungal activities [66] could be developed.

A new route toward the synthesis of chiral morpholines was introduced recently: the epoxy triflates **1** and **2** were allowed to react with an excess of *n*-substituted aminoethanol. Complete control of the off-template chiral center is obtained because of the extremely high reactivity of the triflate group and the stronger nucleophilicity of the nitrogen compared to the oxygen atom. Thus, the first step is the displacement of the triflate by the nitrogen atom to yield the intermediate products **84–87** and **92–95**, respectively. Subsequent opening of the epoxy ring by the oxygen anion leads to the formation of the morpholine derivatives **88–91** and **96–99**, respectively (Scheme 19).

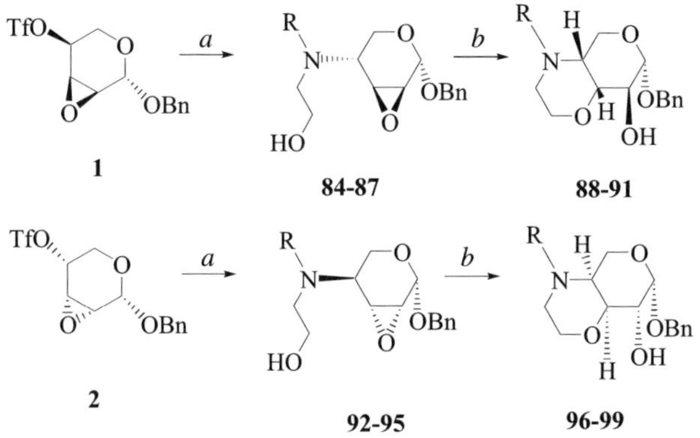

R	Me	Et	Ph	Pr
No.	84,88	85,89	86,90	87,91
	92,96	93,97	94,98	95,99

Scheme 19 Reagents and conditions: *a* RHNCH$_2$CH$_2$OH, THF, rt; *b* NaH, THF, 0 °C [54]

This novel and easy strategy gives access to the preparation of new morpholine derivatives of rich stereochemical complexity and pharmaceutical interest. Analogously to the synthesis of **88–91** and **96–99**, the novel new tricyclic systems **102** and **103** were prepared from **1** and **2**, respectively (Scheme 20).

6
Chiral Tetrahydroquinoxalines and Dihydrobenzodioxins

Quinoxaline derivatives and benzodioxin derivatives [67] possess interesting biological properties [68–85]. From the easily accessible chiral synthons, benzyl 4-*O*-triflyl-2,3-anhydro-β-L-ribopyranoside (**1**) and benzyl 4-*O*-triflyl-

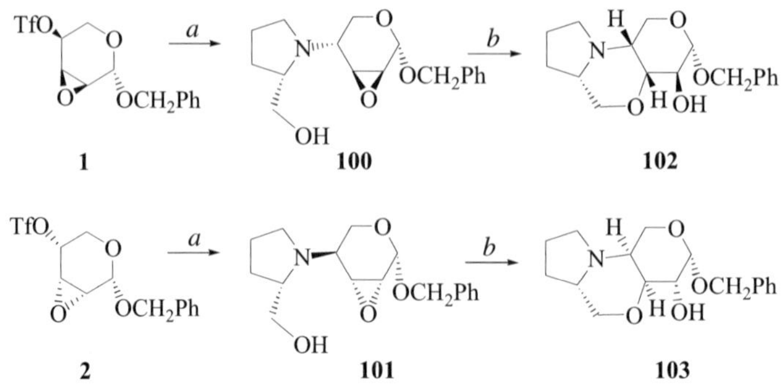

Scheme 20 *a* THF, pyrolidine-2-yl methanol, rt; *b* NaH, THF, 0 °C [54]

2,3-anhydro-α-D-ribopyranoside (**2**), chiral tetrahydroquinoxaline as well as 2,3-dihydro-1,4-benzodioxin derivatives with known absolute configurations can be prepared.

For the preparation of the quinoxaline derivatives, the epoxy triflates **1** and **2** are allowed to react with *o*-phenylenediamine or 4,5-dichloro-1,2-diaminobenzene in THF at room temperature to yield, after conventional workup, the chiral quinoxaline derivatives **104, 105** and **106, 107**, respectively (Scheme 21).

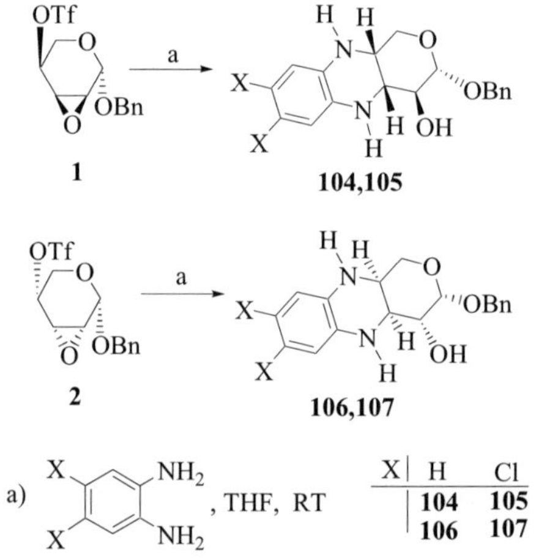

Scheme 21 [67]

Reaction of the epoxy triflates **1** and **2** with the dianion of 1,2-dihydroxybenzene or 2,3-dihydroxynaphthalene at room temperature yields the corresponding chiral benzodioxins **108** and **110** or the naphthodioxins **109** and **111**, respectively (Scheme 22).

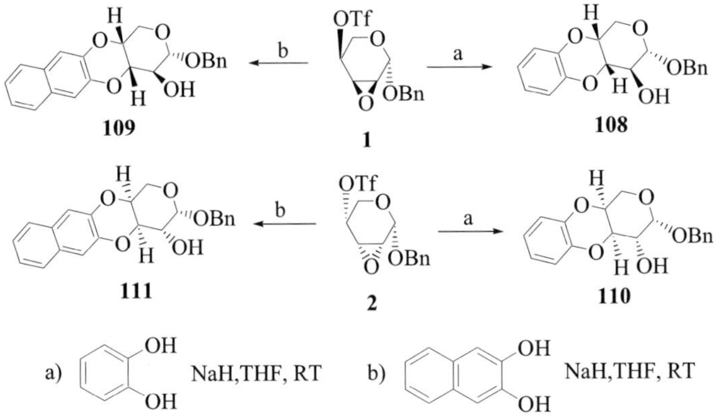

Scheme 22 [67]

7
Oxazolidinones

For the preparation of oxazolidinones [24, 86], treatment of 2,3-anhydrolyxoses **112** and **116** [87, 88] with benzoylisocyanate in CH_2Cl_2 at 0 °C for 30 min affords quantitatively the *N*-benzoylcarbamates **113** and **117**, respectively (Scheme 23). Subsequent treatment with catalytic amounts of NaH in THF leads to the oxazolidinones **114** and **118** in 90 and 95% yields, respectively [89, 90]. In the course of this reaction the benzoyl group migrates from N to O, as indicated by the ^{1}H NMR spectra of **114** and **118**, respectively. Treatment of **114** and **118** with catalytic amounts of NaOMe in CH_2Cl_2 delivered **115** and **119**, respectively, which could also be synthesized directly from **113** and **117** by refluxing in NaH/THF for 1 h.

8
Thiazolines and Episulfides

Thiazolines have been prepared via routes involving complex multistep sequences [91–93]. A strategically different approach to chiral thiazoline uses intermediate thiocyanato derivatives which can produce epithio or thiazoline sugars as the major product, just by varying the reaction conditions [94]. The sugar episulfides themselves are also intermediates of interest, as the nucleophilic opening of a thiirane ring permits introduction of a second heteroatom (such as another sulfur or nitrogen) or carbon nucleophiles, leading to the syntheses of branched-chain mercapto sugar derivatives. Thus, the nucleophilic addition of potassium thiocyanate in acetonitrile to the epoxy triflates **1** and **2** affords the thiocyanato sugars **120** and **121** in 86 and 92% yields, respectively, (Scheme 24).

Scheme 23 *i* PhCONCO, CH$_2$Cl$_2$; *ii* 0.5 equiv. NaH, THF, 0 °C to rt; *iii* 1 equiv. NaH/THF, reflux; *iv* 0.2 equiv. NaOMe/CH$_2$Cl$_2$ [24]

Scheme 24 *a* KSCN, MeOH, rt; *b* NaOMe, hot MeOH [94]

Reaction of the *trans*-oriented thiocyanato compound **120** with catalytic amounts of NaOCH$_3$ in hot methanol gives two products, the episulfide **122** and the thiazoline **124** in 10 and 72% yields, respectively. A similar reaction of **121** affords the episulfide **123** and the thiazoline **125** in 10 and 72% yields, respectively. When the thiocyanato sugars **120** and **121** are treated with an excess amount of $NaOCH_3$ for 30 min at room temperature, the epithio sugars **122** and **123** are obtained in 96 and 97% yields, respectively.

9
Cyclic Trithiocarbonates

Epoxy triflates (e.g., **1**, **2**, **127–129**) are also useful synthons for a simple and efficient route to cyclic trithiocarbonates (e.g., **126–134**) [95]. Formation of these trithiocarbonates involves the addition of a red aqueous solution of Na_2CS_3 [96] to a stirred solution of epoxytriflate. The nucleophilic displace-

Scheme 25 *a* Aliquat 336®, H_2O, 40 °C, 90 min; *b* **126**, ethanol, H_2O, rt, 30 min [95]

ment of the triflyl group at C-4 by sulfur is followed by the simultaneous intramolecular ring opening of the epoxide to afford the required trithiocarbonates (Scheme 25).

References

1. Vethaviyasar N, Kimmich R, Voelter W (1977) Chem Ztg 101:36
2. Kimmich R, Voelter W (1981) Liebigs Ann Chem, p 1100
3. Afza N, Malik A, Voelter W (1983) Chimia 37:422
4. Malik A, Kowollik W, Scheer P, Afza N, Voelter W (1984) J Chem Soc Chem Commun, p 1229
5. Malik A, Afza N, Roosz M, Voelter W (1984) J Chem Soc Chem Commun, p 1530
6. Malik A, Roosz M, Voelter W (1985) Z Naturforsch 40b:559
7. Kowollik W, Malik A, Afza N, Voelter W (1985) J Org Chem 50:3325
8. Afza N, Malik A, Latif F, Voelter W (1985) Liebigs Ann Chem, p 1929
9. Latif F, Malik A, Voelter W (1987) Liebigs Ann Chem, p 717
10. Kowollik W, Janairo G, Voelter W (1988) Liebigs Ann Chem, p 427
11. Kowollik W, Voelter W (1988) Liebigs Ann Chem, p 433
12. Shekhani MS, Latif F, Fatima A, Malik A, Voelter W (1988) J Chem Soc Chem Commun, p 1419
13. Kowollik W, Janairo G, Voelter W (1988) J Org Chem 53:3943
14. Fatima A, Zaman F, Shekhani MS, Malik A, Voelter W (1990) Liebigs Ann Chem, p 389
15. Latif F, Shekhani MS, Voelter W (1990) J Chem Soc Perkin Trans 1, p 1573
16. Kowollik W, Voelter W (1992) Z Naturforsch 47b:589
17. Al-Tel TH, Al-Abed Y, Shekhani MS, Voelter W (1993) Tetrahedron Lett 34:7717
18. Al-Abed Y, Al-Tel TH, Voelter W (1993) Tetrahedron 49:9295
19. Al-Abed Y, Al-Tel TH, Voelter W (1994) Nat Prod Lett 4:273
20. Al-Tel TH, Al-Abed Y, Voelter W (1994) J Chem Soc Chem Commun, p 1735
21. Kazmi SN, Ahmed Z, Malik A, Afza N, Voelter W (1994) Z Naturforsch 50b:294
22. Tschakert J, Voelter W (1994) Z Naturforsch 49b:702
23. Zaman F, Fatima A, Malik A, Voelter W (1994) Z Naturforsch 49b:1434
24. Al-Tel TH, Al-Qawasmeh RA, Schröder C, Voelter W (1995) Tetrahedron 51:3141
25. Abu Zarga MH, Al-Tel TH, Voelter W (1995) Z Naturforsch 50b:697
26. Al-Tel TH, Voelter W (1995) Tetrahedron Lett 36:523
27. Al-Tel TH, Al-Qawasmeh RA, Kaiser T, Voelter W (1995) Tetrahedron Lett 36:4599
28. Al-Tel TH, Meisenbach M, Voelter W (1995) Liebigs Ann Chem, p 689
29. Stang PJ, Hanack M, Subramanian LR (1982) Synthesis, p 85
30. Binkley RW, Ambrose MG (1984) J Carbohydr Chem 3:1
31. Rosowski A (1964) In: Weissberger A (ed) Heterocyclic compounds with three- and four-membered rings, part III. Interscience, New York
32. Baldwin JE, Kruse LI (1977) J Chem Soc Chem Commun, p 233
33. Al-Tel TH, Al-Abed Y Voelter W (1995) J Chem Soc Chem Commun, p 239
34. Baldwin JE, Lusch MJ (1982) Tetrahedron 38:2939
35. Al-Qawasmeh RA, Al-Tel TH, Abdel-Jalil RJ, Thürmer R, Voelter W (1999) Polish J Chem 73:71
36. Al-Abed Y, Voelter W (1997) GIT Labor-Fachzeitschrift, special edition 58–61

37. Al-Abed Y, Zaman F, Shekhani MS, Fatima A, Voelter W (1992) Tetrahedron Lett 33:3305
38. Grieco PA, Miyashita M (1974) J Org Chem 39:120
39. Al-Abed Y, Naz N, Mooto D, Voelter W (1996) Tetrahedron Lett 37:8641
40. Honda T, Kobayashi Y, Tsubuki M (1993) Tetrahedron 49:1211
41. Al-Qawasmeh RA, Al-Tel TH, Abdel-Jalil RJ, Voelter W (1999) Chem Lett, p 541
42. Rong J, Roselt P, Plavec J, Chattopadhyaya J (1994) Tetrahedron 50:4921
43. Hossain N, Papchikhin A, Plavec J, Chattopadhyaya J (1993) Tetrahedron 49:10133
44. Papchikhin A, Chattopadhyaya J (1994) Tetrahedron 50:5279
45. Papchikhin A, Agback P, Plavec J, Chattopadhyaya J (1993) J Org Chem 58:2874
46. Pan S, Amankulor N, Zhao K (1998) Tetrahedron 54:6587
47. Mann J, Thomas A (1985) J Chem Soc Chem Commun, p 737
48. Mancera M, Roffé I, Galbis JA (1995) Tetrahedron 51:6349
49. Saeed M, Abdel-Jalil RJ, Voelter W, El-Abadelah MM (2001) Chem Lett, p 660
50. Trepanier DL (1967) US Patent 3 510 483
51. Trepanier DL (1970) Chem Abstr 73:35415y
52. El-Abadelah MM, Hussein AQ, Thaher BA (1991) Heterocycles 32:1879
53. Butler RN, Scott FL (1970) Chem Ind (London) 1216
54. Abdel-Jalil RJ, Shah STA, Khan KM, Voelter W (2005) Lett Org Chem 2:306
55. Gupta K, Sauder DN, Shear NH (1994) J Am Acad Dermatol 30:911
56. Grieco G, DeAndrade J, Dorflinger E, Kantor T, Saelens J, Sunshine A, Wang R, Wideman G, Zellman V (1989) Clin Pharmacol Ther 45:123
57. Brogdan RN, Speight TM, Avery GS (1975) Drugs 9:164
58. Liu M-C, Lin T-S, Penketh P, Sartorelli AC (1995) J Med Chem 38:4234
59. Acton EM, Tong GL, Mosher CW, Wolgemuth RL (1984) J Med Chem 27:638
60. Özarslan Ö, Ertan M, Sayrac T, Akgün H, Demirdamar R, Gümüsel B (1994) Arch Pharm 327:525
61. Ambra TED, Estep KG, Bell MR, Eissenstat MA, Josef KA, Ward SJ, Haycock LA, Baizman ER, Casiano FM, Beglin NC, Chippari SM, Grego JD, Kullnig RK, Daley GT (1992) J Med Chem 35:124
62. Kato S, Morie T, Kon T, Yoshida N, Karasawa T, Matsumoto J-I (1991) J Med Chem 34:616
63. Dewald HA, Heffner TG, Jaen JC, Lustgarten DM, McPhaill AT, Meltzer LT, Pugsley TA, Wiese LD (1990) J Med Chem 33:445
64. Nozulak J, Vigouret JM, Jaton AL, Hofmann A, Dravid AR, Weber HP, Kalkman HO, Walkinshaw MD (1992) J Med Chem 35:480
65. Tani E, Rekka E, Kourounakis PN (1994) Arzneim Forsch/Drug Res 44:992
66. Bartroli J, Turmo E, Alguero M, Boncompte E, Vericat ML, Rafanell-García J, Forn J (1995) J Med Chem 38:3918
67. Abdel-Jalil RJ, Shah STA, Khan KM, Voelter W (2005) Lett Org Chem 2:238
68. Forlani L, Medici A, Ricci M, Todesco PE (1997) Synthesis, p 230
69. Boido A, Vazzana I, Sparatore F (1994) Farmaco 49:97
70. Monge A, Palop JA, de Cerain AL, Senaedor V, Martinez-Crespo FJ, Sainz Y, Narro S, Garica E, de Miguel C, Gonzalez M, Hamilton E, Barker AJ, Clarke ED, Greenhow DT (1995) J Med Chem 38:1786
71. Glombic H (1990) Offen. DE Patent 3 826 603
72. Glombic H (1990) Chem Abstr 113:9762w
73. Loev B, Musser JH, Brown RE, Jones H, Kahen R, Huang FC, Khandwala A, Sonnino-Goldmann P, Leibowitz MJ (1985) J Med Chem 28:363

74. Khandwala A, Inwegen RV, Coutts S, Dally-meade V, Jariwala N, Huang F, Musser J, Brown R, Loev B, Weinryb I (1984) Int Arch Allergy Appl Immunol 73:56
75. Khandwala A, Inwegen RV, Coutts S, Dally-meade V, Jariwala N, Huang F, Musser J, Brown R, Loev B, Weinryb I (1984) Chem Abstr 100:185318j
76. Kim KS Qian L, Bird JE, Dickinson KEJ, Moreland S, Schaeffer TR, Waldron TL, Delaney CL, Weller HN, Miller AV (1993) J Med Chem 36:2335
77. Catarzi D, Cecchi L, Colotta V, Filacchioni G, Martini C, Tacchi P, Lucacchini A (1995) J Med Chem 38:1330
78. Catarzi D, Cecchi L, Colotta V, Melani F, Filacchioni G, Martini C, Giusti A, Lucacchini A (1994) J Med Chem 37:2846
79. Shin-ya K, Furihata K, Hayakawa Y, Kato Y, Clardy J (1991) Tetrahedron Lett 32:943
80. Salituro FG, Harrison BL, Maron BM, Nyce PL, Stewart KT, McDonald IA (1990) J Med Chem 33:2946
81. Chapleo CB, Myers PL, Butler RCM, Doxey JC, Roach AG, Smith CFM (1983) J Med Chem 26:823
82. Hibert MF, Gittos MW, Middlemiss DN, Mir AK, Fozard JR (1988) J Med Chem 31:1087
83. Werner LH, Barnatt WW (1967) In: Schlitter E (ed) Antihypertensive agents. Academic, New York
84. Fourneau E, Bovet D, Maderni P (1933) J Pharm Clin 18:185
85. Fourneau E, Bovet D (1933) CR Seances Soc Biol Fil 113:388
86. Baker BR (1964) J Org Chem 29:1057
87. Inghardt T, Frejd T, Magnusson G (1988) J Org Chem 53:4542
88. Kowollik W (1987) PhD thesis, Tübingen University, Germany
89. Trost BM, Van Vranken DL, Bingel C (1992) J Am Chem Soc 114:9327
90. Gordon DM, Danishifsky SJ (1991) J Am Chem Soc 56:3713
91. Goodman L, Christensen JE (1961) J Am Chem Soc 83:3823
92. Goodman L, Christensen JE (1961) J Am Chem Soc 83:3827
93. Barker BR, Hullar TL (1965) J Org Chem 30:4038
94. Voelter W, Al-Tel TH, Al-Abed Y, Khan N, Al-Qawasmeh RA, Thürmer R (1998) In: Atta-ur-Rahman, Choudhary MI (eds) New trends in natural product chemistry. Harwood Academic, pp 47–56
95. Saeed M, Abbas M, Abdel-Jalil RJ, Zahid M, Voelter W (2003) Tetrahedron Lett 44:315
96. Sundin A, Frejd T, Magnusson G (1986) J Org Chem 51:3927

Author Index Volumes 1–7

The volume numbers are printed in italics

Almqvist F, see Pemberton N (2006) *1*: 1–30
Appukkuttan P, see Kaval N (2006) *1*: 267–304
Arya DP (2006) Diazo and Diazonium DNA Cleavage Agents: Studies on Model Systems and Natural Product Mechanisms of Action. *2*: 129–152
El Ashry ESH, El Kilany Y, Nahas NM (2007) Manipulation of Carbohydrate Carbon Atoms for the Synthesis of Heterocycles. *7*: 1–30
El Ashry ESH, see El Nemr A (2007) *7*: 249–285

Bagley MC, Lubinu MC (2006) Microwave-Assisted Multicomponent Reactions for the Synthesis of Heterocycles. *1*: 31–58
Bahal R, see Khanna S (2006) *3*: 149–182
Basak SC, Mills D, Gute BD, Natarajan R (2006) Predicting Pharmacological and Toxicological Activity of Heterocyclic Compounds Using QSAR and Molecular Modeling. *3*: 39–80
Benfenati E, see Duchowicz PR (2006) *3*: 1–38
Besson T, Thiéry V (2006) Microwave-Assisted Synthesis of Sulfur and Nitrogen-Containing Heterocycles. *1*: 59–78
Bharatam PV, see Khanna S (2006) *3*: 149–182
Bhhatarai B, see Garg R (2006) *3*: 181–272
Brown T, Holt H Jr, Lee M (2006) Synthesis of Biologically Active Heterocyclic Stilbene and Chalcone Analogs of Combretastatin. *2*: 1–51

Castro EA, see Duchowicz PR (2006) *3*: 1–38
Cavaleiro JAS, Tomé Jã PC, Faustino MAF (2007) Synthesis of Glycoporphyrins. *7*: 179–248
Chmielewski M, see Furman B (2007) *7*: 101–132
Chorell E, see Pemberton N (2006) *1*: 1–30
Crosignani S, Linclau B (2006) Synthesis of Heterocycles Using Polymer-Supported Reagents under Microwave Irradiation. *1*: 129–154

Daneshtalab M (2006) Novel Synthetic Antibacterial Agents. *2*: 153–206
Duchowicz PR, Castro EA, Toropov AA, Benfenati E (2006) Applications of Flexible Molecular Descriptors in the QSPR–QSAR Study of Heterocyclic Drugs. *3*: 1–38

Eguchi S, see Ohno M (2006) *6*: 1–37
Eguchi S (2006) Quinazoline Alkaloids and Related Chemistry. *6*: 113–156
Erdélyi M (2006) Solid-Phase Methods for the Microwave-Assisted Synthesis of Heterocycles. *1*: 79–128
Van der Eycken E, see Kaval N (2006) *1*: 267–304

Printing: Krips bv, Meppel
Binding: Stürtz, Würzburg